HARAMBEE
The spirit of innovation in Africa

Mike Bruton

Published by BestRed, an imprint of HSRC Press
Private Bag X9182, Cape Town, 8000, South Africa
www.bestred.co.za

First published 2022

ISBN (soft cover) 978-1-928246-41-1

Copy-edited by Debbie Rodrigues
Typeset by Firelight Studio
Cover design by Riaan Wilmans
Cover photo by Teun Vonk
Printed by Capitil Press, Paarden Eiland, South Africa

Distributed in Africa by Blue Weaver
Tel: +27 (021) 701 4477; Fax Local: (021) 701 7302; Fax International: 0927865242139
www.blueweaver.co.za

Distributed in Europe and the United Kingdom by Eurospan Distribution Services (EDS)
Tel: +44 (0) 17 6760 4972; Fax: +44 (0) 17 6760 1640
www.eurospanbookstore.com

Distributed in North America by Lynne Rienner Publishers, Inc.
Tel: +1 303-444-6684; Fax: 303-444-0824; Email: cservice@rienner.com
www.rienner.com

Suggested citation: Mike Bruton (2022) *Harambee: the spirit of innovation in Africa*. Cape Town: BestRed

'A winner is a dreamer who never gives up.'

– attributed to Nelson Mandela, late president of South Africa

Map of Africa

Contents

Acknowledgements

I am grateful to Alan Duggan, who has shown a keen interest in my research on African innovation, for writing the inspiring foreword to this book. As an ex-editor of *Popular Mechanics* magazine in South Africa, and a highly respected technology boffin, he is the ideal person to introduce this work to the reader.

This book could not have been written without the assistance and support of many people. I am grateful to the many inventors and innovators throughout Africa (mentioned in this book) who generously provided information on their achievements and life stories. For general information, enthusiastic assistance and inspiration, I also thank Peter Bailey, Phil Beck, Carolynn, Craig, Ryan and Tracey Bruton, Tony Bruton, Tony Budden, Valerie Buhlmann-Strydom, Steve Camp, Shadreck Chirikure, Julie Cleverdon, Andrew Cooke, Chris Cooper, Beverley Damonse, Jason Drew, Alan Duggan, Dirk Durnez, Mike Ellis, Henry Foretia, Theo Govender, Mohamed Hassan, Brian Hogg, Kent Lingeveldt, Ludwig Marishane, James Merron, Shadrack Mkansi, Khotso Mokhele, Haniso Motlhabane, Fredy Mshati, Davis Ndungu, Mthunzi Nxawe, Graeme Murray, Tim Noakes, Habib Noorbhai, Charles Phillips, Dyllan Randall, Sibylle Riedmiller, Melle Smets (and his team), Jeffrey Barbee, Toby Shapshak, Kelly Shibale, Mark Shuttleworth, Sue Swain, Percy Tucker, Joe Tyrrell, Kit Vaughan, Simon von Witt, Caren Vosloo, K. Weaver, John Webb, Rufus Wesi and Dave Woods.

As always, I thank my wife, Carolynn, for the many ways in which she has made it possible for this complex book to come to fruition – often under trying circumstances. I am also grateful to my tech-savvy sons, Craig and Ryan, and daughter, Tracey, for explaining many abstruse modern concepts and ideas to me and for reviewing the chapters on their fields of expertise.

I am extremely grateful to the Human Sciences Research Council (HSRC) for agreeing to publish this book, and especially to Mthunzi Nxawe, Danoline Hanyane and Jeremy Wightman for their enthusiastic support and for shepherding it through the publication process.

Foreword

My long and much-treasured association with Mike Bruton dates back to my early days as the editor of a fledgling science and technology magazine, when I was eager to know how the world worked and why history was inextricably linked to modern science. He helped me to understand just that. Our conversations spanned far and wide, ranging from the idiosyncrasies of quantum mechanics to the wonders and weirdness of the natural world and from unlikely innovations to unexplored educational opportunities; we rarely parted without an undertaking to meet again.

Mike's vast experience across multiple disciplines has equipped him to write with authority and boundless enthusiasm about African innovation, to explain how it has shaped the continent and how it continues to change the lives of billions. In this book, he pulls together disparate threads from many different perspectives to produce a fascinating compendium that the reader will consume in the best possible way. This hardly comes as a surprise: the author's long and distinguished career has spanned scientific research, science administration and science communication.

Using carefully selected examples of innovations and quotes from innovators, with a strong emphasis on the contributions of women and young innovators, and writing in a bright, accessible style, he stitches together the tapestry of African innovation and attempts to answer the compelling question: What is the nature of African innovation? The answer, says Bruton, is both intriguing and encouraging.

In South Africa, the Department of Science and Innovation (DSI) is actively involved in promoting innovation. Like similar government departments throughout Africa, it aims to boost broad-based socioeconomic development through a skilled and innovative workforce. To achieve its goals, the department provides leadership; an enabling environment; and resources for the promotion of science, technology and innovation. The inclusion of 'innovation' in the department's name emphasises the importance that is placed on promoting novelty, transformation and disruption in the science and technology arenas and in furthering its goal of taking advantage of rapid technological change to build a prosperous nation. Similar programmes are being implemented throughout the African continent.

This book showcases the resourcefulness and resilience shown by people in Africa as they search for solutions to the pressing problems they face daily. It serves as a useful introduction to the many significant advances that have been made in science and technology on the continent, at the low-, medium- and high-tech levels, and promises to stimulate innovators – female and male, young and old – to follow their dreams and achieve the extraordinary. In fact, this book is a veritable treasure chest, featuring over 800 inventions and innovations by more than 1 400 innovators from Africa, revealing some surprising facts and figures along the way: From balloons floating in the sky over Kenya to improve internet connectivity and robots patrolling the streets of Tunis City during the Covid-19 lockdown to Africa's leadership in the mobile money domain.

You will read about elephant contraceptives, the heroic efforts of 'locust listeners' in East Africa to combat insect plagues, 'talking gloves' that allow the deaf to communicate and the South African Radio Astronomy Observatory's (Sarao) inspiring array of radio satellite dishes that will soon span the continent. The excitement and prospects of African science and innovation shine through in this remarkable book.

Until a few weeks ago, I would never have imagined an ancient fisherman using a crude form of fishing reel – yet Egyptian tomb paintings dating back over 4 000 years show that very thing. I learned about an entrepreneur named Dayne Levinrad, who was once employed to mop floors in a restaurant and now runs one of South Africa's most popular coffee shops, introducing the concept of Coffee in a Cone to an eager South African audience. Then there is the award-winning children's literacy app Feed the Monster. A joint effort by the MTN SA Foundation, Bellavista School and the American NPO Curious Minds, the application (app) significantly enhances reading fluency and comprehension among children aged six to eight years, allowing them to access reading instructions via a specialised curriculum.

You can expect to say to yourself 'I didn't know that!' at least once for every page, and when you finally put down the book, you will have a lot to talk about. I would encourage anyone with an interest in science and innovation, and the future of Africa in particular, to read this book and share its inspiring message.

Alan Duggan
Founding editor of the sci-tech magazine *Popular Mechanics* (South Africa)

Preface

'It always seems impossible until it's done.'

Nelson Mandela, late president of South Africa (speech 2001)

Harambee: The spirit of innovation in Africa is a celebration of the unique ways in which innovation is expressed in Africa. As innovation does not only take place in the fields of science and technology, but in all fields of human endeavour, topics are included that would not normally be found in a book on innovation, such as agriculture, biodiversity conservation, food and drinks, and clothing and design – all of which contribute to the tapestry of African creativity. It was my intention, when I started writing this book, to include a chapter on African innovations in art and culture, but this topic proved to be so enormous that it had to be excluded. This is a pity, as innovations in African arts and culture have many traits in common with those in other fields and their inclusion would have made the characterisation of African innovation more persuasive. I have not attempted to produce a comprehensive catalogue of African inventions and innovations, as this task would need to be undertaken by a multinational, interdisciplinary team of experts over a period of years, not by a lone scientist and science communicator. Instead, I have selected examples that epitomise the spirit of innovation in Africa to answer the question 'What is the nature of African innovation?'

The cooperative spirit of many ventures in Africa is embodied, in different ways, in the Nguni concept of ubuntu (humanity or 'I am because you are') and in the *siriti* life force that connects individuals in Bitonga communities in Mozambique, but I have chosen to use the word *harambee* in my title because it symbolises the optimistic, cooperative spirit of Africa. *Harambee* literally means 'all pull together' in Swahili and is the official motto of Kenya, one of Africa's most innovative countries (in California, the USA's most innovative state, the state motto is 'Eureka'). Innovation is top-of-mind in Africa, as it is generally agreed that over two-thirds of a country's GDP is derived from innovation, and this percentage is likely to increase during the current Post-Industrial Revolution. By their nature, Africa's leading innovators deal with the big issues, so these issues are discussed in some detail in this book. As the book was partly written during the Covid-19 pandemic, when the novel coronavirus dominated world news, it is inevitable that frequent mention is made of the impact of this health and socioeconomic crisis on Africa.

This book covers inventions, innovations and innovators throughout the African continent and adjacent islands. As far as possible, I have tried to avoid repeating material discussed in my previous books on South African inventions, including *Great South African inventions* (2010) and *What a great idea! Awesome South African inventions* (2017a), although some overlap is inevitable. I have tried to confine my mentions of South African inventions and innovations to those that have had an impact elsewhere on the continent or have the potential to do so.

In my reading on African history, I have been surprised by how few books and other publications discuss the role of science and technology in the development of the continent. Perhaps this is because most histories are written by social scientists, politicians and economists whose interests lie in their fields of specialisation rather than in innovations in science, technology and other creative fields. It cannot be because innovation has not been an important driver of the continent's development. As relatively few African innovations have been described in the formal literature, the information discussed here comes largely from websites, blogs, newspaper and magazine articles, and other informal publications by or about the inventors and/or their employers or investors, or directly from correspondence with the inventors themselves. This information may not be peer reviewed, but it has the advantage of being topical and up to date in this rapidly evolving field.

Any book on innovation is out of date before it is published, and this one is no exception. I nevertheless hope that it will provide a partial snapshot, if not of the status of innovation in Africa, then of the spirit of innovation on the continent. If, on reading this book, you have been as inspired as I have been in writing it, I will have achieved my objective. It is my hope that this book, despite all its inadequacies, will reach every nook and cranny of the innovation space in Africa and encourage innovators – young and old, female and male, of all cultures – to explore the ideas that I share with them, and inspire them to develop their own ideas.

Introduction

'This research has illuminated an aspect of Africa's
past that is often misrepresented or completely
obliterated. Africa has always contributed to global
technological breakthroughs and economic systems.
The continent has an untold history of creativity.'

Abidemi Babatunde Babalola, African heritage scholar (Babalola 2020)

In his groundbreaking book *The wretched of the Earth*, Frantz Fanon (1963), an influential writer on the anticolonial liberation struggle, reflects on the dehumanising effects of colonisation and on Africa's potential contribution to the world as it emerges from the shadow of colonialism. His parting message was that Africa should not try to mimic the West, but should rather focus on contributing its own inventions and perspectives in the service of humanity. He wrote: 'For Europe, for ourselves, and for humanity, comrades, we must turn over a new leaf, we must work out new concepts, and try to set afoot a new man.'

It is a long-established view among most Africans that Africa and its people have much to contribute to the advancement of our world, but we tend to diminish these contributions by dismissing them as 'Third World' solutions. When South African President Cyril Ramaphosa – in his 2019 State of the Nation (Sona) address – spoke of smart cities and bullet trains whizzing across the country, the nation's response to his vision was to ridicule it and tell him to take his head out of the clouds and focus on more pressing issues. The former South African minister of science and technology, the honourable Naledi Pandor, also faced criticism from some quarters when she presented budgets for the development of Sarao's ambitious Square Kilometre Array (Ska) project that is in the process of establishing the largest radio telescope array in the world on African soil.

Their critics forgot that the articulation of a long-term vision is an essential role of any great leader, in Africa or elsewhere. We expect them to plan beyond their term of office, long into the future. Considering the rate of technological change – and the enormous problems that humanity and the other inhabitants of the planet face – it is not enough to think only of the present and the immediate future. Our contributions as Africans need to be future focused, especially because we must carry the undesirable burden of having contributed least to global climate change, yet we will probably be affected the most. Carbon dioxide (CO_2) levels have reached their highest point in 800 000 years and 19 of the 20 warmest years on record have occurred since 2001. If the world warms by an average of 3°C by the end of the century, which we are on track to do, this will mean an average rise of 6°C in southern Africa (Duma 2020). A 6°C rise in average temperature could inter alia result in the failure of maize and other cereal crops – the staple diet of most Africans.

The socioeconomic situation is also not promising, as the latest world population census announced by the BBC on 15 July 2020 (BBC 2020c) predicted that the population of Africa would triple to over 3 billion by 2100 and Nigeria would become the second most populous country in the world after China. Furthermore, the latest UN estimates indicate that over 80 per cent of Africans of employable age are not formally employed; the ILO has estimated that 305 million jobs are likely to be lost due to Covid-19 and not all of them will be recovered after the crisis as a consequence of the austerity measures implemented, similar to those after the 2008 financial crisis (CNN 2020).

We cannot afford to wait for these devastating changes to take their course; we need to act now. We face an array of interconnected crises, including climate change and its threat to food security, out-of-date economic systems that do not cater for modern needs and growing inequality that is widening the poverty gap. Furthermore, it is apparent that a crisis in Mozambique or Zimbabwe is also a crisis in Nigeria, as climate change and its impacts know no political boundaries. In light of these challenges, should we focus on simple bread-and-butter issues or should we, as a continent, participate actively at the cutting edge of science and technology?

I vote for the latter. When Elon Musk boasts about taking people to Mars by the 2030s, most people say that this is a waste of money that would be better spent solving issues on Earth. But, as Coleman (2016) and Duma (2020) point out, this view ignores the historical fact that space travel has been a major enabler of technological advancement and has had a positive effect on our everyday lives.

In 1961, when President John F. Kennedy committed the USA to landing a man on the moon by the end of the decade, the technology to achieve such a feat had yet to be developed. His ambitious goal, mainly driven by Cold War rivalry, forced engineers and scientists to develop the means that would make manned lunar flight possible, with many collateral benefits to Earthlings. In total, the National Aeronautics and Space Administration (Nasa) has produced over 6 300 new technologies in their bid to understand space better, and space agencies in other countries have been equally innovative. Spin-offs from space include more efficient water purification systems, miniaturised digital cameras, space blankets, heat-reflecting materials, nanosatellites, multispectral satellite imagery, thermal imaging, lightweight breathing masks (extensively used during the Covid-19 pandemic), solar panels and cordless drills (Coleman 2016).

The same cycle will be followed with the colonisation of Mars. Many of the technologies required for the colonisation of the Red Planet still have to be invented, but they will be highly relevant to the problems that we face on Earth as we transition to re-usable water, sustainable farming, alternative energy sources and smart cities. The colonisation of another planet will give us the incentive, resources and space to test out new technologies that we can use on Earth. It is therefore essential – considering that Africa is at the forefront of food, water and resource insecurity – that our scientists, engineers and policy-makers are deeply involved in this 'blue sky science'. African scientists already have a proud record of creating innovative solutions to the challenges of living in resource-scarce environments. Now is the time for us to share this expertise and experience with the world, and to participate in the collective ambitions of humankind.

But while embarking on this cutting-edge research, we need to be mindful of the needs of millions of African people who, despite working hard all their lives, are deprived of the opportunity to reach their full potential. This book reveals the extraordinary endeavours of many 'ordinary' people who have achieved amazing goals, often under conditions of great difficulty – from the vanilla farmers of Madagascar who use ingenious means to prevent their valuable crops from being stolen to the fisherman in Angola who made a canoe from the wingtip fuel tank of a downed jet fighter, the librarians in the Sahel who use camels as mobile libraries and the car-makers in Accra (Ghana) who make functional vehicles from scrap parts.

Perhaps more than anywhere else in the world, African innovators are ingenious, enterprising and streetwise. Above all, they are determined to make optimal use of the meagre resources available to them. In fact, a defining character of African innovation may be that deprivation stimulates creativity. Many African innovators (women and men) have experienced failure as well as discrimination and inequality during their lives, but they have learned to cope with it. They have seized the *ithuba* (opportunity), as the Zulu people say, notwithstanding their circumstances, to aim high, to take risks, to fail quickly and often, but – above all – to get up and start again. To the last, they are ordinary people who have chosen to be extraordinary. This is a much better life lesson than being incrementally successful at mediocrity.

1

Precolonial inventions

'What has Africa given to the world, people often ask in a
sneering and derogatory way, as if they already know the
answer – nothing! But Africa has given the world humanity –
and that is no small thing. And secondly, it has given the world
the first human culture. And that is also not a small thing!'

*Prof. Emeritus Philip V. Tobias, University of the Witwatersrand,
Johannesburg, South Africa* (Tobias 1969)

Introduction: In the nearly six decades since the Organisation of African Unity
(now the African Union [AU]) was formed, one of the biggest debates across the
continent has been about the need to find 'African solutions to African problems'.
There have been calls, especially from academia, to do this by drawing from
indigenous knowledge. Communities' knowledge and practices can provide
many answers and have a great deal to teach us, but we have not yet made
optimal use of them. Africa is both the Cradle of Humankind and one of the
most important crucibles of technological innovation. Precolonial innovations
from the continent, especially from ancient Egypt, various West African locations
(including Timbuktu), and Great Zimbabwe and Mapungubwe in southern
Africa are well known. They have been well documented in the literature and will
not be mentioned in detail here except to provide a glimpse of the extraordinary
resourcefulness of ancient African innovators. These early innovations were
in many fields, including fishing, hunting, food preparation, shipbuilding,
medicine, grooming, animal husbandry, metallurgy, town planning, architecture
and military strategies.

The Bantu-speaking people of Africa originated in West Africa near present-day Cameroon and gradually spread out from there about 3 000 years ago. At that time, most of Africa was inhabited by ethnic groups of Stone Age hunter-gatherers (similar to the San people of today) who were constantly on the move, lived off the land in family groups and had few tools, although they were adept at hunting and fishing. The spread of the Bantu people was made possible by three technological advances: agriculture, animal husbandry and metalworking. In contrast to the San, they developed more settled, stable societies that were in control of their food supplies and formed large, increasingly complex communities. Over a period of about 1 000 years, they spread eastwards across Africa to the Indian Ocean and then southwards into southern Africa – one of the greatest migrations in human history. As their economy was based on livestock and cereal crops, they needed grassland for their cattle and summer rainfall for their sorghum and millet, as they had no winter crops. They laid the foundation for the first major technological innovations in Africa (Ehret 2014).

First controlled use of fire: Together with the first use of tools; the agricultural revolution about 12 000 years ago; and the invention of the wheel, paper, printing press and computers, the first controlled use of fire must rank as one of the most transformational innovations in human history – and the earliest evidence for it comes from Africa. Dr Bob Brain of the Transvaal Museum (now the Ditsong Natural History Museum) in Pretoria and Dr Andrew Sillen of the University of Cape Town in South Africa discovered evidence of this important event at Swartkrans, a farm near the famous archaeological site of Sterkfontein, in the form of burnt bones dated between 2 and 1.8 million years before present (ybp) (Brain & Sillen 1988). The controlled use of fire was a major breakthrough, as it allowed humans to produce heat to keep warm and cook their food, create light at night and chase away predators. Later, fire was also used to produce baked clay pots, melt glue, prepare chemical concoctions, and eventually melt metal to make iron tools and weapons. Fire was also used to herd prey, encourage the growth of fresh grazing and clear away bush. Because of its mystical nature, fire was given magical and religious powers; it was a living entity that had to be carefully tended. The first controlled use of fire placed early African hominids at the forefront of human innovation, as it enabled them to start controlling their environment, which led to major changes in their habits and opportunities.

Ancient inventions: Blombos Cave and Klasies River Cave on the southern coast of South Africa have provided remarkable insights into the cognitive development of early humans. Blombos Cave near Stilbaai has yielded the oldest known abstract artwork – a piece of ochre with abstract patterns engraved in it, hashtaq style, dated at about 87 000 ybp, which may suggest that geometrical thinking predated numerical thinking (Webb 2019). Blombos Cave also yielded perforated shell beads covered with ochre (evidence of the use of heat to flake stones) and the 'first chemistry set', a perlemoen shell containing a mixture of ochre and fat, dated at about 100 000 ybp. Recently it yielded the earliest known drawing by *Homo sapiens*. This 73 000-year-old silcrete stone flake features six straight lines crossed obliquely by three other lines and was made with an ochre crayon. It is further evidence that early humans were able to produce and use symbolic material to communicate and store information (Henshilwood & Van Niekerk 2018).

A delicately made bone arrowhead found in Klasies River Cave, which would have been hafted and glued into a reed shaft and probably dipped in poison before being used to hunt, has been dated at about 60 000 ybp – at least 20 000 years earlier than bone arrowheads found elsewhere. It is like the arrowheads used by indigenous San hunter-gatherers from the 18th to the 20th centuries (Bradfield et al. 2020). Bow-and-arrow technology gave hunters a unique advantage, as it allowed them to hunt from a distance and from concealed positions, which added to their success. Furthermore, a bow and arrow consist of multiple parts, each with a function but operating together.

These findings demonstrate that, from at least 100 000 ybp, the early inhabitants of southern Africa were combining multiple ingredients to make coloured pastes, possibly for decoration or skin protection. By 70 000 ybp, they were making glues and other adhesives from a wide range of ingredients in a series of complex steps and by 60 000 ybp, they were crafting finely honed bone arrowheads – all of which signal advanced cognitive ability, including notions of abstract thought, analogical reasoning, multitasking and the ability to 'think outside the box' (Bradfield et al. 2020).

Ancient mathematics: Written mathematics was born in Africa with the calculations of the surveyors ('rope stretchers') and pyramid builders of ancient Egypt. There are papyruses with mathematical writing from 1750 BC on display in the British Museum, which suggests that the recording of numbers predated writing in some cultures (Webb 2019). Counting the days of the lunar month, or solar year, would have been important for early agriculturists so that they could predict when the rains would come and when to plant. Hunter-gatherers would also have benefited from keeping track of the seasons, or when the wildebeest migrated or the impala foaled. Coastal communities living off fish and shellfish would have tracked the phases of the moon and the rhythm of the tides to optimise their catches.

The earliest known mathematical counting device (known as the 'lunar stick') was found in Border Cave on the eastern slopes of the Lebombo Mountains in Maputaland (KwaZulu-Natal, South Africa). It is a 7.7 cm-long fibula of a baboon, dated at about 35 000 ybp, that has 29 notches carved into it and resembles calendar sticks still used by the San in Namibia (Bruton 2017a). An engraving tool with a bone handle found in the Congo, which has groups of notches cut into it, has been dated at 22 000 ybp. Do these notches represent prime numbers or lunar calendars? We do not know (Webb 2019).

The games played in Africa provide fascinating examples of the everyday use of mathematical principles. *Mankala* is the generic name for a family of games with many variations throughout the continent: it is called *bao* in East Africa, *wari* in West Africa, and *ncuvo* and *mufuvha* in South Africa (Bruton 2010; Webb 2019). It is a two-person strategic game played with a board comprising two rows of six holes and counters, usually stones or seeds, that are placed in the holes. The game starts with an initial number of stones placed in each hole. Players then scoop up stones and distribute them in the holes. Depending on where the last stone lands, the player can capture the opponent's stones and the player capturing the most stones wins. *Mancala* boards carved into flat rock have been found at the Mapungubwe World Heritage Site in Limpopo province and there is even one carved into a paving stone in the Castle of Good Hope in Cape Town.

Morabaraba is another two-person strategic game that is played throughout Africa. The goal is to place pieces ('cows') on intersections of the board and move them around to form a 'mill' (three in a row), similar to noughts and crosses but far more subtle. The player who forms a mill can then remove ('shoot') one of the opponent's cows, with the aim of 'shooting' all of them. In Sesotho, *morabaraba* means 'to go round in a circle'. Other names for the game include *mlabalaba* (isiXhosa), *mmela* (Setswana) and *mororova* (Sepedi). Variations of *morabaraba* can be traced back thousands of years – the poet Ovid (43 BC–18 AD) mentions it in Roman times, when it was known as *latrones*. Similar games are played in England (nine men's morris), India (*nao-guti*) and the Philippines (*tapatan*). In South Africa, *morabaraba* is recognised as an official indigenous game and national colours are awarded to players who compete in international tournaments (Bruton 2010; Webb 2019).

In 1925, the Belgian ethnologist Emil Torday discovered a game played by Bushongo children in Central Africa that involved drawing, in the sand, mesh-like diagrams resembling fishing nets. The challenge is to draw the complete diagram without lifting your finger or going over a line twice. The study of such networks is a first step in an important field of modern mathematics called graph theory (Webb 2019). Sotho girls in South Africa make intricate string patterns between their fingers in a game called *uzamanyeka*, which means 'it sways of itself'. In Lesotho, the favourite pattern gives the game the name 'fowl's foot'. In Tanzania, teams of four children play a strategic thinking game called *tarumbeta*, with a pattern of 45 beans laid out in a triangular array. The aim of the game is to memorise the number of the beans as they are gradually removed from the array. Kpelle children in Liberia play a variation of the river crossing game with a leopard, goat and pile of cassava leaves. *Mbuti* pygmies in the Congo play a game called *panda* in which beans are thrown onto the ground and one of the players scoops up a handful of beans; the other players must then estimate how many beans are needed to make a multiple of four (Webb 2019).

Ancient musical instruments: Musical instruments are not preserved well in the archaeological record, but there are some examples dating back about 10 000 years, from the Late Stone Age to the Iron Age. They fall into two groups: aerophones (which produce sound by vibrating air) and idiophones (which vibrate solid materials). They include spinning disks; bullroarers; bird bone tubes used as single-tone flutes; clay whistles; wooden keys from thumb pianos (*mbiras*); metal musical bells; and trumpets made from shells, bone and ivory that have been found in South Africa, Zimbabwe, Zambia, Mozambique, Ghana and elsewhere. Of interest are musical bells recovered from Great Zimbabwe and clay whistles from Mapungubwe. Some traditional African musical instruments are still in use today, including *mbiras* (widespread), conch shell trumpets (Tanzania), clay whistles (by Basuto herders), reed flutes (by San hunters in Botswana) and bone flutes (by the Venda in South Africa while performing their *tshikona* dance) (Kumbani 2020).

Traditional fishing methods: African fishers have developed an astounding variety of methods to catch fish and shellfish in marine and freshwaters. They include poisons, spears, gaffs, bows and arrows, handlines, barriers, baskets, traps, nets, artificial reefs and fish aggregation devices, as well as rafts, bark and dugout canoes, dhows and other craft used to set the gear and recover the catch. Spearing fish has been carried out in Africa for aeons. A rock painting in a cave on Mpongweni Mountain in KwaZulu-Natal, dated to the Late Stone Age (about 45 000 ybp), depicts nine canoes arranged in a semicircle with a man in each canoe spearing a fish. Tomb paintings over 4 000 years old in Egypt show fishermen using spears with retrieval lines and wooden spools on which to store the line. These spools precede fishing reels in the West by over 3 500 years (Bruton 2016)!

Spear points were made from bone, horn or ivory – and later copper – and were used for fishing as well as hunting crocodiles and hippopotami (hippos). A rock painting at least 7 000 years old at Aouanrhet in the central Sahara depicts a hippo hunting scene. In Lake Itasy in Madagascar, divers dig holes underwater to create hiding places for eels and then catch them using spears. Gwembe children in Zambia make simple spears by inserting thorns into the ends of reed shafts, and the Ovimbundu in Angola use 10-pronged spears to catch fast-swimming river fish (Bruton 2016).

The Twa living on the Kafue Flats in Zambia previously used a 'dark hut' method of fishing (*mbulu*) that was unique. They built a reed platform with a wide, angled tube in the centre that extended into the water and enabled them to see fish swimming past. They speared the fish with two-pronged spears (*namako*) or three-pronged spears (*muingo*) up to 6 m long. In the 1950s, there were at least 250 of these 'dark huts' in the Twa region but today there are none, as this and many other innovative traditional fishing methods have gone extinct and been replaced by monofilament gillnets (Maclaren 1958). Bow-and-arrow fishing is less common in Africa than in South America, but Liberia has issued postage stamps depicting bow fishermen. Bow fishing has also been recorded on the Luangwa River in Zambia and on the Lundi and Sabi Rivers in Mozambique, and the use of crossbows for fishing has been reported in West Africa (Bruton 2016).

Fish poisons derived from plants are widely used in Africa and represent the earliest examples of humans using chemicals to collect food. African people have an intimate knowledge of the roots, bark, wood, sap, leaves, flowers and seeds of plants, and use at least 29 species to make fish poisons. Derris root extracts, from which rotenone is made, are widely used because it stuns but does not kill the fish, whereas extracts from *Tephrosia vogelii* (fish-poison-bean) kill fish outright and have been banned in Kenya, Malawi and elsewhere. The toxic sap of *Euphorbia* (milkbush) is used as a fish poison (and arrow-tip poison) throughout East Africa, as is the sap of the violet tree and latex from tamboti trees. Poisons from several sources may be used to create toxic concoctions. For example, powdered dogwane tree wood is mixed with milkbush sap, crushed scorpions, toxic caterpillars, snake venom and toad secretions; or the saliva of rock lizards is combined with crushed meloid beetles (Bruton 2016).

Rod fishing without a hook (bobbing) is practiced by the Lovale in Zambia who fish with a bunch of cords and the worm bait tied directly onto the line. Small bags of bait may also be cast into the water to catch crabs, crayfish and octopi along marine shores. Gorges – straight pieces of wood, bone or shell (or flint in Egypt) sharpened at both ends and tied (usually off-centre) onto a line – are widely used (baited or unbaited) to catch fish. They are swallowed lengthwise by the fish and then lodge crosswise in its gullet. Three-pointed acacia thorns and porcupine quills are also used as gorges. The Bakongo on the Lower Congo River use wooden gorges, whereas the Ovimbundu in Angola use pieces of stiff grass. Unusual spring gorges made from flexible pieces of wood that are tied to the bait and spring open when they are swallowed are used on the Congo River (Bruton 2016).

Barbless fishing hooks made from bone, horn, shell and hippo ivory have been used in Egypt for over 5 000 years; copper hooks were already in use there about 4 700 ybp. Barbed hooks with eyes or flanges for attaching the line appeared in Egypt about 3 900 ybp, but they did not reach Central Africa until much later (Bates 1917; Brewer & Friedman 1989). In precolonial Africa, fishhooks were initially made from any suitably shaped objects such as curved thorns or spines, eagle's beaks or talons, octopus beaks and rhinoceros beetle horns, and were also fashioned from stones, wood, bones and shells.

Early Khoi fishers in southern Africa used hooks made from mammal teeth and bone tied onto wood. The sharp spines of raffia palm and ilala palms are widely used in Mozambique and Maputaland (South Africa) as fishhooks. Young fishers at Vilanculos make their fishing rods almost entirely from the ilala palm (*Hyphaene coriacea*), with the rod fashioned from the flexible leaf stem, the eyes from loops of dried leaf, the twine from leaf fibres and the hook (with three barbs) from the spines on the leaf stem; only the reel is made from a three-pronged stick cut from another plant. The Venda in southern Africa use a unique device for catching fish comprising a palm tree leaf stem from which all the thorns, except the last one, have been removed. The combined rod and baited hook with no line is dipped into the water.

In Nigeria, fishers use modified rodent snares to catch fish as well as rods ingeniously equipped with automatic trigger releases that pull it up sharply when a fish bites (Bruton 2016). On the African Great Lakes, Bangweulu Swamps and Upper Zambezi River, barbless iron hooks were in use long before the colonial era. These hooks were often strongly recurved so that they could hold a struggling fish even if the line were not recovered immediately. There is no evidence that barbed metal hooks were made in sub-Saharan Africa until imported barbed hooks from Europe could be copied (Goodwin 1946; Hurum 1976; Maclaren 1958). The earliest fish traps were almost certainly crescent-shaped rock barriers arranged in the intertidal zone of the marine coast to trap fish at low tide, or rock or log barriers in rivers that trapped fish when the water level receded after a flood. Numerous rock-barrier traps or vywers are found along the southern Cape coast in South Africa at Stilbaai, Rooi Els and Skipskop, and they have also been recorded in freshwater in South Africa, Botswana, Malawi and Namibia.

The use of reed barriers (*cisasa*) in combination with fish traps has been recorded in 5 000-year-old paintings in the Faiyum Oasis in Egypt (Bates 1917) and they are still widely used throughout the continent. There are two main kinds of traps. Constriction traps are trumpet shaped but have no valve; fish swim into them (usually down current) and become trapped in the narrow funnel. Valve traps have a valve comprising interlocking sticks that allow fish to swim in but not out; they are used throughout Africa in marine and freshwaters, are set singly or in groups, and may be baited or unbaited. The most famous examples are those set on high fences that straddle the rapids on the Congo River and at Malepo Pool (between the Republic of Congo and the Democratic Republic of the Congo [DRC]), and the centuries-old barricade fish traps in the Kosi Bay estuary in Maputaland (South Africa). In the Cabinda enclave near the Congo River mouth, and in the DRC, Nigeria, Zanzibar and Mozambique, more sophisticated valve traps with trigger-release doors are used.

Thrust baskets (which are pushed over fishes in shallow, turbid water) are commonly used in coastal wetlands and floodplains in Maputaland and Mozambique as well as in the Upper Zambezi River and Okavango Swamps and on Lake Kyoga in Uganda; they are known variously as *isiFonya, ivumbu, chongo* or *shiranga* (Bruton 2016). On the Phongolo floodplain in Maputaland and on the Buzi River in Mozambique, thrust basket fishing drives previously involved hundreds of people (Tinley 1964); however, today fewer than a dozen people participate and this innovative method of fishing is threatened with extinction.

Seine nets made from plant fibres have been used in Egypt for thousands of years. A scene showing fishermen catching snoutfishes (family Mormyridae) and Nile tilapia (Cichlidae) adorns a wall in the 3 000-year-old tomb of Ipouy in Egypt. The net fishery at Malebo Pool on the Congo River is hundreds of years old and features the earliest known netting knot, the 'lake-dweller's knot' (Gabriel et al. 2005). Other types of traditional fishing gear developed in Africa (as well as on other continents) include handmade fishing baskets, scoop nets, drag nets (*mkwau*), drift nets, lift nets (*lituwa*), gillnets (*amakonde*), tangle nets, drop nets (*imbwa*), purse seines (*chirimila*), trawl nets and lampara nets – all drawn in by hand. Cast nets, synthetic fibre multifilament and monofilament gillnets, and trawl nets were all introduced during the colonial era (Bruton 2016).

By far the most harmful gear used in African marine and freshwaters is monofilament gillnets, which are ruthlessly efficient and have replaced many traditional fishing nets and traps that harvest resources more sustainably. On the East African coast and in Madagascar, huge large-mesh gillnets called *jarife* are set in deep water to catch sharks for the shark fin and oil trade, but also catch sawfishes and large bony fishes (including coelacanths) as well as turtles, dugongs and dolphins (Cooke, Bruton & Ravololoharinjara 2020). Traditional fishing methods represent the collective intellectual property of generations of African men and women who have a close affinity with the natural environment and have developed techniques to catch fish sustainably. They should not be regarded as relics from the past, but as part of our sustainable future.

Traditional hunting methods: Precolonial hunting methods in Africa were similar to those used on other continents and included shaped stones; hand axes; battle axes; stone, bone and later metal knives; spears (with fire-hardened points, some with retrieval strings); long bows; crossbows; fall traps; and snares. Fire was also used to herd game over cliffs and into holes. The main weapon of the Venda in southern Africa was the bow (*vhura*), which was used with iron- or wood-headed arrows (*masevha*). Featherless iron-headed arrows were used for hunting big game, while feathered wood-headed arrows were preferred for shooting small game and birds. Iron-headed arrows were sometimes tipped with poison made from a mixture of dried mouse (*Lutema matanda*) and the powdered seeds of the mudulu tree (*Balanites maughamii*). The San use arrows tipped with poison extracted from the larva of a beetle (*ka/ngwa*), poisonous plants such as euphorbias and snake venom (Goitsemodimo 2019).

DeGeorges (2012) reports that the Waliangulu/Wata of Kenya hunted elephants using 68-kg long bows and arrows tipped with poison from the bark, wood and roots of the arrow-poison-tree (*muriju; Acokanthera schimperi*), whereas the San in Botswana use small poison-tipped arrows shot from delicate bows. Knobkerries or clubs made from the hard wood of tree roots and shields fashioned from hippo, rhinoceros (rhino), buffalo and giraffe hide were also used in hunting, as were straight- or curve-bladed daggers.

Later, in the 19th century, after they had first seen Western guns, crude 'blackpowder muskets' were made by village blacksmiths and used in several West African countries (including Mali, Senegal and Guinea-Bissau) for hunting and fighting. The hunters made their own gunpowder from ashes and saltpetre taken from the ground, and crafted firing caps from moulded goat skin and crushed match heads. In the 20th century, they even fashioned barrels from the steering columns of Land Rovers! As bullets were scarce, hunters frequently used rocks or jagged pieces of metal as projectiles; however, these would sometimes clog the barrel and cause the gun to explode, with disastrous consequences. Modified single-shot Russian Baikal shotguns and later AK-47s were also popular in several West African countries (DeGeorges 2012).

Early glass-making: Babalola (2020) found archaeological evidence of the earliest glass-making technology in sub-Saharan Africa at Ile-Ife in Nigeria, dated to about 1 000 ybp. His findings show that the people of Ile-Ife were not just consumers of glass made elsewhere but also contributed to technological innovation, and suggest that glass beads were mass produced and traded there as early as 600 to 400 BC. The earliest evidence of glass-making dates to 2500 BC in the Middle East, Mediterranean and Levant. Babalola's discoveries include ancient furnaces, over 20 000 glass beads, 1 500 crucible fragments (ceramic vessels used in glass production) and glass waste. Chemical analyses revealed that it is 'high lime, high alumina glass', which is not known from anywhere else in the world. The composition of the glass is consistent with the geology of the region, which indicates that the glass-makers invented their own glass recipe using locally available resources. They used feldspar-rich granitic sand and pegmatite as a source of silica and added snail shell to reduce the melting temperature and improve the quality of the glass, which was as good as glass from other ancient societies (Babalola 2020).

Mapungubwe: The ancient city of Mapungubwe ('hill of the jackal') is an Iron Age archaeological site in Limpopo province near the border between South Africa, Zimbabwe and Botswana, close to the confluence of the Limpopo and Shashe Rivers. One thousand years ago, Mapungubwe appears to have been the centre of the largest known kingdom in southern Africa, with thriving trade in gold and ivory with China, India and Egypt from about 1200 to 1300 CE. Research at the site has revealed that gold smelting took place there, as the most famous artefact from the site is the Golden Rhino, a symbol of authority made from a single sheet of gold. In addition, a female skeleton was found with at least 100 gold wire bangles around her ankles and over 1 000 gold beads in her grave. Evidence of expertise in ceramics, especially pottery, was also recovered. Mapungubwe was declared a World Heritage Site by Unesco in July 2003.

Great Zimbabwe: The ancient city of Great Zimbabwe is located in the south-eastern hills of Zimbabwe near Lake Mutirikwe. It was the capital of the Kingdom of Zimbabwe during the Late Iron Age, with construction by ancestors of the Shona people beginning in the 11th century and ending in the 15th century. Its most formidable edifice, the Great Enclosure, has walls 11 m high extending for 250 m, making it the largest ancient structure south of the Sahara. The level of technological advancement of Great Zimbabwe is shown by artefacts found on the site, including meticulously carved 2 m tall soapstone carvings of the 'Zimbabwe bird' (probably a bateleur); other soapstone figurines; baked pottery; iron gongs; elaborately worked ivory; iron and copper wire; iron hoes; bronze spearheads; copper ingots and crucibles; and gold beads, bracelets, pendants and sheaths. Gold and ivory were traded from the site; Chinese pottery, coins from Arabia and glass beads were imported. Great Zimbabwe gave its name to the modern-day country and was declared a Unesco World Heritage Site in 1986.

Timbuktu: Timbuktu is a city in Mali that is 20 km north of the Niger River and became a permanent settlement early in the 12th century. The city flourished from the trade in salt from the Sahara Desert that was exchanged for gold, ivory and slaves from the Sahel. In its Golden Age, Timbuktu (with its numerous Islamic scholars and extensive trading networks) became a nexus for trade in books, which led to it becoming a major centre of learning in Africa from the 13th to the 17th centuries. Hundreds of thousands of manuscripts were collected there over the centuries, some written locally and others (including exclusive copies of the Quran for wealthy families) imported through the lively book trade. Many of these manuscripts – buried, hidden in cellars or between the mud walls of mosques, or safeguarded by their patrons – survived the city's subsequent decline and now form the collections of many libraries in the city that hold over 700 000 manuscripts. Although rebel forces destroyed some of the manuscripts in January 2013, most have survived. The ancient city continues to be a source of information and inspiration for scholars worldwide.

Discussion: Discoveries of technological innovations in ancient African societies have challenged Western theories that Africa did not contribute to early innovation. Some Western scholars have even suggested that Mapungubwe and Great Zimbabwe could only have been built with outside help, but this view is rejected (Babalola 2020). The evidence suggests rather that Africa was a major contributor to early technological development, but has an untold history of creativity.

There is also increasing appreciation of the value of indigenous knowledge systems (IKS) in stimulating modern inventions and innovations. IKS refers to the understandings, skills and philosophies developed by societies with long histories of interaction with their natural surroundings. For rural and indigenous peoples, this local knowledge informs decision-making about fundamental aspects of day-to-day life and encompasses not only information but also language, classification systems, resource-use practices, social interactions, rituals and spirituality.

These unique ways of knowing are important facets of the world's cultural diversity and provide a foundation for locally appropriate sustainable development. Unesco's Local and Indigenous Knowledge Systems (LINKS) programme has been influential in ensuring that local and indigenous knowledge holders and their knowledge are included in contemporary science–policy–society discussions on issues such as biodiversity management, climate change assessment and adaptation, natural disaster preparedness and sustainable development. Working at local, national and global levels, LINKS strives to strengthen indigenous peoples and local communities, foster transdisciplinary engagements with scientists and policy-makers and pilot novel methodologies (https://en.unesco.org/links).

2

Agriculture, stock farming, apiculture and rural living

'It is time to change the way we think. Farmers are not the cause of Africa's poverty; they are a potential solution. They are key to creating the future envisioned by the SDGs.'

Kofi Annan, former UN secretary-general (Annan 2015)

Introduction: About 12 000 years ago, in the 'Fertile Crescent' of west Asia, humans began to abandon their hunter-gatherer lifestyle and started to domesticate animals and cultivate grains for the first time. This secure source of storable food allowed for the growth of stable communities living in towns and villages. Many people in Africa today eat starchy food made from plantain, yam, rice, maize and other grains, but is it good for them? Human evolutionists such as Jared Diamond (1987) have suggested that, from the perspective of the future of the planet, this was 'the worst mistake in the history of the human race' because it produced a number of serious disadvantages, including periodic starvation, epidemic diseases and malnutrition. Yuval Noah Harari (2014), in his epochal book *Sapiens: A brief history of humankind*, calls it 'history's biggest fraud' and argues that wheat domesticated humans (not the reverse) and kept 'more people alive under worse conditions'. Agriculture also produced deep class divisions, as it allowed some people to accumulate wealth by storing food.

Over the course of the first 4 million years of human evolution, eating high-fat, animal protein foods made us remarkably healthy and robust. However, when we began eating cereals after the Agricultural Revolution, especially the processed foods of modern commerce, human health began to deteriorate and the so-called 'modern lifestyle diseases' started to take their toll. Many recent studies (reviewed in Noakes & Sboros 2017) reveal that humans are naturally carnivorous and that the structure and function of their digestive systems are better designed for processing fats and animal protein than carbohydrate-rich plant foods. Of course, feeding 7.6 billion people on animal protein would be an ecologically expensive exercise, as livestock is responsible for up to 14 per cent of all greenhouse emissions from human activities.

Alliance for a Green Revolution in Africa (Agra): Agnes Kalibata's ambitions know no bounds. As the president of Agra, this Rwandan woman is working to increase the incomes and improve the food security of over 40 million farming households in 18 African countries. Agra (a partnership-driven institution that is African led and farmer centred) was formed in 2006 in response to a call of the former UN secretary-general, the late Kofi Annan, who stated that the time had come for African farmers to wage a 'uniquely African Green Revolution'. Annan served on Agra's board of directors together with internationally respected experts in agriculture and policy management, including Kalibata, who became Agra's president in 2014. The founding partners included the Bill & Melinda Gates Foundation and the Rockefeller Foundation. For over 15 years, Agra has transformed smallholder farming in Africa from a solitary struggle for survival into thriving businesses by focusing on distinct problems related to seed production, soil health, policy development, international collaboration and agricultural marketing.

Kalibata's immense contribution is reflected in the many international honours she has received. In 2019, she was awarded an honorary doctorate by McGill University and won the US National Academy of Sciences' Public Welfare Medal. She was appointed by UN Secretary-General António Guterres as his special envoy for the 2021 UN Food Systems Summit, which had been jointly requested by the UN Food and Agriculture Organisation (FAO), International Fund for Agricultural Development, UN World Food Programme (WFP) and World Economic Forum (WEF). The goals of this summit are to expand knowledge and encourage sharing experience to help countries unleash the benefits of food systems for all their people. Kalibata's roles at this summit will include providing leadership and strategic direction, developing outreach programmes and fostering cooperation among key leaders (Ayene et al. 2020).

Great Green Wall of Africa: In August 2019, 399 volunteers from 27 countries arrived in northern Senegal to plant over 150 000 acacia, baobab and moringa shoots in kilometre-long furrows ploughed in the ground. They were participants in one of the most audacious efforts ever to combat the effects of climate change – an US$8-billion plan to reforest 100 million ha of degraded land across the width of Africa, stretching from Dakar to Djibouti.

The Great Green Wall project (which is spearheaded by the AU and funded by the World Bank, EU and UN) was launched in 2007 to halt the southward expansion of the Sahara Desert by planting a barrier of trees running 7 746 km along its southern edge. But it will also fulfil additional roles by recharging the water table, creating microclimates that increase local rainfall, restoring agricultural land, providing food and fuel, stemming conflict and discouraging migration. When the project is completed in 2030, the restored land is expected to absorb over 250 million metric tonnes of CO_2 from the atmosphere (Baker 2019a).

Volunteers working from the village of Mbar Toubab will turn 494 acres of barren land into another forested brick in the Great Green Wall. But the project has been hindered by late rains. El Hadj Goudiaby (who has spent the past nine years overseeing the project in Senegal) asks, 'How is it possible to grow trees to combat climate change if climate change is making it impossible to grow trees?' He recognises that the problem is not so much that the Sahara Desert is expanding but that the Sahel is shrinking, destroyed by decades of overgrazing, climate-change-induced drought and poor farming practices that have stripped the once lush grasslands of the fertile topsoil needed for it to regenerate (Baker 2019a).

Vanilla farming in Madagascar: Vanilla farmer Lydia Soa, from Sahabevava on the east coast of Madagascar, has had to resort to stamping her green vanilla pods with her unique producer number (MK021) to avoid them being stolen off her vines. If a thief steals her crop, it can be traced at the local market. Vanilla theft is a big problem in Madagascar, as climate change, crime and speculation have led to skyrocketing prices – from US$20 per kilogram in 2013 to US$515 per kilogram in June 2018. Furthermore, vanilla is not easy to cultivate because the plants only bloom once a year for one day, must be pollinated by hand and the fruit takes nine months to mature.

To stabilise the supply of vanilla, leading fragrance companies have invested in the Livelihoods Fund for Family Farming programme that provides Malagasy farmers with vanilla seedlings, trains them in sustainable farming practices, organises neighbourhood vanilla-watch programmes and supplies the antitheft stamps. Vanilla thieves now face up to four years in prison but, to Soa, that is not enough: 'You invest all your life in growing the vanilla. Stealing it is the same thing as killing someone' (Baker 2018).

In the Sava region of Madagascar, about 20 000 children work on the vanilla plantations but their families receive little income from the lucrative US$192-million industry. Although vanilla is the second most expensive spice in the world after saffron, farmers receive only US$6.00 per kilogram, which locks traditional vanilla farmers into a cycle of exploitation, poverty and child labour (Lind 2016) that has been exacerbated by Covid-19.

Sorghum: *Striga* (a genus of parasitic weed) is one of the major constraints limiting the production of cereal crops such as sorghum, maize, pearl millet and rice, with yield losses of 30 to 90 per cent being reported regularly in the semi-arid regions of sub-Saharan Africa. One species, *Striga hermonthica* (purple witchweed), has had devastating impacts on sorghum and pearl millet cultivation from Senegal eastwards to Ethiopia and in the DRC, Tanzania, Angola and Namibia. Many remedial measures have been proposed by international experts, but few have been adopted by the affected farmers due to the high cost of implementation or the ineffectiveness of the proposed methods. Scientists at the University of Zululand in South Africa have now developed sorghum varieties that are resistant to *Striga* and will help to solve the problem throughout the continent.

Carbon-storing wonder plant: *Spekboom* (*Portulacaria afra*, a soft-wooded, semi-evergreen shrub or small tree that occurs naturally in South Africa) is being hailed as the new 'wonder plant'. It lives in both moist and dry climates, although favouring the latter, and often grows with succulents such as euphorbias. *Spekboom* (also known as 'elephant food', as the pachyderms like to eat it) has a number of amazing properties that should attract the attention of horticulturalists and environmentalists throughout Africa. Most interestingly, it has enormous carbon-storing properties and could therefore become an important 'weapon' in the fight against climate change. It can isolate and store more than 4 tonnes of CO_2 per year per hectare planted, making it more effective than the equivalent area of Amazon Rainforest at sucking CO_2 out of the air.

One of the reasons for this is that it can switch its photosynthetic mechanism to optimise performance in wet conditions (when they behave like normal plants) and in drier conditions (when they mimic desert cacti), which allows them to grow and capture CO_2 much faster than other plants. *Spekboom* is also fire and drought resistant, and its fleshy, succulent leaves have a lemon-like taste and can be used in salads. The leaves can also be chewed to treat exhaustion, dehydration and heatstroke, as well as sore throats and mouth infections. They grow to an age of over 200 years and have been shown to improve soil fertility and combat soil erosion (Anon 2020a).

In South Africa, an 11-year-old boy with autism, Chris-Tiaan Nortjé, has taken the lead by planting 4 918 *spekboom* plants and distributing over 1 000 cuttings to others to plant. He has inspired over 100 000 people on Facebook to each plant at least 10 of the 'carbon sponge' plants per year and plans to persuade every South African to do the same. The Grade 4 learner from KwaZulu-Natal, who radiates positivity despite being orphaned, views his autism as a 'superpower' that drives his passion to raise awareness of global warming and do something about it (Dzakwa 2020). African horticulturalists and climate warriors should take note – perhaps they have an indigenous plant in their country that could be used for the same purposes?

The Great Labyrinth Project featuring a large *spekboom* labyrinth has been created at East Hill Farm in Stellenbosch. Over 6 000 trees had been planted by the start of the Covid-19 lockdown in South Africa in March 2020 and another 84 000 will be planted after the lockdown – all by volunteers. According to Peter Shrimpton, the creator of the project, the campaign has been launched to raise awareness of the threat of climate change and the steps that everyone can take to mitigate the damage. 'It's about inspiring people to plant *spekboom* and reduce their carbon emissions. We hope that other African countries will be inspired to launch their own *spekboom* labyrinth projects,' he said (Dzakwa 2020). The 13-circuit labyrinth will have the same dimensions as the base of one of the Giza pyramids in Egypt (230 m × 230 m) and will replicate the design of the Chartres Cathedral Labyrinth in France.

Miracle tree: We do not know the names of the traditional healers and agriculturalists who first developed the multitudinous traditional medicinal remedies, foods, drinks and grooming products derived from Africa's rich inventory of indigenous plants. Over the centuries, they would have experimented through trial and error with wood, leaves, roots, bark, flowers, saps and fruits to find useful products, and then passed their knowledge on to the next generation by word of mouth. A plant that would have attracted their attention is the moringa tree (*Moringa oleifera*), also known as the drumstick or miracle tree. It is indigenous to Africa and Asia, and has been prized by traditional societies for centuries for its nutritional and medicinal properties.

Every part of the moringa tree is useful: the seeds can purify water; the seed pods provide a nutritious food; the bark, roots and flowers all have medicinal properties; the flowers can be used to make tea or added to salads; and the leaves are one of the most nutrient-dense food sources available. Dried flowers can be fried in moringa oil and added as flavourants to lasagna, omelettes, stir fries, pastas or pizzas. The pods and leaves provide a wide range of essential proteins, vitamins and minerals, and are rich in amino acids that are the building blocks of proteins the human body cannot make on its own. They also contain antioxidants and substances that lower blood pressure and reduce asthero-sclerotic plaque build-up. Moringa trees are also drought resistant, have exceptionally long roots that penetrate into deep water tables and can grow in a variety of soils. They are fast growing, reaching maturity in just nine months, and their leaves can be harvested six to eight weeks after planting and again six to eight weeks later.

Canna culture: Hemp has been used by humans for thousands of years – evidence of cannabis use dated at 2 700 ybp has been found in a temple in the Middle East. Industrial hemp, a variety of *Cannabis sativa*, has been hailed as a miracle crop that can generate a range of useful products, create jobs and be used to build low-cost houses. However, its 'ugly sister' marijuana (another variety of *C. sativa*), considered by some horticulturalists as a different species (*C. indica*), is hampering the development of the industrial hemp industry. Hemp and marijuana contain the psycho-active agent tetrahydrocannabinol (THC) but hemp has much lower concentrations of THC than marijuana and higher concentrations of cannabidiol (CBD), which decreases or eliminates the pyscho-active effects of THC. Despite these differences, politicians in several African countries have been hesitant to legalise industrial hemp, although its cultivation in rural areas is tolerated (Bruton 2017a). Internationally, Uruguay and Canada were among the first countries to legalise cannabis, while the use of recreational and medicinal cannabis have been legalised in several US states. In Spain, the Netherlands and Denmark, marijuana can be smoked in designated clubs. Surprisingly, in arguably the most authoritarian state in the world, North Korea, growing and smoking marijuana are widely tolerated and are even considered prestigious among young people!

In 2020, China was the world's leading producer of hemp (about 70 per cent), followed by France (25 per cent). In 2018, retail hemp sales in Africa reached about US$15 million. It is estimated that by 2022, retail sales in African markets will increase to a combined US$133 million, with Malawi, South Africa and Zimbabwe being the leading hemp producers in Africa (New Frontier Data 2018). In 2015, the Malawian government approved hemp cultivation for export on a trial basis. The company Invegrow has had success in producing hemp seed varieties and has worked with Hemporium in South Africa to bring hemp-based clothing and cosmetics to the market. Since 1994, Hemporium has also been conducting feasibility studies on hemp cultivation but has been reticent to embrace a full-blown hemp industry because of restrictive legislation.

In 2018, the South African Constitutional Court unanimously ruled to decriminalise the private consumption, cultivation and possession of cannabis. The government was given until September 2020 to bring legislation on cannabis in line with the Constitution. By February 2020, the possession of up to 600 g of dried cannabis in private homes was approved but legislation permitting the culture and sale of cannabis has not yet been passed (Harper 2020a). In early 2020, the Eastern Cape government announced that it would establish the first cannabis college in Lusikisiki, where the rich soil has supported the traditional culture of hemp for centuries. According to the spokesperson for the

Department of Rural Development and Agrarian Reform, Ayongenzwa Lungisa (Sokanyile 2020):

> The Eastern Cape has identified cannabis as one of the critical subsectors of agriculture that can be exploited to create economic growth and … employment. This college is intended to be a centre of knowledge on all aspects of cannabis. Unlike alcohol and cigarettes, cannabis has great health benefits such as treating cancer, Alzheimer's and insomnia.

In 2017, the Zimbabwean government legalised hemp production for research purposes; however, despite taking in more than US\$7 million in application fees, their programme was put on hold until feasibility studies were completed. In 2019, Zimbabwe announced that it would legalise commercial hemp production in the hope that it would serve as a substitute for tobacco, its biggest agricultural export (www.newfrontierdata.com/cannabis).

Krithi Thaver is among the 50 per cent of South African university students who has dropped out of tertiary programmes, but nevertheless he has made a name for himself by selling cannabis-infused ayurvedic oils at South Africa's first cannabis dispensary in Durban North. As it is still illegal to grow cannabis in South Africa for commercial purposes, he imports the oils from the USA. Thaver chairs the KwaZulu-Natal branch of the Cannabis Development Council of South Africa and has helped to shape the council's vision to develop a cannabis industry employing people from the country's most impoverished rural communities. He plans to grow hemp, not just for recreational and medicinal purposes but also to make biodiesel, biodegradable plastics and even sanitary pads (Viljoen 2018).

Other cannabis activists in South Africa include Myrtle Clark and Julian Stobbs (the 'Dagga Couple'); Dr Thandeka Kunene from the House of Hemp; Jason Law, the founder of the Zubenathi Trust in the Eastern Cape; Sheldon Cramer ('Bobby Greenbush') from Richards Bay, who sells his harvest through Swaziland; and Tony Budden, the co-owner of Hemporium in Cape Town that has been importing hemp fabrics and other material since 1996 and makes hemp clothing and health and nutrition products. Budden built South Africa's first hemp house using hempcrete, a concrete made from hemp fibre (Bruton 2017a).

Genetically modified (GM) cowpeas: Cowpea is an important source of protein in West Africa, and Nigeria is the world's largest producer of this crop. Yields are, however, greatly reduced by an insect pest, the cowpea pod borer (*Maruca vitrata*). Although farmers have used chemical insecticides to control the pest, these are expensive and dangerous to human and environmental health. Now an international team led by the African Agricultural Technology

Foundation in collaboration with the Commonwealth Scientific and Industrial Research Organisation in Australia and scientists in the USA have developed a GM variety of cowpea called Sampea 20-T that carries a microbial insecticidal gene from the bacterium *Bacillus thuriniensis* (Bt) (Barrero & Higgins 2020).

Bt is a soil bacterium that can produce a natural insecticide. It has been used in organic insecticides for decades following the increased resistance of many insect pests to inorganic (chemical) insecticides. Gene technology has made it possible to produce GM cowpea plants by introducing into them the genes that encode for the insect toxin from Bt. In 1995, the US Environmental Protection Agency approved the commercial production and distribution of Bt crops (corn, cotton and potato), and most of those grown around the world now are Bt varieties. In 2019, the GM cowpea variety was approved for release to smallholder farmers in Nigeria, making it the second GM crop commercialised in the country (after Bt Cotton) and the first GM food crop. This is an important step, as many African countries are still wary of GM crops even though they provide significantly increased yields and could play an increasingly important role in reducing poverty. Nevertheless, several African countries (including Nigeria, South Africa, Burkino Faso, Egypt and Sudan) have introduced GM crops, and Ghana and Niger have expressed an interest in growing Sampea-20-T. Other African countries have passed laws that make it legal to cultivate them (Barrero & Higgins 2020).

Biofertilisers and the *youkoulef* seeder: The use of symbiotic micro-organisms as biofertilisers has been shown to improve crop yields in Senegal, where soils are notoriously poor or saline, and the involvement of subsistence farmers in the development of more efficient farming methods has also reaped rich dividends (Le Quéré & Wade 2020). Researchers from the French Development Research Institute, Senegalese Agricultural Research Institute and Cheikh Anta Diop University in Dakar have combined their expertise to teach farmers how to apply biofertilisers to improve yields of cowpea, groundnut, onion, tomato, okra, watermelon, sorghum, millet, fonio, chilli and cassava. Furthermore, the production of mycorrhizal fungal inoculants has been outsourced to farmers, who earn an additional income. The farmers supply the fungal 'starters' to the Shared Microbiology Laboratory for checking and they are then sent to the inoculant producers. The inoculated seeds are planted using a *yookoutef* seeder, developed by Senegalese artisans, which streamlines the planting process and promotes the use of inoculation technology.

Locust plagues: Insect pests have decimated African agriculture and horticulture for centuries, and still pose a severe threat. In January 2020, Kenya's food production and grazing land were under threat again from a huge desert locust invasion that originated in Saudi Arabia, Yemen and Oman (Allison 2020a, 2020b). The massive swarms of desert locusts were the result of the heavy rains precipitated by three cyclones – Mekunu (in May 2018), Luban (in October 2018) and Pawan (in December 2019) – each of which caused a renewed cycle of locust breeding and growth. The swarm over Kenya alone was estimated to contain several billion insects and covered an area of over 100 km². According to Piou (2020), a swarm of 1 billion locusts eats about 2 000 metric tonnes of vegetation each day. What has made this invasion particularly tragic is that Kenya experienced its first good rains in 60 years in 2019, following cyclone Luban, and a good maize harvest was expected in 2020.

Desert locusts are among the most dangerous of all migratory pests, as they develop from hoppers (which cannot fly) into flying adults and form cohesive swarms that can cross seas and deserts. The 2020 swarm migrated from Yemen through Djibouti, Somalia and Ethiopia to South Sudan, Uganda, Kenya and Tanzania, causing the worst locust plague in 70 years. In Kenya alone, they consumed the same amount of food per day as the country's human population, and the FAO estimated that 13 million people in East Africa could go hungry as a result of the plague (Allison 2020a, 2020b). The main victims were the small subsistence farmers who stood to lose their entire crops.

To fight these pests, the Kenyan government used aerially sprayed chemical pesticides, previously used to curb the spread of invasive fall army worms in Kenya, Malawi and Ghana and tsetse flies in Botswana. Many African countries (including Botswana, Cameroon, Ethiopia, Ghana, Kenya and Nigeria) have a history of making extensive use of chemical insecticides; in 2017 alone, Nigeria spent over US$400 million on this pest control method (Ngumbi 2020). But using pesticides is a short-term solution, as the locusts can become resistant to them. They may also have severe environmental side effects because they are indiscriminate killers of all insects, including beneficial species such as bees and wasps, and may have negative effects on the health of farmers and consumers. In 2017, the UN announced that over 200 000 people (mostly from developing countries) die every year from pesticide poisoning and many European countries have banned the use of the pesticides commonly used in Africa (Ngumbi 2020). Innovative alternative solutions need to be found.

Integrated pest management is the answer. This approach includes the limited use of pesticides combined with biocontrol (using natural enemies to control pests), biopesticides (naturally occurring pesticides that have limited side effects) and cultural control (traditional practices that have been developed over time through trial and error). African countries also need to work towards preventing invasions before they take place. For example, the Horizon Scanning

Tool enables countries to generate a list of insect species that might invade from neighbouring countries so that the control authorities can prepare an action plan before the invasion takes place. African countries also need to improve their pest surveillance routines.

Piou (2020) supports a preventative management strategy whereby solitary locusts are sprayed before they create swarms. Soil moisture and the appearance of new vegetation after rain in previously dry areas, both of which can be monitored by satellites, are used to forecast the swarming of locusts. Unfortunately, countries suffering from civil war and political instability, such as Yemen, are often unable to carry out preventative control measures and are less able to react quickly when swarms develop or arrive from neighbouring countries. Countries in the Horn of Africa (Djibouti, Eritrea, Somalia and Ethiopia) are most at risk, but East African countries such as Uganda, Kenya, South Sudan and Tanzania – and even West African countries – are increasingly prone to locust invasions, depending on wind and other conditions. Countries that do not have proper preventative management systems in place need to rely on the expertise of those that do. For example, despite its economic woes, Mauritania has built an effective antilocust centre that sends dozens of field teams out every year to look for early signs of swarming and sprays an average of 20 000 ha per year (Piou 2020).

Keith Cressman has an unusual job. His official title is senior locust forecasting officer for the FAO in Rome, but unofficially he is the 'locust whisperer' responsible for predicting when and where the next plague will take place. He predicted the 2020 swarms and issued a warning to Kenya about them in mid-2019, but he did not anticipate the sheer scale of the invasion. Africa has its own 'locust whisperer', Baldwyn Torto, a chemist at the Nairobi-based International Centre of Insect Physiology and Ecology (ICIPE). About 25 years ago, he and his team taught themselves how to 'talk' to locusts through chemical cues. They analysed the chemical signals locusts use to communicate with one another and were able to decode their chemical language. They found that hoppers use different chemical cues for the adults and were able to synthesise the chemicals concerned. When they presented their mimic chemical signals to the locusts, the insects responded naturally – which gave Torto the opportunity to communicate with the locusts and modify their behaviour.

Now the ICIPE team can disrupt the locusts' development and breeding cycle by overwhelming them with adult chemical signals, which disorientate them and make them vulnerable to predators. But this remedy only works if you know where the locusts are breeding and get there in time. So Torto is working on another solution. His hypothesis is that locust eggs release a chemical scent that is the signal for them to hatch. If he can isolate and synthesise this scent, he may be able to disrupt the breeding cycle (Allison 2020b). But Cressman, Torto and other locust researchers realise that the problem is huge and may be out of human control, as global warming is likely to cause locust swarms to form more often and on a larger scale than before as the oceans warm and the frequency of cyclones and other extreme weather events increases. UN Secretary-General António Guterres has warned that there is a link between climate change and the unprecedented locust crisis plaguing East Africa. Cressman is also pessimistic about our ability to control the pests: 'These guys are professional survivors. They have so many tricks to survive the harshest weather. They've been through climate change several times and survived. How we fare is another story' (Allison 2020b). Cressman has nevertheless initiated a programme to use drones to spray insecticides, monitor locust swarms and carry out post-disaster mapping.

John Karongo, a regional agronomist for the International Committee of the Red Cross, warns that the plague of desert locusts cannot be forgotten in the race against Covid-19 because about 19 million people in East Africa are already battling hunger (the FAO has a less optimistic estimate of 25 million people) (Allison 2020b). In April 2020, another brood of hoppers hatched and soon matured into flying adults. Alarmingly, the new swarm was estimated to be 400 times larger than the original one (Byaruhanga 2020). Furthermore, the Covid-19 crisis hampered the delivery of essential expertise and supplies to the most affected areas.

Fall army worm: Caterpillars of the fall army worm (*Spodoptera frugiperda*), an invasive moth, are serious pests in North America, where they devastate crops such as maize, rice, sorghum, sugarcane, peanuts and soybean, as well as non-food crops like cotton. In 2016, this destructive pest arrived on the São Tomé and Príncipe islands and in Nigeria, and then spread into 38 African countries within just two years (Niassy 2018). The fall army worm represents a threat to food security and to the livelihoods of millions of farmers, and the potential loss it causes to agriculture in Africa could reach US$4.6 billion per year (Bateman 2019).

As is the case with desert locusts, mass campaigns to eradicate fall army worms using insecticides are a short-term and potentially harmful solution and, once again, biocontrol appears to be the better long-term solution. An evaluation of 23 possible biocontrol agents revealed that products containing neem plant extracts and the bacterium *Bacillus thuringiensis* (two of the most widely used bio-insecticides) showed the most promise. But for these bio-insecticides to work, they have to be supported by other practices such as intercropping with edible legumes, manual removal of pests, the development of insect-resistant crop varieties and traditional insect control methods such as applying ash.

Furthermore, governments have to assist farmers by subsidising biopesticides, as is already happening in Ghana (Bateman 2019). The development of early warning and surveillance systems and pheromone traps, which use the smell of a female army worm to attract males, is also showing promise (Niassy 2018). Sevgan Subramarian, the principal scientist at the ICIPE in Kenya, recently discovered a parasitoid (that is, an insect that parasitises other insects) that can be used to control the fall army worm. During extensive field trials, the parasitoid *Cotesia icipe* was found to successfully parasitise 45 per cent of fall army worms (Calatayud & Subramarian 2020).

Trapping stable flies: Stable flies (*Stomoxys calcitrans*) are serious pests in Africa – and elsewhere – that feed on the blood of their hosts (including cattle, camels, horses, donkeys, sheep, dogs and humans) and may transmit viruses and bacteria that cause diseases such as West Nile Fever, Rift Valley Fever and anthrax. Because the females lay their eggs in dung, researchers at the ICIPE decided to determine which dung they preferred in order to explore ways to manage their populations. They found that the flies favoured donkey and sheep dung, which has larger quantities of nitrogen, potassium and zinc than other dung. They isolated the chemicals characterising these dung types (β-citronellen and carvone) and added them to traps designed to catch the flies. This led to a 400 per cent increase in the number of flies trapped and a greatly improved method of controlling the pests (Weldon 2020).

Agriprotein Fly Farm: AgriProtein Fly Farm, established by Jason and David Drew north of Cape Town in 2012, won the UN's Innovation Prize for Africa in 2013 and has thrived ever since. They reckoned that chicken feed naturally on fly larvae, so why not culture them commercially for chicken (and fish) farmers? Using blood and guts scavenged from slaughterhouses, and natural cycles of organic decomposition and energy flow, they culture fly larvae and sell the protein-rich product to farmers who would normally have to use more expensive soya protein or marine fishmeal for feed. The fly larvae are not only wholesome and nutritious to chickens and fish, but also take the pressure off land and sea. Agriprotein technology also helps to tackle local waste disposal problems by rearing insects on organic waste that would otherwise go to landfill – an example of the classic win–win situation that entrepreneurs like to embrace.

AgriProtein (a Cleantech Global Top 100 company and the largest fly farming company in the world) is expanding rapidly and recently signed a deal with Saudi Arabia, where three fly farms are being built in collaboration with the Saudi technology hub Sajt. The Saudi deal follows similar agreements in the USA, South America and Africa, and a lucrative arrangement with the Twynam Group, to build 20 fly farming factories in Australasia. AgriProtein has also announced a partnership with Austrian engineers Christof Industries that will enable it to roll out its fly farm blueprint on a turnkey basis anywhere in the world.

> We're delighted that Sajt shares our vision. Fishmeal production is destroying the marine environment. Replacing it with insect meal leaves more fish in the sea for human consumption, allows the oceans to heal and reduces greenhouse gases at every stage of the supply chain from point-of-catch to point-of-sale. We estimate an environmental cost saving of USD 2000 per tonne in reduced CO_2.
>
> *Jason Drew, co-founder and CEO of AgriProtein* (Agriprotein 2017: para. 5)

Each Agripotein fly farm has a population of about 8.5 billion black soldier flies (*Hermetia illucens*), which are able to convert 250 tonnes of organic waste per day (91 000 tonnes per year) into protein that is easily assimilated by fish and poultry, with a similar amino acid and protein profile to fishmeal. Unlike houseflies, black soldier flies avoid human habitations and do not pose a threat as pests or disease vectors. In December 2016, Agriprotein won an award for its industrially scalable solution to the depletion of fish stocks in the Indian Ocean Blue Economy Challenge backed by the Australian government.

Beekeeping: Apitherapy is a branch of alternative medicine dating back thousands of years in which honeybee products such as pollen, propolis (a resin-like substance made by bees), honey and venom are used to treat illnesses. Apiculture is the business of beekeeping. Aiming to become Africa's biggest beekeeper is no easy feat but Mokgadi Mabela is well on her way to realising her ambition, as her company Native Nosi (established in 2015) expanded from 140 hives in 2017 to 650 by 2018. She launched her business through crowdfunding in 2016 when she realised that there was a need for new entrants into apiculture to ensure South Africa's food security. Her crowdfunding campaign, hosted by The People's Fund, required minimum contributions of R1 200 (the cost of one new beehive) and raised R1.2 million.

Beekeepers earn revenue by selling honey as well as by marketing pollination services, which may make up 50 per cent of their income in South Africa. More than 50 different crops are dependent on insect pollination, which is as important to horticulture as water, land and air. According to Tlou Masehela, the chairperson of the Western Cape Bee Industry Association, honey production in South Africa has dropped by 40 per cent since the 1980s due to a lack of forage, pollution, fires and droughts. A detailed plan has been put in place to transform the honey and pollination industry so that it meets consumer demands (Booysen 2018a). Wicander and Coad (2018) found that in several West and Central

African countries, beekeeping provided the fastest returns on various alternative, small-scale farming livelihoods that had been tried.

In Mauritius, the African Leadership College has embarked on an ambitious project to optimise honey production and in Uganda, the non-profit company ApiTrade Africa (which registered in July 2008) aims to develop the African honey and beeswax trade. Their mission is to lobby and advocate for a competitive apiculture sector and promote the growth and supply of quality bee products from Africa. SmartXChange (a company in Port Shepstone, South Africa) has developed the Digital Apiary Optimisation Beedale, a 'smart beehive' that offers a fully integrated hardware and software solution for beekeepers. Its management system tracks vital health and productivity metrics of a beehive and offers remote monitoring capabilities to beekeepers who want to optimise their operations and decrease their costs. Furthermore, Beedale incorporates commercial elements related to the bee and honey industry that will benefit beekeepers at any scale.

SaveTheChicken app: Poultry farmers in Cameroon now have an app (SaveTheChicken) that uses AI, big data and big data analytics to produce predictive models that function in real time and facilitate the rapid and accurate detection and diagnosis of poultry diseases (H.N. Foretia, personal communication, April 2020). The app (developed by Henry Foretia and his colleagues) is accessed via a smartphone and makes it possible for farmers to respond quickly when a disease is detected, which decreases the risk of further transmissions. The app uses data captured from different locations to provide farmers with the means to monitor and control both the birds and their environment. This is achieved in three simple steps: a farmer takes a photograph or video of the chicken or chicken's faeces and submits it via the app; the app scans, analyses and evaluates the condition; and it then provides the results and treatment advice to the farmer. Farmers can also consult a veterinarian using the platform and receive updates on best practices and vaccination reminders.

Donkey work: Donkeys are the backbone of many agrarian economies in Africa, but these hardy animals are now threatened by the illegal donkey skin trade (especially in Kenya). It is estimated that over 4.8 million donkeys are killed annually to sustain traditional 'medicine', food and grooming markets in the Far East. In particular, a traditional Chinese medicine (*ejiao*) is made from a gelatine-like substance extracted from boiled donkey skin and allegedly has life-prolonging and aphrodisiac properties. Donkey meat is also used for making snacks and beauty and anti-ageing products (Maichomo 2020). China's donkey population has decreased by 76 per cent since 1992 and they are now importing donkey skins on a large scale from other parts of the world, including Africa (Lockwood 2020a).

Kenya has become a focal point of the donkey trade with the opening of four government-licenced slaughterhouses in 2016 that have the capacity to slaughter 1 260 donkeys per day (Maichomo 2020). The increase in trade has escalated the price of donkeys from Ksh4 000 to Ksh13 000, which has also stimulated illegal trade. Trade in donkeys was legalised in Kenya in 2012, but they soon regretted the decision. In 2018, about 160 000 donkeys (8.1 per cent of the total population) were slaughtered and between 2016 and 2018, 16 544 tonnes of donkey skin and meat were exported, according to Lyne Iyadi (the information and communications officer at Brooke East Africa, an NGO that protects working horses, donkeys and mules). A report compiled by the Kenya Agriculture and Livestock Research Organisation suggests that at the current rate of slaughter, donkeys will become extinct in Kenya by 2023 (Lockwood 2020a). In late February 2020, the Kenyan government reversed its decision and decided to ban the commercial slaughter of donkeys.

The illegal trade is spilling over into adjacent countries. Donkeys are smuggled into Kenya from neighbouring Tanzania, Somalia and Ethiopia, where regulations and local outcries prevent the governments from setting up slaughterhouses. Ethiopia has the largest donkey population in Africa (about 8.8 million animals) and Kenya's latest abattoir (financed by the Chinese) is on the Ethiopian border, which fuels the illicit trade. The international smuggling of donkeys also led to an outbreak of equine flu in Niger that resulted in the death of over 60 000 donkeys in 2019. Subsequently, Botswana, Burkina Faso, Mali, Niger and Senegal all banned the export of donkey products and shut down all the donkey slaughterhouses (Lockwood 2020a).

Donkey conservation is an African problem that needs an African solution. Perhaps donkeys should be regarded as part of a people's cultural heritage rather than its natural heritage, as they are all domesticated and are so entwined in the daily lives of millions of rural people that they are an integral part of their culture. This would give people the right to fight for the well-being of their donkeys without fear of reprisal. In the Comoros, there was concern for many years over the survival of the ancient 'living fossil' fish, the coelacanth (*Latimeria chalumnae*). It was only when the Comorian people themselves declared the fish sacred, and engaged actively in its conservation, that the species was saved from extinction (Bruton 2016).

Donkeys have been domesticated for over 5 000 years, having evolved from the African wild ass (*Equus africanus*). They have been part of African culture for millennia and are an integral component of millions of households, where they play a vital role in the transport of farm produce, water, firewood and people (especially the young and aged). Now foreign greed is driving them to extinction, and the wild ass may be next on the list (Lockwood 2020a). Monicah Maichomo (the director of the Veterinary Sciences Research Institute at the Kenya Agricultural and Livestock Research Organisation) has recommended that if the donkey trade is further developed, pregnant females should not be slaughtered (27 per cent of females currently slaughtered are pregnant) and that there should be a greater focus on research on donkey breeding and on the size of animals produced. The dominant breeds of donkeys are descendants and crosses of Nubian and Somali wild asses, which are both relatively small animals. Larger breeds would provide more skin and meat if they are farmed professionally (Maichomo 2020).

Indigenous cattle: Sanga (the collective name for the indigenous cattle of sub-Saharan Africa) are now regarded as a separate subspecies, *Bos taurus africanus* (Grigson 1991). They descended initially from cattle domesticated in the Near East and, after their introduction into Africa through Egypt about 8 000 years ago, spread throughout the Sahara (which was green at the time) to West and then East Africa. In the last few hundred years, they have hybridised with local humped Zebu cattle (Grigson 1991). There are 11 breeds of Sanga cattle in Africa, including Ankole (the original African type), of which the most famous strain is the Watusi (the long-horned cattle from Burundi, Rwanda and the Congo). Now the famous long-horned cattle, which are very well-adapted to semi-arid conditions, are listed by the FAO as threatened with extinction by the introduction of exotic breeds such as Friesians.

There is also concern that crossbreeding Ankoles with Friesians may weaken their gene pool but, according to Dr Nicholas Kauta (commissioner in charge of livestock in Uganda), the sperm of Ankole bulls has been preserved in the Entebbe-based National Animal Genetic Resource Centre and Data Bank to safeguard the integrity of the species (Timothy 2018). In South Africa, there are several indigenous breeds of domesticated animals, including sheep (BaPedi, Zulu, Damara and Namaqua Afrikaner), goats (Boer, Kalahari Red, Mbuzi, Savannah White, Speckled and Lob-eared) (Kruger & Ramukhithi 2019) and an indigenous breed of dog, the Africanis (Lawson 2017a) – all of whose genetic integrity is carefully preserved.

Cattle with hindsight: An interesting experiment carried out by researchers from Australia in collaboration with Dr Weldon McNutt and Tshepo Ditlhabang of Botswana Predator Control in the Okavango Delta may solve the problem of reducing predation on cattle by lions while conserving the predators. The experiment involved cattle from 124 herds that had recently suffered lion attacks. The study covered a total of 2 061 cattle, with about one-third having a large eyespot painted on their rumps, one-third with simple cross marks on the rump and the rest unmarked. The cattle foraged in the same area and were exposed to a similar risk of predation (Jordan, Radford & Rogers 2020).

They found that the cattle painted with eyespots were significantly more likely to survive than those that were cross-painted or unmarked. In fact, none of the 683 'painted eye cows' were killed by ambush predators during the four-year study, while 15 (out of 835) unpainted and four (out of 543) cross-painted cattle were killed. These results confirmed their hunch that creating the perception that the predator had been seen by the prey could cause it to abandon the hunt. They also found that cattle marked with simple crosses were significantly less likely to be killed than unmarked individuals. Although eye patterns are widely used by animals such as butterflies, fish, frogs and birds to deter predators, no mammals are known to use this technique (Jordan et al. 2020).

Agri-fintech pioneer: Onyeka Akuma wanted to make it easy for people to invest in agriculture from the comfort of their living rooms, so he founded Farmcrowdy (a digital agritech platform that connects small-scale farmers with potential investors). Farmcrowdy gives Nigerians the opportunity to participate in farming by selecting the kind of farms they want to sponsor. Farmers use the sponsored funds to secure land, buy and plant seeds, insure their assets, train staff and sell the harvest. Farm sponsors are paid a return on their investment and keep track of the farming cycle by receiving regular updates in text, images and videos. Akuma's plan to reshape the agricultural value chain in Nigeria attracted immediate attention and he received his first angel investment of US$60 000 just a month after launching! Then, in 2019, the Oyo state government announced a partnership with Farmcrowdy that would potentially benefit over 50 000 farmers in Nigeria (Ayene et al. 2020). Akuma's idea has taken root in his home country and is likely to spread across the continent.

Female smallholder farmers: Olayinka Adegbite obtained her MSc degree in Agricultural Economics from the University of Ibadan in Nigeria and is now a PhD candidate at the University of Pretoria in South Africa. Her innovative research on women smallholder farmers in Nigeria has revealed that their lack of access to financial resources holds back their development. She found that most women farmers did not have bank accounts or own cellphones, which meant that they could not access basic financial services such as payments, remittances, transfers, savings, credit and insurance. Other factors that contributed to their financial exclusion included poverty, illiteracy, limited education and few assets.

Adegbite recommends that the gender gap in financial inclusion could be closed by identifying key performance indicators to allow women smallholder farmers to access financial services and form mutually beneficial partnerships. She also recommends that financial institutions should define options other than land (which few women own) that can be used as collateral when securing loans and that women farmers need to be integrated into agricultural value chain financial models (Adegbite, Mkandawire & Machethe 2020). Proof that this approach works can be found in Limpopo province (South Africa), where a group of women farmers is producing 27 000 boxes of table grapes per year for the European market on tribal trust land. These women, who belong to the Peace Table Grapes cooperative, won the export category of the Female Farmer of the Year award in Limpopo province in 2019. They have been empowered by a provincial agricultural development programme that includes improved finance and partnership models, better access to land, capacity building and training.

Want to hire a tractor? Kamal Yakub's idea is so obvious that it is one of those 'Why didn't I think of that?' innovations. The Ghanaian noted that, as in many other African countries, agriculture in Ghana is dominated by smallholder farmers who cannot afford to buy their own tractors so they laboriously cultivate their fields using ploughs. He established Trotro Tractors to connect farmers to tractor hirers via a smartphone app. Now a single farmer can plough an acre of land in 45 minutes instead of using four labourers to do the same job in two days, which leaves the farmer more time for other tasks and greatly increases his or her productivity and profits. Yakub is now expanding Trotro Tractors into Zimbabwe and other markets. He is a strong advocate for attracting more young people into agriculture, saying that modern technology makes it more exciting and profitable for them (Ayene et al. 2020).

Drycard: Scientists at the University of California-Davis in the USA have developed an affordable device that allows small-scale farmers in Africa to measure whether dried food is dry enough to prevent mould growth during storage. The device does not need batteries or electricity, as the farmer places the Drycard and a sample of dried product in a sealed plastic bag and, after 30 to 60 minutes, reads the card's indicator colour and compares this colour with the scale on a card. Pink indicates that the product is too wet for safe storage, and blue or grey that it is adequately dried. Drycard is already widely used in Rwanda and is likely to be adopted in other African countries.

ITIKI drought predictor: Muthoni Masinde grew up in the Mbeere farming community in rural Kenya, where she learned that farmers predict the weather (especially the threat of droughts) by relying on indigenous knowledge about natural cycles – when flowers bloom, the behaviour of mammals, or changes in the numbers of migrating insects and birds. She realised that while these natural signals are useful, they are not accurate. She therefore decided to develop a tool that uses modern science and indigenous knowledge to improve the accuracy of weather predictions for farmers who do not have access to modern, Web-based forecasts.

Her PhD at the University of Cape Town entitled 'ITIKI: The bridge between African indigenous knowledge and modern science on drought prediction' set the tone for her project. Now the head of the Department of Information Technology at the Central University of Technology in the Free State, she has developed a smartphone app and website (ITIKI) that combines indigenous knowledge and accurate weather prediction data, and acts as an early warning system for small-scale farmers. The app was officially launched at the Central University of Technology in June 2019 and is being used in South Africa, Mozambique and Kenya, where it has yielded promising results.

In each location, the ITIKI app is populated with indigenous knowledge on possible weather-related indicators which are supplemented by data from wireless sensor networks that serve as automatic weather stations, recording rainfall, temperature, humidity, atmospheric pressure, wind and soil moisture. The information is fed into computer models that use AI to downscale the forecasts to the local level, which are then sent to farmers via SMS. ITIKI is a wonderful example of two knowledge systems being combined into one device. In 2016, Masinde was the first recipient of the DSI's Distinguished Young Woman Scientist: Research and Innovation award.

Waste-to-food systems: Khepri Innovation (a 100 per cent black-owned biotechnology company in Johannesburg) has designed an insect-based bioconversion system that converts abattoir and food production waste into protein-rich animal feed, vermicompost or worm tea using black soldier fly larvae. As the system is modular, it can be rolled out at any scale. Likewise, the Nambu Group in South Africa has developed a system that turns food waste into high-quality insect protein that is a cost-effective alternative to soya- and fishmeal-based feeds in the poultry, pig and aquaculture industries.

Emerging innovators: Ntuthuko Shezi (the founder of the South African 'crowdfarming' company Livestock Wealth) has launched an app, MyFarmbook, that allows city dwellers to invest in farming using their cellphones. They can, for example, buy shares in a cow for as little as R576 and receive interest on their investment ranging from 5 to 14 per cent per year, depending on market fluctuations and other factors. Launched in 2015 with 26 cows, the project had over 2 000 cows on its books by July 2019, with 10 per cent of the investors coming from outside South Africa. Groups of investors can buy a whole cow, while individuals can buy shares in a pregnant cow or calf. According to Wandile Sihlobo, an economist with the agribusiness association Agbiz, many people in cities who have an interest in agriculture but cannot participate in it can now do so through MyFarmbook. Nontokoozo Sabela, a small business consultant, brought the term 'cash cow' to life when she bought her first cow through

MyFarmbook in 2016 and earned R6 000 from it within two years. Shezi is now planning to expand his business into the produce market after launching a vegetable growing scheme in which urbanites can invest (*The Citizen* 2019).

Emerging innovations: The Cape Town-based start-up Aerobotics is set to save the agriculture industry millions of dollars using satellite imagery and drone technology. Aerobotics was launched in 2015 by James Paterson and Benji Meltzer, and focuses on analysing data on tree crops such as citrus, other fruits and nuts, and reducing crop losses due to pests and diseases. They have developed software that uses layers of data from satellite and drone imagery to map individual trees to identify those suffering from a lack of water or food, or from diseases or pests. The software includes a Google Map-type app that directs farmers to individual trees requiring attention.

While many drone services have been offered to farmers to provide an aerial overview of their crops, Aerobotics claims to be ahead of the game globally by developing unique software that uses aerial data to individually identify and analyse every tree on the farm. The potential of Aerobotics has been recognised by Nedbank Corporate Investment Banking's (CIB) Venture Capital Fund, which has bought a stake in the start-up. According to Nedbank CIB's Head of Disruption and Innovation Stuart van der Veen, this new division was launched to 'experiment with frontier technology' and to 'create alternative futures' for their clients. From Paterson's perspective, farming has moved from the age of mechanisation to the age of digitisation in which data will be analysed to optimise yields. He predicts that the next step will be to enable work at leaf level so that pests and diseases can be identified before they affect the whole tree (Kretzmann 2018).

Discussion: Today multinational giants such as Corteva Agriscience offer a wide variety of cutting-edge technologies to African farmers, from high-quality seeds to crop protection solutions, environmentally-friendly insecticides and herbicides, and digital services that use data and analytics to develop cropping models, improve teamwork efficiency and communication, maximise yields and quantify profits. It is widely anticipated that inexpensive agricultural robots will soon do the hard work in the fields, which will release farmers to be managers. Laboratory-grown meat could become mainstream within 10 years and insect protein will be more widely consumed, possibly disguised behind innovative pseudonyms.

There is increasing evidence that politicians and policy-makers throughout Africa are showing renewed respect for the key roles that farmers play in ensuring food security, providing jobs and supporting the economy while protecting the natural environment. The attempts in 2019 by then Australian Home Affairs Minister Peter Dutton to attract South African and Zimbabwean farmers to his country through fast-track visas were a below-the-belt move that threatened to steal vital skills from Africa at a time when we needed them most.

There are many initiatives in Africa that promote the interests of farmers. The Future Farmers Foundation, based in Howick (KwaZulu-Natal, South Africa), provides platforms from which young men and women in agriculture can become successful commercial farm managers or farmers. Its apprenticeship system places learners on farms where they gain experience and learn skills such as driving tractors, operating milking machines, controlling irrigation systems, and dairy herd management and leadership. The apprentices who excel are selected to do a yearlong internship in Europe, Australia or the USA, where they are exposed to high working standards and a different work ethic. They then take their new skills and experience back to the countries they represent.

3

Terrestrial biodiversity conservation

'We cannot afford to have the same mode of recovery, the same mode of doing business, the same mode of economic activity.'

Juliet Kabera, director-general of the Rwanda Environment Management Authority (Worland 2020)

Introduction: If 2019 was the year in which climate change action was factored into global financial planning, 2020 was the year when biodiversity claimed a seat at the global 'green finance' table. Biodiversity loss was flagged as one of the top five global threats at the 2020 World Economic Forum Annual Meeting in Davos (Switzerland), an indication that investors were moving 'beyond carbon' to assess the impact of companies and countries on global biodiversity. But biodiversity is difficult to shoehorn into a market-based financial framework, and assigning monetary values to the resources and services provided by nature is problematic and controversial. Financial models are poorly suited to capturing the non-linear tipping points, feedback loops, and intricate and everchanging systemic connections characterising natural ecosystems and biomes. Also, biodiversity-related projects may not yield returns that are monetisable. For example, wetland restoration delivers real economic and social benefits (flood defences, carbon sequestration, water purification and storage, wildlife habitat and recreation), but these benefits are not easily translated into an income stream. Biodiversity gains may also take decades to be realised, which is an unappealing risk–return profile for most investors and politicians, and the small and localised nature of many projects makes them unattractive to big global investors.

All the above point to the importance of involvement by local people and economic systems in the management of scarce biological resources. Many indigenous people, especially in Africa and South America, have developed their own commons-based governance systems to sustainably use and manage natural resources *in the long term*. Instead of the top-down, data-orientated methods advocated by the West, such approaches use a sophisticated combination of indigenous knowledge, local rules and adaptive management to steward complex ecosystems. This multistakeholder approach has merit, as many of the products and services of nature are public goods that are part of the commons and should not be privatised. Furthermore, the public sector can play a vital role by fostering commons governance and being a reliable, long-term source of capital and policy support (Kedward 2020). Incorporating ecological concerns into mission-orientated public policy can also force the private sector to take the issue seriously.

Early human threats to biodiversity: Humans have been a threat to biodiversity for aeons. The 'overkill theory' proposed by US palaeo-biologist Paul Martin (1967) posits that early humans were mainly responsible for the extinction of the African megafauna. He suggested that early *Homo sapiens* as well as its predecessors, *H. erectus* and *Ardipithecus* species, played a major role in the disappearance of mammoths, giant sloths, wolves and hyenas, woolly rhinos, sabre-tooth cats, and the okapi-like *Sivatherium* from Africa and elsewhere during the Pleistocene era from about 2.58 million ybp until just over 11 700 ybp. Their destructive ability was made possible by their controlled use of fire, coordinated tactics by groups of hunters and the use of tools, and was aided by non-human interventions such as lethal solar flares, disease outbreaks, volcanic eruptions and meteorite impacts.

However, a new generation of scientists with more accurate chronological data is questioning this theory and suggesting that climate change (not caused by humans) was the main cause of early megafaunal extinctions (e.g. Du Plessis 2020; Faith et al. 2018; Nagaoka Rick & Wolverton 2018). They have proposed that early megaherbivore extinctions may rather have been caused by loss of habitat brought about by the shift from forested terrain to grassy savannah beginning about 4.6 million ybp due to natural climate change. Today, the situation is different because destructive interventions by humans (including habitat loss, the impact of invasive species, overexploitation and pollution), combined with the climate changes that they have brought about, are causing the most catastrophic extinction event since the Mesozoic era. But this time, we can do something about it.

The first African game reserves: The Hluhluwe-Imfolozi Park, originally a royal hunting ground for the Zulu kingdom, was established in 1895 and is the oldest proclaimed terrestrial nature reserve in Africa. It was also the forerunner for the proclamation of several other protected areas in the region. Then, from

1896 onwards, the twin disasters of rinderpest and nagana (diseases in domestic livestock that are spread by wild game) caused the South African government to launch a campaign to eliminate big game in Zululand (Minnaar 1989). Predictably, the policy failed but the wildlife of Zululand was decimated.

The Kgalagadi Transfrontier Park, which was brokered by the Peace Parks Foundation and established in 2000, was the first transfrontier park in Africa. Its 3.6 million ha join the Kalahari Gemsbok National Park in South Africa to the Gemsbok National Park in Botswana, a vast conservation area larger than Belgium. An even more ambitious venture is the Kavango–Zambezi Transfrontier Conservation Area, which is centred at the confluence of the Zambezi and Chobe Rivers where the borders of Botswana, Namibia, Zambia and Zimbabwe meet, and is being brokered by the Peace Parks Foundation and WWF International. It includes a major part of the Upper Zambezi River and Okavango Delta and catchment, the Zambezi Region of Namibia, south-east Angola, south-west Zambia, northern Botswana and western Zimbabwe. In November 2006, a memorandum of understanding was signed by the five partner countries to work towards the establishment of the park, which will be Africa's biggest conservation area and the world's largest transfrontier park.

Gorongosa National Park: During the transition from colonialism to democracy, many African game reserves suffered at the hands of poachers, with the majestic Gorongosa National Park in Mozambique being a classic example. It flipped from a wildlife haven into a killing ground and then back to a sacred area for peace-making and reconciliation. Battles raged in Gorongosa during Mozambique's civil war (1977–92), leaving large numbers of people dead or wounded. The battles also ruined the natural ecosystem and resulted in the killing of 90 per cent of the elephants, buffalo, zebra and wildebeest as soldiers poached them for money or meat. At the end of the war, just 15 buffalo, about 100 hippos and a few lions survived (Desalegn 2020).

With peace came the opportunity to rebuild the natural environment and the lives of over 100 000 people. The reserve was re-envisioned in 2004 as a 'human rights park' (i.e. one that protects wildlife while investing in local communities). Its rebirth is the result of a co-management agreement between the Carr Foundation, an American NGO and the Mozambican government whereby about one-third of its budget goes to community development programmes. By 2018, grasslands, shrublands and forest were recovering rapidly, over 1 000 buffalo could be seen roaming the area and the hippo population had increased fivefold. By the time cyclone Idai struck in 2019, healthy wetlands and terrestrial ecosystems had developed that could absorb the resulting floods and the wildlife had increased by 700 per cent (Matthews 2019).

Programmes initiated by the Gorongosa Restoration Project have restored the balance between nature and people by improving the welfare and health of local human communities, restoring essential ecological processes and ecosystems, and increasing work opportunities through ecotourism. The restoration of Gorongosa has demonstrated that the protection of the natural environment is central to sustainable development, the mitigation of climate change, and the development of peaceful and healthy human societies in Africa.

Hailemariam Desalegn (a former prime minister of Ethiopia and now a leading African conservationist) points out that despite the success at Gorongosa, natural areas are being devastated at an alarming rate throughout Africa regardless of the fact that the continent generates 62 per cent of its GDP from industries such as agriculture and ecotourism that are highly or moderately dependent on nature. He further notes that the disintegration of natural systems leads to the breakdown of human societies and to conflict. Ecosystem degradation, biodiversity loss, climate change and other human-caused environmental changes now affect more than 458 million people in Africa and cost an estimated US$9.3 billion annually (Desalegn 2020). It is therefore imperative that African people take the lead in environmental conservation efforts and that they partner with international campaigns that address this critical global issue.

Women rangers: Sergeant Vimbai Kumire and her all-female wildlife ranger team, the *Akashinga* ('brave ones' in Shona), have been trained to protect the wildlife in a vulnerable part of the Zambezi Valley in Zimbabwe. The rangers are an arm of the non-profit International Anti-Poaching Foundation that manages the Phundundu Wildlife Area, a 300-km² former trophy-hunting tract which has lost thousands of elephants and other wildlife to poachers over the last two decades. The Akashinga patrol Phundundu, which borders 29 communities, with rifles because they must be prepared to face heavily armed poachers, although some conservationists argue that arming them increases the threat of violence. The Akashinga's founder, Damien Mander, disagrees: 'With the women, [the rifle] is more of a tool. With the men, it's more of a toy' (Smith 2019).

Mander, an Australian and former Special Forces soldier who has trained game rangers in Zimbabwe for more than a decade, believes that change – be it peace among humans or attitudes towards wildlife – cannot happen without the buy-in of the local community. With that local-first mentality, he turned to women in the villages surrounding Phundundu to fill the ranks of the Akashinga. After years of training male rangers, he concluded that, in some ways, women are better suited for the job because they are less susceptible to bribery from poachers and more adept at de-escalating potentially violent situations.

He knew that working women invest 90 per cent of their income in their families, compared with only 35 per cent by men, and therefore demonstrate a key conservation principle – wildlife is worth more to the community alive than it is dead at the hands of poachers (Smith 2020). Mander sought women who had suffered trauma: AIDS orphans and victims of sexual assault or domestic abuse. Who better to task with protecting exploited animals, he reasoned, than women who had suffered from exploitation themselves? He modelled his selection course on Special Forces training and the women have responded magnificently. A National Geographic documentary film produced by David Cameron, *Akashinga: The Brave Ones*, has immortalised their contribution to biodiversity conservation.

The wildlife in the locally-owned Mara Naboisha Conservancy at Leopard Hill in Kenya is stunning but the most interesting feature of the reserve is that the all-Kenyan guide staff include three trailblazing Maasai women who attended guiding school and learned to drive 4×4s with the help of Basecamp Explorer, Leopard Hill's parent company and a staunch supporter of female guiding. In South Africa, the all-women, unarmed Black Mambas Anti-Poaching Unit was founded in 2013 by Transfrontier Africa NPC to protect the Olifants West Region of the Balule Nature Reserve adjacent to the Kruger National Park. Within a year, they were invited to expand their activities into other regions and now protect all the boundaries of the reserve, which is part of the Greater Kruger Area, where their patrolling has reduced snaring and poaching by 76 per cent (Goyanes 2017). The strategy of Craig Spencer, Black Mambas founder, is to saturate the landscape with rangers, make them visible wearing badges, practice early detection and then call armed response if necessary. The rangers endeavour to make their area the most difficult and least profitable place for poaching, and aim to address the social and moral decay within communities brought about by the false economy of poaching (Goyanes 2017).

> I've been doing this for 24 years and have never had to raise a weapon to a wild animal. The poachers would have to consider defending themselves against these women. Creating orphans and widows is not the answer to this problem. You can't shoot this problem away.
>
> *Craig Spencer, Black Mambas founder and head warden at Balule Nature Reserve*
> (Goyanes 2017)

African climate change warriors: While Greta Thunberg of Sweden is the leading young eco-warrior internationally and Klaus Schwab (founder and executive chairperson of the WEF) is even talking about the 'Greta Effect', several young

African women are also leading the charge (especially since they were snubbed by the Davos conference in January 2020). They include Vanessa Nakate of Uganda, Makena Muigui of Kenya, and Ayakha Melithafa and Ndoni Mcunu of South Africa (Lundahl 2020). Mcunu, a PhD student at the University of the Witwatersrand (Wits University) in Johannesburg, has pointed out that almost 20 million people have fled the African continent due to the repercussions of climate change and major droughts have caused almost 52 million others to become food insecure. She has emphasised that Africans have begun to adapt using indigenous knowledge combined with international research on climate change. Nakate, a business administration graduate, stated during a video conference from Kampala to Davos in January 2020, 'This is the time for the world to listen to the activists from Africa and to pay attention to their stories. This is an opportunity for media to actually do some justice to the climate issues in Africa' (Lundahl 2020). Muigui, Nakate, Melithafa and Thunberg have all emphasised that Africa is essentially blameless when it comes to climate change. Although the continent is home to 17 per cent of the world's population and over a quarter of its nations, it accounts for only 5 per cent (some say only 2 to 3 per cent) of greenhouse gas emissions; yet it is most impacted by climate change (Lundahl 2020). According to the World Resources Institute, in 2000 Africa's per capita emissions of CO_2 were 0.8 metric tonnes compared to the global figure of 3.9 metric tonnes (United Nations 2006).

Campaign for Nature: The Campaign for Nature offers a science-driven, ambitious new deal for nature that calls on world leaders to protect at least 30 per cent of the planet (land and water) by 2030. The so-called 30×30 campaign is a partnership of the Wyss Campaign for Nature, the National Geographic Society, and a growing coalition of more than 100 conservation and indigenous peoples' organisations around the world. The campaign has also launched a High Ambition Coalition for Nature and People, composed of government leaders, to drive high-level action on the 30×30 campaign. It calls on world leaders to mobilise financial resources to effectively manage protected areas and to fully integrate and respect indigenous leadership and rights in conservation work.

Many African countries have already committed to conserving their natural heritage as an integral component of sustainable development. For example, the governments of Rwanda and Uganda have resolved to protect 30 per cent of their natural lands and wetlands by 2030, and Namibia has designated its entire coastline a national park. Ethiopia has written environmental protection into its constitution so that every political party that comes into power must act accordingly (Desalegn 2020). The Gorongosa National Park has shown us that nature can recover quickly if it is given a chance. It is our responsibility to ensure that sufficient natural areas in Africa are given an equal chance to recover, for the benefit of all its inhabitants.

Conservation outside protected areas: In Kenya, about 19 per cent of the land area is protected in national parks, reserves and conservancies; however, recent research has shown that only 16 per cent of amphibians, 45 per cent of birds and 41 per cent of mammals are adequately protected in these controlled areas. Furthermore, 80 of the 1 535 animal species surveyed were not protected at all, including the critically endangered Taita warty frog (*Callulina dawida*) (Tyrell 2019). As in many African countries, a disproportionate amount of conservation attention is focused on large mammals in protected areas. To offer more protection to smaller wildlife outside protected areas, Kenya is embarking on a new strategy to identify and conserve biodiversity hotspots in rangelands, urban environments and forest patches on farmland. The Kenyan conservation authorities are also putting more effort into compiling accurate national Red Lists of Threatened Species, as has been done in Uganda and South Africa, so that conservation efforts can be based on accurate scientific data. A landscape-based approach to conservation, pioneered by McHarg (1969), is being adopted to balance competing land uses in a way that is best for both humans and wildlife (Tyrell 2019).

Africa's tree crusader: Kenyan Wangari Maathai was the first woman in Central and East Africa to earn a PhD (in 1971) but it was her work among rural people that earned her international accolades, including the Nobel Peace Prize in 2004 (Maathai 2010). In 1977, she founded the Green Belt Movement to teach peasant women how to earn a living by planting trees. This project, which she initiated in response to the wholesale deforestation of Kenya, was seen as a threat by land-grabbing politicians and she spent International Women's Day in 2001 in prison. Undeterred, she actively pursued the links between environmentalism, poverty reduction and democratic rights, and won a seat in the Kenyan parliament in 2002 with 98 per cent of the vote (Vick 2020).

In recognition of her deep commitment to the environment, the UN secretary-general named Maathai a UN messenger of peace in December 2009, with a focus on the environment and climate change. In 2010, she was appointed to the Millennium Development Goals Advocacy Group, a panel of political leaders, businesspeople and activists established to galvanise worldwide support for the achievement of the Millennium Development Goals. In 2010, in partnership with the University of Nairobi, she founded the Wangari Maathai Institute for Peace and Environmental Studies to bring together academic research and peace studies. By the time of her death in 2011, her Green Belt Movement had launched branches in 30 countries, planted over 50 million trees and changed the course of forest conservation in Africa.

CO₂ absorbing forests: While the world's attention is mainly focused on the CO_2 absorbing properties of the Amazon Rainforest, little attention is given to the role of Central African forests in this regard. Due to land reclamation, deforestation, agricultural expansion, road building and other interventions, the amount of CO_2 absorbed by the Amazon Rainforest has halved since the 1990s and 20 per cent of its area (that which has been deforested) is now a carbon source rather than a carbon sink. Recent research in Bolivia by Iokine Rodriguez and Mirna Inturias (2020) revealed that those areas of the Amazon Rainforest that are under the control of indigenous people (52 per cent of the total area) store 58 per cent of its carbon and that 90 per cent of net CO_2 emissions come from outside these protected lands. Now a range of threats to the autonomy of indigenous people, and their right to manage their own land, in Bolivia and elsewhere in Amazonia have further reduced the rainforest's ability to act as a significant carbon sink (Rodriguez & Inturias 2020), which places even more importance on the carbon sequestrating power of African forests.

The tree that bleeds: Haidor el Ali's passion is protecting the rosewood trees (genus *Dalbergia*) of Senegal, but he is fighting a lone battle. When the rosewood tree is cut, it bleeds a bloodred sap – which is appropriate, as it is one of the most trafficked plants on Earth even though it is internationally protected. The Fafa Kourou Forest near Pata Village in southern Senegal is one of the last strongholds of this valuable tree, but it is being exploited ruthlessly by Chinese traders who (having exhausted the rosewood stocks in Madagascar and The Gambia) have turned to Senegal, where the trees are felled and then smuggled into neighbouring The Gambia and shipped to China. Ironically, The Gambia is now one of the fifth largest exporters of rosewood trees, although its own forests have been decimated. The million-dollar trade in rosewood is based on the tree's dark brown timber that is used for making luxury furniture, guitar fretboards, marimbas, recorders, billiard cues and flooring.

Aeumumbo Diedhiou, the head of Women of the Sacred Forests in Pata, is another avid campaigner for rosewood conservation. In an interview with the BBC in March 2020, she stated that her organisation recognised that trees play an ecological role because they bring rain. She also pointed out that they are important for the spiritual well-being of her people, yet she had witnessed her own people chopping them down for cash.

In vitro cultivation of endangered plants: The artificial cultivation of endangered plants is becoming an increasingly important biodiversity conservation tool, as many plants are threatened by urban and industrial development; overgrazing; and illegal harvesting for horticulture, trading and medicine. Scientists at the University of Limpopo in South Africa have developed techniques for the in vitro culture of *Euphorbia groenewaldii* and *E. clivicola*, which are indigenous to Limpopo province. In vitro technology provides a complete protocol for shoot multiplication, rooting and acclimatisation of plantlets under controlled conditions, and has the potential to facilitate the mass production of economically valuable succulents that are rare and/or difficult to propagate without posing a threat to natural plant populations.

Snake whisperer: We have heard of horse, elephant and locust whisperers, but Africa also has a snake whisperer in Toy Bodbiji (consulting ecologist for ECO-ED Environmental Education and Training in KwaZulu-Natal, South Africa). Bodbiji's main concern is for gaboon adders (*Bitis gabonica*), which are threatened by dune mining, land invasions and forestry plantations, and are also vulnerable to sugarcane burning operations. He rescues injured snakes, rehabilitates them and releases them back into the wild. Gaboon adders occur in 21 African countries from Benin to South Africa, and are particularly common in Gabon. They are ambush predators that take large prey. Bodbiji has seen a 2 kg male snake take a 2.17 kg large-spotted genet (Lockwood 2020b). In South Africa, iSimangaliso Wetland Park in northern Zululand provides protection for the species, which occur in the coastal dune forest and in Dukuduku Forest near Mtubatuba (Bruton & Haacke 1980), but they are threatened elsewhere in Africa. Although they are deadly, with 5-cm-long fangs that store a large amount of cytotoxic venom, gaboon adders are iconic African animals that deserve to be protected.

Studying snakes via Facebook: We know that all snakes are predators, but it is difficult to study their secretive, infrequent and unpredictable feeding behaviour in the wild. Knowing what snakes eat helps scientists to understand ecological connections between them and other species, and leads to a better understanding of how ecosystems work. Bryan and Robin Maritz, who work at the University of the Western Cape in South Africa, decided to use Facebook groups to learn about what snakes eat because they found that this was the quickest and most accurate way to collect this elusive data. They have already gathered over 1 100 feeding records, which represent 27 per cent of all scientifically documented snake feeding records in southern Africa, with more than 70 per cent of their snake victims not having been recorded previously in the scientific literature.

Their findings highlight the remarkable power of citizen science to reveal undocumented details about the natural world. In the case of snake diets, harnessing thousands of social media users facilitates data collection. Social media, and the widespread use of smartphones with cameras, means that even difficult-to-observe events can be recorded in large numbers and across wide geographic zones. They are planning to extend their research beyond southern Africa into the rest of Africa (Maritz & Maritz 2020).

Vulture conservation: Vultures are among the most endangered birds in Africa, which is especially problematic because as scavengers they play a vital role in natural ecosystems. Scientists at the University of Cape Town studied 143 vulture feeding sites across South Africa and found that these 'restaurants' provide over 3 300 tonnes of meat per year – enough to feed almost the entire regional vulture population. Vulture restaurants were started in the 1970s in Europe, Africa and Asia as a conservation measure to save the endangered birds because it was hoped that the feeding sites would limit their exposure to poison-laced carcasses put out by farmers or poachers intent on killing them. The sites are mainly used by species with wide home ranges, such as African white-backed and Cape vultures, but appear to play an important role in sustaining populations. Feeding sites have also been used in vulture conservation in Botswana, Zambia and Kenya.

All six vulture species that occur in West Africa, and all vultures elsewhere in Africa except the palm-nut vulture, are endangered or critically endangered. Research by Beckie Garbett of the Fitzpatrick Institute of African Ornithology at the University of Cape Town has revealed that most species have decreased by over 90 per cent in recent years and over 4 000 critically endangered hooded vultures have died from poisoning by agricultural pesticides in Guinea-Bissau since 2017 (Garbett 2020). This is the largest incidence of mass vulture deaths in the world and represents a loss of about 20 per cent of the population of hooded vultures in Guinea-Bissau, which is 22 per cent of the global population. Losses of this magnitude of an endangered species are a huge setback for conservation in Africa. Many of the dead vultures were beheaded, which suggests that their body parts were harvested for 'medicinal' or belief-based purposes. In June 2019, 530 endangered vultures were killed in Botswana after feeding on an elephant carcass poisoned by poachers (De Greef 2019).

Addressing the complex threats to African vultures requires a multipronged approach. In Nigeria, Birdlife International and the Nigerian Conservation Foundation are working with traditional healer associations, hunters and wildlife traders to advocate for the use of plant-based alternatives to vultures in traditional medicine. Nature Kenya has established a model that involves creating antipoisoning networks within local communities to rapidly detect and respond to poisoning incidents. In January 2019, Kenya made wildlife poisoning a crime punishable with a fine of about US$50 000 or five years in prison, which is a major deterrent (Garbett 2020). If these measures were introduced throughout Africa, vultures may be able to continue playing their vital role in nature. If not, they are doomed.

Ground hornbills in South Africa: Dr Lucy Kemp (the director of the Mabula Ground Hornbill Project in South Africa) adopts a strictly scientific approach to the conservation of these iconic birds, which are endangered in South Africa and threatened throughout their range in sub-Saharan Africa. She uses scientific evidence to formulate on-the-ground conservation action that takes into account socioeconomic realities, cultural sensitivities and conservation biology. Previously Kemp worked on conservation projects for black rhino, African wild dog and cheetah; community-based natural resource management; and food security for communities living in national parks in Namibia and South Africa. Southern ground hornbills have always been part of her life because her parents, Alan and Meg Kemp, did much of the early research on this species in the Kruger National Park and her childhood was filled with adventure while she accompanied them on their research trips. Now, as a professional conservation biologist, she believes that this flagship species is an excellent candidate for testing conservation tools in the savanna biome and for connecting people through a common conservation goal.

Pangolins in danger: Pangolins are nocturnal, insectivorous mammals that are highly susceptible to stress and generally die in human care. They are hunted for their large scales that are made of keratin, the same substance in rhino horn and human fingernails. Between 2017 and 2019, seizures of pangolin scales tripled in volume, with 97 tonnes of scales (equivalent to about 150 000 animals) being intercepted leaving Africa. Their scales are used in cultural rituals in Africa and traditional medicines in Asia; they are also trafficked for food. The giant pangolin (the largest of the eight pangolin species) is native to Central Africa and is classified as endangered by the International Union for Conservation of Nature (IUCN), with more than 1 million animals poached between 2004 and 2014 (Braczkowski et al. 2020).

The intense global trafficking of pangolins (the only mammals covered in scales) means that the entire order Pholidota is threatened with extinction. Their survival is dependent on the ingenuity and determination of African conservationists, such as those associated with the Sangha Pangolin Project in the Central African Republic and the African Pangolin Working Group (APWG) in South Africa. Each year, the APWG retrieves between 20 and 40 pangolins, which are admitted to the Johannesburg Wildlife Veterinary Hospital for medical treatment and rehabilitation before they are released. In 2019, seven rescued Temminck's pangolins were re-introduced into the Phinda Private Game Reserve in KwaZulu-Natal.

Okapis in the DRC: The Okapi Wildlife Reserve in the Ituri Rainforest is home to about 5 000 of the estimated 30 000 okapis that survive in the wild – all in the DRC. Although *Okapia johnstoni* (a forest giraffe) only came to the world's attention when the explorer Harry Johnston collected the first specimen in 1901, it is well known to the Congolese (especially the Mbuli pygmies who share the forest with them) and is to the Congo what the giant panda is to China or the kangaroo is to Australia. Many innovative and determined people have contributed to the struggle to conserve the okapi, including John Lukas, who co-founded the Okapi Conservation Project in 1987 with Karl Ruf, Jean Nlamba and conservation biologist Kambali Sambili. Tragically, the latter three were all killed in a car accident in Uganda in 2003 while returning from negotiating with a rebel group that was in control of Epulu at the time.

Okapi conservationists (including rangers from the Congolese Institute for Nature Conservation) have had to deal with attacks by armed Mai Mai rebels, poachers, bushmeat hunters, illegal miners and loggers, armed militias, slash-and-burn farmers and a six-year civil war (1997–2003) in their efforts to save from extinction the 'ghost of the forest' and the many other inhabitants of the Okapi Wildlife Reserve (including gorillas, chimpanzees and forest elephants). The population of okapis has decreased by over 50 per cent in the last 25 years, and they are now listed as endangered by the IUCN (Dasygupta 2017).

Hyena men: One of the more bizarre wildlife experiences in Africa is witnessing the nightly show of the 'hyena men' near Harar in Ethiopia. As night falls, they take up positions outside the city and throw meat, offal and bones to packs of wild hyenas that slink down from the surrounding hills, apparently responding to names that the hyena men chant. Slowly the wild beasts appear out of the darkness, their eyes glowing and jaws slavering, until they are close enough to the hyena men to take food from their hands. Occasionally fights break out and throaty growls mingle with their eerie, demented giggles. This is a uniquely African experience (Hancock, Pankhurst & Willets 1997).

Controlling elephants with bees: African elephants can inflict damage on farm crops, infrastructure and the habitats of other species, and their movements sometimes need to be controlled. Culling is not practical or desirable in many circumstances, so non-lethal methods of control have been developed. As elephants are strongly deterred by African honeybees (*Apis mellifera scutellata*), techniques have been developed to protect farms and trees from elephants using beehives as well as recordings of honeybee sounds. However, both methods are difficult to implement on a large scale. Researchers in the Jejane and Maseke Nature Reserves adjacent to the Kruger National Park in South Africa have taken this method one step further by developing an elephant repellent based on the volatile alarm pheromones released by honeybees. Their research showed that the method, which uses natural cues to which elephants have an evolved response, holds potential for the development of new options for controlling elephant movements. They found that the elephants were not stimulated to bolt in response to the pheromones but showed a calm response, slowly moving away from the affected sites, which is a desirable degree of deterrence for large animals (Wright et al. 2018).

Elephant contraceptive: Elephants are keystone species that transform their savanna habitats and influence ecosystem function. They can be destructive if their populations are left unchecked, as they may decimate woodlands such as mopane forests in Botswana and make ecosystems uninhabitable for other species. Historically, culling has been the standard method of population control, but increasing awareness of the intelligence of elephants – and their complex social systems – has led to public pressure to have culling banned in some African countries (it was banned in South Africa in 1995). The search for an alternate method of population control has been difficult but in 1999, researchers in South Africa revealed that the *porcine zona pellucida* (pZP) vaccine could be used to stop elephant cows from conceiving. The drug provokes an immune reaction in cows, causing them to produce antibodies that bind to the surface of their eggs, which prevents the sperm from binding onto and fertilising the egg. In contrast to hormonal contraceptives, pZP is efficient, reversible, safe, remotely deliverable, does not require animal immobilisation and has minimal impact on elephant social behaviour.

The application of pZP has been tested in the Addo Elephant National Park and Welgevonden Game Reserve in South Africa, and found to be highly effective. The vaccine is administered from helicopters, with each cow being darted with a biodegradable dart that falls out once the contraceptive has been injected into the blood stream. The intelligent dart ensures that no cow receives a double dose by spraying a coloured dye onto the rump of each darted elephant (Oosthuyse 2020). It is hoped that the pZP contraceptive will soon be used throughout Africa.

Elephant conservation in Chad: The Zakouma National Park in Chad is home to one of the most stunning conservation success stories in Africa. During a period of political unrest, poachers took advantage of the country's lawless state and massacred 90 per cent of the park's elephants as well as some of the rangers. As the park manager from 2001 to 2007, Luis Arranz lost seven rangers killed by poachers and was twice visited by the rebels. Now Zakouma has one of the largest herds of elephants in Africa and even black rhinos are making a comeback.

After taking over Zakouma's management in 2011, Rian Labuschagne and his team from African Parks (a South Africa-based public–private consortium specialising in managing and rehabilitating failing protected areas) transformed it into a haven for Africa's imperilled elephants. The Labuschagnes, who moved to Tanzania in February 2017 after six years at Zakouma, laid the foundation for visitors to return to the park by creating Camp Nomade, a luxury safari camp, the first of its kind in Chad. As its name suggests, the camp's aesthetics match those of the surrounding nomadic communities and it is also mobile, which allows it to move with the wildlife on the Rigueik floodplain (Nuwer 2018).

Elephant rebound: Elephant populations in the Amboseli National Park in Kenya are experiencing a boom, with more than 170 calves born in 2020 (to mid-August), including two sets of rare twins; 113 new calves were born in 2018. Furthermore, the country's total elephant population increased from 16 000 in 1989 to 34 800 in 2019. The recent boom is attributed to heavy rains, which makes more forage available, and increased antipoaching efforts during the Covid-19 lockdown. According to Kenya's Cabinet Secretary for Tourism and Wildlife Najib Balala, 80 elephants were poached in 2018 and 34 in 2019. 'Our number of poached elephants from January to today [14 August 2020] has been seven,' Balala said. 'We regret it has been seven' (Saldivia 2020).

Madagascar's lemurs: There are 111 species of lemurs in Madagascar, but many of these highly unusual primates are threatened with extinction through habitat loss, logging, slash-and-burn farming, wildlife trafficking and poaching in this island country's severely depleted forests. Today, only about 10 per cent of Madagascar's original natural vegetation remains intact and 90 per cent of its lemur species are threatened with extinction (Mittermeier & Mittermeier 2019). One of Madagascar's most innovative and passionate lemur conservationists, Jonah Ratsimbazafy, launched the World Lemur Festival to focus international attention on the plight of these primates – the most endangered group of land mammals in the world. Every October, communities across Madagascar celebrate with a week-long festival that is connected internationally through zoos, the IUCN's Save our Species initiative and the Lemur Conservation Network (LCN), which unites over 50 conservation organisations working to protect lemurs in Madagascar. The LCN works closely with local communities to implement a lemur survival action plan and protect lemur habitats. Prominent lemur conservationists and researchers include Joel Ratsirarson at the University of Tana who, together with Alison Richard, has studied lemurs at Beza Mahafaly Scientific Reserve for over 40 years; Jurg Ganzhorn (University of Hamburg in Germany); Russell Mittermeyer (formerly of Conservation International); Patricia Wright at Stony Brook; and the late Alison Jolly (primatologist and children's book author).

Silverback mountain gorillas in Uganda: In June 2020, poachers killed a 25-year-old silverback mountain gorilla, Rafiki, in Uganda's Bwindi Impenetrable Forest National Park. Rafiki's death was a major setback, as he had been the leader of a group of 17 gorillas since 2008. The poacher claimed that he had stabbed the gorilla in self-defence but, under Uganda's stringent laws, he will face life imprisonment or a fine of US$5.4 million if he is found guilty of killing an endangered species. The last time a mountain gorilla died at the hands of humans in Uganda was in 2011. Mountain gorillas have undergone an astonishing revival in recent years, following decades of a devastating civil war and uncontrolled poaching that reduced their population to about 350 animals in the 1980s. The apes, which now number over 1 000 individuals, occur in Bwindi and in a network of parks in the Virunga range of extinct volcanoes.

> Rafiki's family regularly foraged beyond the park boundaries, making it a symbolic group in regard to co-existence with people. Rafiki's death, and the circumstances surrounding it, are significant. He was the only mature male in this iconic group.
>
> *Anna Masozera, director of the International Gorilla Conservation Programme in Uganda* (Losh 2020)

Chimpanzees in Liberia: Only 20 per cent of the original population of critically endangered West African chimpanzees still exists. They have already been eliminated from Benin, Togo and Burkina Faso, but Liberia (where they still occur) has taken the lead in their conservation. Confiscated and orphaned chimpanzees can be relocated to the Liberia Chimpanzee Rescue and Protection (LCRP) Centre, a 40-ha forest sanctuary in the Marshall Wetlands. The LCRP was co-founded by Jenny and Jimmy Desmond (animal welfare and conservation consultants), and currently has 40 orphaned chimpanzees in its care – all victims of the illegal bushmeat and pet trades. The LCRP, which moved to its new sanctuary in 2017, supports chimpanzee conservation as well as broad-based public awareness campaigns, vaccination and rabies control programmes, disease research, legislative development and conservation action planning. Fauna & Flora International (FFI) is working with partners in Liberia and elsewhere in West Africa to develop a chimpanzee action plan to safeguard the beleaguered great apes and their shrinking forest habitat. The LCRP is one of the FFI's partners in an ambitious project supported by the UK government's Illegal Wildlife Trade Challenge Fund in Liberia aimed at developing national capacity to respond to the challenge of illegal wildlife trade and encourage cross-border collaboration (Knight 2020).

Covid-19 and the great apes: All species of great apes (gorillas, bonobos, chimpanzees and orangutans) are at risk of being infected by Covid-19, as they share between 97 and 99 per cent of their DNA with humans and may even experience a higher mortality rate than us (Gibbons 2020). In 2013, a respiratory virus swept through the 56 chimpanzees in the Kanyawara community at Kibale National Park in Uganda, where research has been carried out for 33 years; more than 40 apes were sickened and five died. The culprit was *Rhinovirus C*, a human common-cold virus. Human respiratory viruses are also the leading cause of death in chimpanzees at the Gombe Stream National Park in Tanzania, where Jane Goodall worked. Since 1999, researchers have detected repeated outbreaks of viral and strep diseases among chimpanzees in the Taï National Park in Côte d'Ivoire, including of a human coronavirus in early 2017. Each time a respiratory virus sweeps through the population, about one-quarter of the chimpanzees die. Those deaths, coupled with poaching and habitat loss, have shrunk the Taï Forest chimpanzee population from 3 000 in 1999 to less than 400 today (Gibbons 2020).

In mountain gorillas, respiratory viruses cause up to 20 per cent of sudden deaths. Only 1 063 mountain gorillas are left, of which half live in the Bwindi Impenetrable National Park in Uganda, which is visited by 40 000 tourists each year (under normal circumstances). Bwindi veterinarian Gladys Kalema-Zikusoka, who is at the forefront of great ape conservation, recently held a Covid-19 training session with 130 Ugandan Wildlife Authority rangers on how to keep the virus away from the gorillas and monitor them for signs of sickness. To protect them, researchers and veterinarians have closed great ape tourism and research sites, worked with local villagers to reduce contact with apes, and wear masks and observe social distancing (between themselves and the apes) in the forest. So far, there have been no confirmed cases of Covid-19 in the apes (Weber, Kalema-Zikusoka & Stevens 2020).

Animal Demography Unit (ADU): The ADU (formerly the Avian Demography Unit) was established in December 1991 as a research unit of the University of Cape Town, initially as part of the South African Bird Ringing Unit and the Southern African Bird Atlas Project. Since then, the ADU's scope has expanded to include insects (butterflies, dung beetles, lacewings and dragonflies), spiders, scorpions, echinoderms, fish, frogs, reptiles and mammals, while maintaining a strong interest in birds. Its mandate has even been extended to include mushrooms, orchids and trees. The records are now housed in a virtual museum at the world-renowned Fitzpatrick Institute of African Ornithology at the University of Cape Town.

The ADU's mission is to contribute to the understanding of animal populations, especially population dynamics, and provide input into their conservation. This is achieved through mass participation, citizen-science projects, long-term monitoring, innovative statistical modelling and the population-level interpretation of results, with a strong emphasis on the curation, analysis, publication and dissemination of data. The concept behind the ADU, which is now housed in the Department of Biological Sciences, can be traced back to 1983 when it was realised that modern technology allows scientists and non-scientists to collect geo-referenced field data that can be used to construct distribution maps and other analyses of animal demography. Each data point that the ADU's citizen scientists collect is a piece in the jigsaw puzzle of biodiversity, and the ADU's role is to fit together all the puzzle pieces so that it can map South Africa's biodiversity through time. The myriad bits of raw data are thus turned into information on which conservation decisions can be made.

The Internet of Things (IoT) and conservation: We live in the Anthropocene Age, the first era in evolution during which the survival of plants and animals – and essential ecological processes – is imperiled by a living species rather than a catastrophic physical event such as a meteorite strike. We also live in the information age, which provides us with the opportunity to mitigate human impacts by monitoring, modelling and responding to climate change and its effects. The rise of the IoT provides us with an opportunity to continuously monitor the pulse of the natural world. The IoT has taken off thanks to cost-effective production at scale of high-tech sensors that enable scientists to capture more comprehensive data more frequently and accurately, and in more circumstances, than ever before. IoT sensors also facilitate automated high-resolution monitoring in places that were previously inaccessible or dangerous, and they require minimal maintenance.

In southern Africa, the Sigfox Foundation has partnered with Dark Fibre Africa to launch SqwidNet, an IoT network that uses connected sensors to help conservationists monitor white rhino populations. According to the CEO of SqwidNet, Rashaad Sha, the sensors are implanted in the rhino's horns and reliably send out three GPS signals per day via the Sigfox network. By knowing the exact location of the animals, conservationists are better able to look after them. The eventual aim of the project is to GPS-tag all 29 000 remaining southern white rhino individuals (Sha 2017). Other examples of IoT helping scientists to study and conserve wildlife include the use of drones to catch poachers and illegal fishers; the GPS-tagging of sharks, dolphins, whales, turtles and elephant seals (like Yoshi and Ziggy, see the next chapter) to track their movements and monitor their environments; and the use of automatic, remotely-located 'camera traps' to record the presence and behaviour of rare animals.

Biomimicry: In contrast to humans, plants and animals do not pollute the planet nor deplete its resources. In 3.8 billion years of 'research and development' through evolution, they have evolved into consummate architects, engineers and designers that have already solved many of the problems with which humans still grapple. Furthermore, they perform their functions and meet their needs while helping one another to create conditions that are conducive to life. What better model could there be for human development? Biomimicry involves the design and production of materials, structures and systems by humans that are modelled on biological species and processes. The concept has been recognised as one of the breakthrough developments in science and technology, as it represents a novel approach to innovation that seeks solutions to the problems created by humans by turning to nature for her time-tested strategies for success.

> Biomimicry is both an ethos and a blueprint presenting our species with the opportunity to really 'find our way' and begin functioning like a species that is a welcome member of the interconnected web of life.
>
> *Sue Swain, founder of BioWise* (Swain 2020)

BioWise, a non-profit company that promotes the implementation of biomimicry solutions, was established by Sue Swain in South Africa in 2007. Her mission is to establish the coastal town of Knysna and the adjacent Garden Route as the biomimicry hub of Africa and to facilitate the widespread uptake of the concept throughout the continent. Initially BioWise planned to establish the world's first Biomimicry Discovery Park (consisting of a public ecotourism destination, a biomimicry education hub, and a cutting-edge biomimicry research and innovation centre), but financial circumstances prevented this initial dream from being realised. Now Sue is pursuing an alternative vision in the form of a decentralised discovery park experience in partnership with existing ecotourism destinations along the Garden Route, including the Pezula Resort Hotel & Spa and Village n Life. Her aim is to establish a 'community of practice' in biomimicry, with the Garden Route serving as a living laboratory and inspirational flagship for designers, planners and developers in Africa.

Chinese puzzle: Chinese traders have been implicated in activities that exploit many endangered African wildlife taxa (including perlemoen, pangolins, rhinos, elephants and donkeys), but there are also signs that their mindset on environmental conservation is changing. A chance meeting between a Chinese visitor with a deep commitment to the environment and Anton Lategan, the managing director of EcoTraining, has led to over 700 Chinese students attending a seven-day EcoTraining EcoQuest course in South Africa and Kenya. The course is aimed at creating a generation of environmentally conscious travellers whose new knowledge and experiences will percolate through their social and business circles when they return to China (Lategan 2020). 'It is imperative that we inspire sustainable behaviour in the most populous countries in the world. If you mobilise an entire population, you can achieve incredible things,' Lategan says. The signs are promising. In 2018, China announced major plans to rehabilitate the Yangtze River and its estuary; in January 2020, the Chinese government announced that it would ban all single-use plastic by the end of the year.

Before we completely dismiss the usefulness of Chinese traditional medicines, we need to remind ourselves of the contribution made by Dr Tu Youyou, the Chinese pharmaceutical chemist who developed the first treatment for malaria. History will remember her for the role she played in discovering artemisinin, a drug derived from sweet wormwood that is used in traditional Chinese remedies, which formed the basis for a commercialised treatment that has prevented millions of deaths from malaria. In 1979, Youyou described her team's findings as 'a gift from traditional Chinese medicine to the world'. The discovery earned her the Nobel Prize in Physiology or Medicine in 2015, and won humanity important ground in the battle against one of history's deadliest diseases (Gates 2020).

Emerging innovation: Fighting fires in remote areas and over difficult terrain is a challenge throughout Africa. Tshwane University of Technology in South Africa has developed a cost-effective, lightweight firefighting cart that has three times the water capacity of a standard backpack firefighting system but is small enough to penetrate into areas that larger units cannot reach. As the unit is pulled from the waist, rather than carried or pushed, the firefighter's hands are free to pump water or clear a path. The cart has a quick-release belt in case it needs to be jettisoned in a hurry.

Discussion: An increasing number of people believe that the only way in which the environmental crisis will be reversed is to totally rethink the ways in which capitalist and communist societies live and work, but their ideas are widely rejected because they are too disruptive of current socioeconomic models. Yet, during the Covid-19 pandemic, humans have been prepared to endure major economic and social disruptions to overcome the deadly virus. So why can we not adopt the same extreme measures in confronting the environmental crisis that is potentially far worse? Greta Thunberg (2020), *Time*'s Person of the Year in 2019, points out that even if countries meet their stated emission reduction targets, we would still be heading for a catastrophic global temperature rise of at least 3°C above pre-industrial levels. She notes that by the year 2030, the world's planned fossil-fuel production alone will account for 120 per cent more than what would be consistent with a targeted 1.5°C temperature rise. The maths does not add up.

From a sustainability point of view, all political and economic systems have failed. But humanity has not yet failed … Nature doesn't bargain, and you cannot compromise with the laws of physics. Either we accept and understand the reality as it is, or we don't. Either we go on as a civilization, or we don't. Doing our best is no longer good enough. We must now do the seemingly impossible.

Greta Thunberg, Swedish climate activist and founder of Fridays for Future
(Thunberg 2020)

The most outspoken proponent of this extreme view is probably the radical British NGO Extinction Rebellion (XR), which was founded in May 2018 by Roger Hallam and Gail Bradbrook and now has a presence in 75 countries. Initially, their ideas were so extreme that they did not gain traction but now, in an increasingly radicalised world, they are having an impact through their disruptive, non-violent civil disobedience campaigns. Within weeks of their first two-week mass mobilisation in London, the UK government declared a climate emergency and announced a legally binding target for net-zero carbon emissions by 2050 (Nugent 2020). Perhaps the world's leaders need to realise that the sacrifices that people have made worldwide to 'flatten the curve' of Covid-19 infections are the kinds of lifestyle changes that are needed to 'flatten the extinction curve' and allow the planet's essential ecological processes and life-support systems to recover and operate normally again?

While the Covid-19 lockdowns globally have undoubtedly saved hundreds of thousands of lives, the stakes are much higher in relation to the environmental crisis – which will affect hundreds of millions of people. There also fears that in the rush to save ailing economies after the pandemic, governments will abandon their climate change and biodiversity conservation goals and may even revert to supporting the fossil-fuel industry. The optimists are hoping that the Covid-19, climate change and racial injustice crises that are sweeping the planet simultaneously will force governments to rethink how their economies are structured and prioritised, and develop new, greener models.

Mordecai Ogada (Kenyan writer, conservationist and co-author of the book *The big conservation lie*) has advised for many years that tourism should be treated as a by-product of conservation, rather than the basis for it, because of the fickle nature of the tourism business. He believes that Kenya's relentless pursuit of tourism at the expense of more resilient indigenous livelihoods, like pastoralism, is a fallacy. Now that tourism has crashed for the foreseeable future, the weakness of this model has been brutally exposed, with tourism interests begging for donations and bailouts while livestock production continues unimpeded.

4

Fishing, aquaculture and aquatic biodiversity conservation

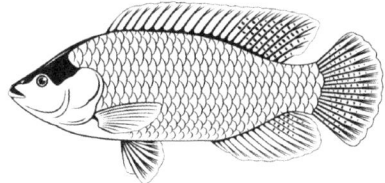

'Give me a fish and I will eat today, teach me
to fish and I will eat for a lifetime.'

African proverb

Introduction: Which came first: hunting or fishing? They probably developed simultaneously, depending on the location of the early inhabitants of Africa, as a wide variety of aquatic and terrestrial animals has been harvested by humans since time immemorial. Some of the earliest tools were made for procuring, preserving and storing food, including plants, shellfish, fish, birds and mammals that live in water. The first fish farms in Africa, where fishes were grown under controlled conditions, were probably the brush parks or *acadjas* in the shallow coastal lagoons of West Africa in Benin, Cameroon and Nigeria. They are made by planting vegetation or placing branches in rows in shallow water (1 to 1.5 m deep), with brushwood around the margins, and then left to 'lie fallow' for six to 12 months before they are stocked with fish. The *acadjas* serve as shelter from predators and provide food in the form of algae, diatoms and bacteria that grow on the branches. After a few months, the fish can be harvested using scoop or cast nets, and yields as high as 7 to 20 tonnes can be obtained (Bruton 2016). Brush parks are also made in shallow water in Lake Malawi (*zivilundu*), Lake Albert and lakes of Madagascar.

A related early form of 'fish farming' is the use of fish aggregation devices (FADs) and artificial reefs (ARs) using locally available materials such as bundles of brush, tubes of bark, palm or coconut leaves, logs and rocks. Over time, these devices create an environment in which fish congregate in large numbers and can be caught more easily. FADs and ARs have been deployed along the marine coast as well as in freshwater lakes. Fisherfolk from Nkata Bay, Lake Malawi, have created over 500 ARs by sinking logs and branches attached to sand bags in water 15 to 20 m deep and then harvesting the plankton-feeding *utaka* fishes that congregate around them using open-water seine nets. In Uganda, Nigeria, Malawi, Mozambique and the DRC, clumps of grass are thrown onto the water surface and then (after a few days) they are surrounded by nets or reed screens and the fishes are caught. In Côte d'Ivoire, submerged coconut leaves are used to attract shrimps in shallow water (Bruton 2016).

Although the establishment of formally protected areas in marine and freshwater ecosystems lagged behind those on land, this deficiency has largely been rectified in recent decades. The oldest marine national park in Africa is the Tsitsikama Coastal National Park in South Africa, proclaimed in 1946, which was amalgamated with the Wilderness National Park to form the Garden Route National Park in 2009. Since then, many marine protected areas (MPAs) have been proclaimed throughout the continent (see below).

Traditional boatbuilding: Africa has a rich tradition of boatbuilding dating back millennia, ranging from simple palm-frond, reed, papyrus or log rafts to bark canoes, wooden dugout canoes and majestic ocean-going dhows. Papyrus rafts have been used in Egypt for over 4 000 years (Bates 1917), raffia palm rafts are still in use on the Kosi lake system in Maputaland, and wooden dugout canoes (with or without outriggers, some with square or triangular lateen sails) are extensively used in West and East Africa and around the islands of the western Indian Ocean, historically as far south as Durban (Bruton 2016). In many places, these craft are still built in the time-honoured way using natural materials and simple, handmade tools (adze and bow drill).

An essential component of traditionally built boats is longlasting, water-resistant timber from hardwood trees such as mango, sausage tree, lucky bean tree, *umdoni*, *trichilia*, wild teak, red mahogony, kapok and breadfruit, which are becoming increasingly scarce in coastal areas. Furthermore, mangrove trunks and branches (which are widely used for making boat ribs, spars and masts, and are very resistant to sea water and boring animals) are also becoming scarce in some areas or are protected in marine reserves. These trends are particularly acute in small island nations such as the Comoros, where fewer traditional dugout canoes and dhows are being built and the skills to make them are being lost. Modern boatbuilding materials such as fibreglass and aluminium for hulls, brass, copper and galvanised iron for fittings, and synthetic materials for ropes and sails, can rarely be afforded by traditional fisherfolk. The adoption of Western technology for canoe-making was taken to the extreme by a traditional fisherman in northern Angola who fashioned a canoe from the wing-tip fuel tank of a crashed F-84 Thunder Jet flown by the Portuguese Air Force during the Angolan civil war (Bruton 2016)!

As a result of these trends, a rich aspect of Africa's heritage will soon be lost, which will spell the end of an era in our maritime history when serene, environmentally friendly, handmade craft plied its waters (Bruton 2016; Stobbs & Bruton 1991). Attempts to replace paddled wooden dugout canoes with inboard- or outboard-powered fibreglass craft in various African maritime states have generally failed, as the skills to build, maintain and use these craft have not been passed on to local fisherfolk. Furthermore, these craft require infrastructure (such as concrete slipways, harbours and moorings) that is often not available or secure.

Most new boats include elements of both traditional African and Arabic design combined with Western technology. The dhows that are typically used off the coasts of Kenya and Tanzania today (*mashua*) have planked (clinker-built) hulls up to 10 m long with no outriggers; many are powered by outboard motors, and are commonly used for laying and retrieving gillnets and fish traps far offshore. The small dhows used in inshore waters in northern Madagascar (*boutre* or *pirogues*) that are usually less than 10 m long have a higher stern and freeboard, possibly to cope with the rougher seas. The inshore dhows used in Mozambique (*bote*), which are usually less than 9 m long, are similar to the *boutre* but a greater part of the sail above the mast pulley is fixed to metal rings that allow it to be dropped quickly in changeable winds (Bruton 2016).

Wetland conservation: The world's largest mangrove restoration project, located in Senegal, has produced amazing results. Not only did the mangroves and their associated plants and animals recover, but the restoration led to higher rice yields in adjacent paddies; increased fish, oyster and shrimp catches that generated valuable income for local fishers; and improved food security for local residents. However, despite these and other significant wetland conservation gains, wetlands are the most endangered ecosystems on the planet and are disappearing three times faster than forests. In West Africa, the vast inland sea of Lake Chad contracted by a massive 95 per cent between 1963 and 2001. Globally, 87 per cent of wetlands have been lost in the past 300 years and 35 per cent since 1970. This is catastrophic, as 35 per cent of wetland plants and animals (which comprise 40 per cent of all the world's species) are at risk of extinction, and stocks of the remaining species are declining rapidly. Martha Urrego, the secretary-general of the Ramsar Convention on Wetlands, estimates that worldwide these ecosystems provide an estimated US$47 trillion worth of 'free' services annually and a livelihood for about 1 billion people (Urrego 2019).

Wetlands are particularly important in Africa because they store, cleanse and make available clean water on a continent where many millions of people have limited access to safe drinking water and water-related conflict is on the rise. Wetlands also provide a livelihood for millions of artisanal fishers and food gatherers. On a broader scale, wetlands are important carbon sinks that play a vital role in climate regulation. The international community has therefore set wetland conservation goals that include no net loss of wetlands by 2030 and a 20 per cent increase in protected wetlands by 2050 (Urrego 2019).

Saving seagrass: Seagrasses are flowering plants that have adapted to survive in the sea and are mostly found in shallow sandy habitats where they form extensive meadows. They are an important source of food, shelter and nurseries for commercially important fish as well as endangered species such as dugongs, sea turtles and seahorses. Seagrass beds support artisanal fisheries and are important carbon sinks, trapping carbon from the atmosphere 40 times faster than tropical rainforests and storing it for hundreds of years.

There are 12 species of seagrass along the Kenyan coast, covering an area of about 317 km². However, between 1986 and 2016, about 21 per cent of this cover was lost due to human disturbance and overgrazing by sea urchins (due mainly to overfishing of their main grazers, parrotfish). Lillian Daudi of the Kenya Marine and Fisheries Research Institute is actively involved in conserving seagrass beds and has found that the most effective way of restoring them is to plant seedlings in punched holes in large bags on the sea floor. She recommends the use of remotely operated vehicles to map seagrass beds and measure the effectiveness of conservation methods (Daudi 2020). Although seagrass does not have the emotional appeal of tropical forests, its conservation is equally important.

Saving riverside forests: Riparian forests are a vital link between terrestrial and aquatic systems, and provide valuable products and services (including shade, shelter, food, bank stabilisation and erosion control). They absorb CO_2 and regulate the microclimate, and their presence prevents humans from carrying out industrial activities so close to a river that they pollute the waterway. The incessant demand for more space (e.g. Nigeria's population is growing at a rate of over 2.5 per cent per year) places enormous pressure on these ecosystems, as most towns and villages are located near water.

Emmanuel Akindele at Obafemi Awolowo University in Ile-Ife is passionate about conserving riparian forests in Nigeria. His research focuses on Opa Stream near Ile-Ife where he has found high levels of heavy metal pollution (lead, cadmium, iron, arsenic and copper) in the water as a result of forests being replaced by carwashing and brick-making activities on the river edge. In addition, pesticides used on adjacent farmlands wash into the stream if the riparian forest has been removed, and the toxins accumulate up the food chain and kill aquatic life. He recommends that riparian forests should be managed as an integral component of river ecosystems and need to be restored so that they can continue to play their vital roles in aquatic ecosystem functioning (Akindele 2020).

Biocontrol of water weeds: Invasive alien plants (IAPs) are among the main threats to wetlands and, as is the case with plagues of invasive animals such as locusts on land, biocontrol appears to offer the best long-term solution for their control. Biocontrol uses the alien plants' natural enemies (including insects, mites, fungi and pathogens) to control them, but the process must be implemented under strict control to avoid damage to indigenous plants. According to Kim Weaver (the community engagement officer at the Centre for Biological Control at Rhodes University in Makhanda, South Africa), the biocontrol organism is brought over from the home range of the invasive plant and then tested over a period of up to five years to confirm that it will only attack the targeted IAP. Only then can application be made for its release into the wild from the quarantine facility. Five of the worst aquatic plant invaders in Africa currently have biocontrol agents deployed against them: water hyacinth (*Pontederia crassipes*), Kariba weed (*Salvinia molesta*), water lettuce (*Pistia stratiotes*), parrot's feather (*Myriophyllum aquaticum*) and Brazilian waterweed (*Egeria densa*).

What makes the biocontrol project in Africa unique is that high-school learners are at the forefront of the control measures, as they are raising millions of tiny insects that are used to attack IAPs. For example, learners at Mountain Cambridge School and Pecanwood College in Gauteng (South Africa) have raised large numbers of weevils that are used to control water hyacinth in the Hartebeespoort Dam (Montgomery 2020; K. Weaver, personal communication, January 2020), and learners at Merrifield College in East London are raising and releasing biocontrol agents into the Nahoon River. This highly successful citizen science project needs to be rolled out in other parts of Africa such as Egypt, Kenya, Uganda, Rwanda, Tanzania, Zimbabwe and Zambia, and in the Senegal, Niger and Volta River basins where water hyacinth, water lettuce and Kariba weed have caused serious problems.

Coral reef protection: David Obura is a founding director of Cordio East Africa, a knowledge organisation supporting the sustainable management and conservation of coral reefs and other marine ecosystems in East Africa and the western Indian Ocean. His primary research has been on the distribution and conservation status of Indian Ocean corals and on coral reef resilience to climate change. Although coral reefs are among the best-studied marine ecosystems globally, humanmade changes are decimating them. Obura is applying his knowledge in practical ways to respond to local needs and national and intergovernmental goals on coral reef conservation, and has worked with multiple forums (including online newsletters and newspapers, and TV and documentary films such as *Vamizi: Cradle of Coral*, *Chasing Coral* and *Blue in Focus*) to bring his message to scientists and non-scientists alike (Obura 2020).

Obura strongly emphasises science communication and advocates that science must be made understandable to everyone and framed in socially relevant terms. For example, while scientists are endeavouring to influence policy so that we do not exceed CO_2 emissions of more than 370 parts per million in the atmosphere, it is easier for laypeople to understand that this would involve keeping global warming below 2°C (preferably below 1.5°C). He is also a participant in an international team that is setting new global goals as part of the Global Biodiversity Framework for the decade 2020 to 2030. These goals include ensuring that there is no net loss in the area and integrity of natural ecosystems, reducing the percentage of species threatened with extinction and increasing the abundance of species, enhancing or maintaining genetic diversity, securing and expanding the benefits that nature provides to people, and increasing the benefits shared equitably from the use of genetic resources and associated traditional knowledge.

Chumbe Island Coral Park: The proclamation of the Chumbe Island Coral Park off the coast of Zanzibar in Tanzania is an example of what one determined individual can achieve if they have the right skills, knowledge and mindset. After two decades of working on educational aid programmes in Latin America and Africa, Sibylle Riedmiller arrived in Tanzania in the 1980s. She was a keen diver and soon became entranced by the magnificent coral reefs around Zanzibar, but also deeply distressed by the rate at which they were being destroyed. She found that the local people had little appreciation for the value of coral, known as *matumbawe* ('rocks and stones') in Swahili, in marine ecosystems and had no hesitation in breaking it up and heating it to form lime to make mortar for building. She realised that there was an urgent need to create awareness of the ecological importance of coral, so she made proposals to the government for a small educational marine park that would be funded by tourism. Together with local fishers, she travelled around Zanzibar searching for a suitable site and finally found Chumbe Island, 13 km south-west of Zanzibar Town, bordered along its western shore by a beautiful fringing coral reef of exceptional biodiversity.

For four years she lobbied the government, including a decisive meeting with the president, before they approved the establishment of the park in 1993. The park comprises a 30-ha no-take marine reef sanctuary and a coral-rag forest reserve covering most of the island's 22 ha. It took another four years for her to develop the ecotourism infrastructure before the park could finally open with seven rustic guest bungalows in 1998. By mid-2015 over 6 500 schoolchildren, 1 100 teachers and 700 community members had benefitted from the educational programmes at Chumbe Park. In 2016, Riedmiller was awarded the Cross of the Order of Merit of the Federal Republic of Germany as 'an outstanding conservationist and activist working for the environment and natural resources'. In retirement, she remains involved in marine conservation in Tanzania.

> Now, after over 20 years of protection, we have more than 450 fish species, from more than 50 fish families, more than 200 species of hard coral representing ninety percent of all coral species found in Africa on only one kilometre stretch of reef, plus turtles permanently resident because they are protected and find enough to feed on in the no-take zone.
>
> *Sibylle Riedmiller, founder of the Chumbe Island Coral Reserve*
> (personal communication, August 2020)

South African Sustainable Seafood Initiative (Sassi): Fourteen-year-old Kerry Sink kickstarted her glittering career in marine science while working as a volunteer at the Oceanographic Research Institute in Durban during the school holidays. She later worked at the Monterey Bay Aquarium in California and then returned to Ezemvelo KZN Wildlife in South Africa in 2000. There she was shocked to find that local shopkeepers and restaurants openly sold protected fish to their customers. In 2002, her research revealed that 92 per cent of retailers were contravening the Marine Living Resources Act (18 of 1998) in some way, mostly out of ignorance, so she launched a campaign to raise awareness among suppliers, retailers and shoppers on which fish should and should not be caught, distributed, sold and eaten.

In 2004, at Kerry's initiative, WWF South Africa established Sassi to inform and educate wholesalers, restaurateurs and seafood lovers about sustainable seafood. To use Sassi, you need to know what fish you are about to eat, how it was caught and where it came from. Fishes on the Sassi list are divided into three categories: Green = best choice sustainable fish food; Orange = think twice, as it is increasingly rare; and Red = do not buy, as it is threatened. Sassi's SMS service (079 499 8795) allows diners to make on-the-spot choices about what to eat and the free app (sassiapp.co.za) encourages them to check the sustainability of their seafood choice. Fish on the 'red list' that should not be eaten include *poenskop* (direct translation 'skinhead'), *dageraad* (direct translation 'predawn'), red *steenbras* (sea bream), Miss Lucy, Scotsman, *galjoen* (black bream), potato bass and kingfish.

Abalobi: Serge Raemaekers and his team at the University of Cape Town, in collaboration with government officials and fishers, have developed a suite of apps (Abalobi) that not only allow them to harvest the knowledge and data of small-scale fishers along the west coast of South Africa but also help fishers to manage their fishery and connect with the chefs who most value their catch. The app, named after the isiXhosa phrase *abalobi bentlanzi* ('someone who fishes'), allows small-scale fishers to upload their daily catch and catch location onto a digital marketplace to which chefs have access. The Abalobi team then helps to facilitate the delivery of orders, with each delivery being accompanied by a quick response code that directs users to the app as well as to information on where, when and by whom the fish were caught.

The app also provides fishers with information on weather and sea conditions so that they can decide when best to fish. It makes it possible for them to share ideas and information with one another, develop and revise budgets, and even obtain bank loans. The Abalobi initiative is an open-source, transdisciplinary, social learning endeavour that has the potential to completely change the way in which artisanal fisheries are managed because it allows fishers to be both harvesters and custodians of their resource. As a substantial proportion of the fish caught in Africa are harvested by subsistence fishers, this is an important initiative.

As Abalobi's core programme operations are entirely cloud based, and its staff is experienced at working remotely, it was able to continue to offer its services throughout the Covid-19 lockdown in South Africa. Furthermore, the Abalobi team supports the maxim that fishery managers should be 'glocal' (i.e. act local, think global) and believes that the future of ethical sourcing and food security starts with being hyperlocal and then expanding internationally with customised variations on the original theme. The project has expanded into the Seychelles under the name *Lansiv* (the local name for the triton conch shell) and is also being rolled out in other African countries. It is likely to be a game changer in the development and management of African artisanal fisheries, and has the potential to be used in the small-scale farming sector in a similar way to the Haller Farmers app launched by Godan Partners and the Haller Foundation to promote sustainable agricultural development in East Africa. Abalobi is also a founding partner of the global ICT4Fisheries network that is pioneering innovative ICTs for small-scale fisheries.

Protecting West African fish stocks: Dr Ifesinachi Okafor-Yarwood at the Centre for Strategic Research and Studies of the National Defence College in Abuja (Nigeria) is a strong advocate for the protection of West Africa's marine fish stocks. He argues that many of the agreements that West African countries such as Guinea-Bissau, Côte d'Ivoire, Liberia, Cape Verde, Mauritania, São Tomé and Príncipe, Senegal and The Gambia have reached with the EU, China and Russia have contributed to the overexploitation of their marine fish stocks.

His research has revealed that EU vessels undermine local food security and provoke conflict with artisanal fishers, partly because they target fragile fish species such as European anchovy, bigeye grunt, sardinellas, bigeye tuna, yellowfin tuna and swordfish. He and Dyhia Belhabib from Ecotrust Canada recommend that the EU should review the implementation of the provisions of its Common Fisheries Policy and that West African countries should negotiate more robust fishing agreements. In 2019, Guinea-Bissau negotiated an improved deal with the EU whereby it was paid €15.6 million per year for providing five years of access to 50 EU fishing vessels compared to €9.2 million per year in the previous agreement. The EU was also required to invest more heavily in marine governance and law enforcement (Okafor-Yarwood 2019).

Errant mosquito nets: One of the main means of defence against malaria-carrying mosquitoes is insecticide-impregnated bed nets, with the WHO (2018) reporting that bed net distribution campaigns over the previous 15 years had reduced malaria cases by about 40 per cent. But scientists in Rwanda, Kenya and other African countries have shown that the wily insects have developed resistance to the insecticides in the nets and frequent washing of the nets reduces their potency. Pyrethroids have historically been the only insecticides that the WHO has approved for use in bed nets but the global chemicals producer BASF has added another insecticide (chlorfenapyr) into the mix to produce a next-generation Interceptor G2 bed net, which has been approved for use by the WHO. A 2018 trial in Côte d'Ivoire pitted Interceptor G2 nets against their pyrethroids-only predecessor and found them to be far more effective. Rwanda is now handing out the Interceptor G2 nets as part of its regular bed net distribution campaigns, which happen every two to three years (Schreiber & González 2020).

But insecticide-impregnated mosquito nets have a dark side. From Nigeria, Mozambique and Madagascar to the Congo, rural people use them as seine nets to catch marine and freshwater fish – which is having a devastating impact on aquatic life. The insecticides in the net dissolve in the water and kill aquatic insects and fish. In addition, their fine mesh (smaller than a mosquito) causes them to catch small fish and the aquatic stages of insects, which are of no interest to fishers (Bruton 2016). Hundreds of millions of insecticide-impregnated mosquito nets have been distributed in Africa in recent decades and they are imperiling a vital food source.

> The nets go straight out of the bag into the sea. That's why the incidence of malaria here is so high. The people don't use the mosquito nets for mosquitoes. They use them to fish.
>
> *Isabel Marques da Silva, marine biologist at the Universidade Lúrio in Mozambique* (Gettleman 2015)

A study at Lake Tanganyika revealed that 87.2 per cent of local households used mosquito nets for fishing. In Zambia, Harris Phiri (a fisheries manager) blames mosquito nets, as well as deforestation and human population growth, for the dramatic decline of fish stocks in the Bangweulu Swamps. In August 2014, Uganda's President Yoweri Museveni threatened to jail anyone who fished with them. The Malagasy name for the nets is *ramikaoko* ('the thing that takes all things together') and the government of Madagascar has banned their use in Antongil Bay, where the industrial shrimp catch has plummeted by 36 per cent in recent years (Gettleman 2015).

Saving 'old fourlegs': Coelacanths are a group of fish whose fossil record stretches back 420 million years but ended about 65 million ybp during the Cretaceous extinction; it was logical to assume that they had gone extinct with the dinosaurs. It was therefore a great surprise when a living specimen was caught off East London in the Eastern Cape (South Africa) in December 1938. The South African scientist Prof. JLB Smith named it *Latimeria chalumnae*, after Marjorie Courtenay-Latimer (the curator of the East London Museum who had saved it for science) and the Eastern Cape river off which it was caught. Smith predicted that the first specimen was a stray from tropical waters and mounted a series of fish-collecting expeditions up the East African coast in the 1940s and 1950s to find another specimen. He and his wife, Margaret, suffered tremendous hardships and survived human-eating lions, giant snakes and treacherous seas; however, their efforts were eventually rewarded when an artisanal fisherman on Anjoaun Island in the Comoros, Achmed Hussein, caught a second coelacanth in December 1952. Smith famously flew from South Africa to the Comoros in a military aircraft to fetch the specimen from French territory (Bruton 2018).

Research has since revealed that Comorian fishers have been catching coelacanths for centuries (Stobbs & Bruton 1991). Three-hundred-and-thirty individual specimens of the iconic fish have since been caught by traditional and semi-commercial fishers in South Africa, Mozambique, Tanzania, Kenya, the

Comoros and Madagascar (Bruton et al. 2020; Nulens, Scott & Herbin 2011), and another living species (*L. menadoensis*) has been found off Indonesia. Using research submersibles, the German diver/scientist Prof. Hans Fricke and his team have conducted detailed research on the ecology and demography of the ancient fish in the Comoros. They also teamed up with the African Coelacanth Ecosystem Programme to study the fish and its environs in southern and East Africa (Hissmann et al. 2006).

The coelacanth is the ultimate survivor and has become a symbol of marine conservation, the 'panda of the seas'. Furthermore, it carries important messages in its DNA on the critical evolutionary transition of backboned animals from water onto land about 320 million ybp (Bruton 2017b), and it is imperative that we take advantage of every opportunity to study and conserve them. In Madagascar, a novel conservation measure has been proposed whereby traditional fishers who catch coelacanths as a bycatch will be incentivised to photograph them with a smartphone, tag and release them in exchange for a reward (Cooke et al. 2020). The *Conseil pour la Conservation du Coelacanthe* (Coelacanth Conservation Council) was established in Moroni, Grand Comore, in April 1987 and the Association for the Protection of Gombessa (the local name for the coelacanth) and Coelacanth Centre in Dzahadjou in 1995. In 2009, Tanzania established the Tanga Coelacanth Marine Park along 100 km of its northern coastline. The colony of coelacanths living in Maputaland is protected in the iSimangaliso Wetland Park proclaimed in 2017.

Turtle conservation: Tomas Diagne is a leading African marine turtle specialist who has been working to save threatened and endangered turtles throughout West Africa for the past 22 years. He co-founded and built the *Village des Tortues* in Rufisque (Senegal). In 2009, he began developing the African Chelonian Institute to promote turtle research, captive breeding, re-introduction to the wild, conservation and education on all African species. Tomas is also a co-founder of the African Aquatic Conservation Fund that promotes turtle and manatee conservation and research.

In the early 1960s, the Natal Parks Board in South Africa received complaints that turtles nesting on the beaches near Kosi Bay in northern Zululand (Maputaland) were being killed. In response, they launched a turtle monitoring initiative in the summer of 1963/4 that has since developed into one of the longest continuous data collection series for any vertebrate conservation project in the world. The first surveys revealed that loggerhead (*Caretta caretta*) and leatherback turtles (*Dermochelys coriacea*) were nesting on the beaches; in 2014, green turtle nests were also found. Since then, an intensive programme of turtle monitoring and conservation has been carried out by temporary *amaThonga* patrolling staff supported by permanent field officers and rangers from the Natal Parks Board (now the KwaZulu-Natal Nature Conservation Services).

The main driver of this turtle conservation campaign has been Dr George Hughes, who joined the Natal Parks Board as a fisheries officer in 1961 and subsequently conducted research for his PhD on the Maputaland turtles through the Oceanographic Research Institute in Durban. He was elected as a founding member of the Marine Turtle Specialist Group of the IUCN in 1969 and appointed as the director of the then Natal Parks Board in 1988. In 1999, he became the CEO of the KwaZulu-Natal Nature Conservation Services and in 2002, the CEO of the KZN Conservation Trust. In 2009, he received a Lifetime Achievement Award from the International Sea Turtle Society. The research and conservation programmes on sea turtles carried out by Hughes and others have had a broad ripple effect and resulted in the initiation of similar projects in Madagascar, Réunion, Mozambique, the Comoros and the Seychelles. These programmes have all had a positive impact on turtle conservation and have improved our understanding of their biology and ecology.

Travels of Yoshi and Ziggy: The journeys undertaken by marine animals such as turtles, sharks, whales and elephant seals were relatively unknown until satellite-linked tags became available. Now we can track their movements with great accuracy as well as monitor the ocean conditions that they experience. As a public aquarium cannot justify its existence through displays alone, it must become actively involved in marine education, conservation and research. The major aquaria in South Africa – Two Oceans Aquarium (Cape Town), Bayworld (Port Elizabeth/Gqeberha), uShaka Marine World (Durban) and National Aquarium (Pretoria) – have excelled in this regard.

For example: Scientists at the Two Oceans Aquarium tagged and released a loggerhead turtle (Yoshi) in December 2017. Her journey was tracked up the west coast of southern Africa, back to Cape Town and then eastwards across the Indian Ocean to Australia – a distance of over 37 000 km in her first 24 months at sea. According to Maryke Musson (the head of the Two Oceans Aquarium Education Foundation), Yoshi guided Australian scientists to a previously unknown loggerhead turtle feeding ground near Port Samson. There they found about 40 other loggerheads (juveniles and adults) gathered at an ideal, protected feeding ground. By late March 2020, Yoshi moved further north towards Port Hedland, travelling about 300 km in two weeks in 32°C water with a gentle south-west current. She spent most of the winter of 2020 in the Eighty Mile Beach Marine Park in Western Australia, a Ramsar wetland site. By June 2020, she had travelled over 40 000 km at an average of 44 km per day across two continents and four countries, and her satellite-linked tag made 23 167 transmissions (Maryke Musson, personal communication, 1 September 2020).

In January 2020, Bayworld released a tagged young elephant seal that had been born near Port Elizabeth. Within a month, Ziggy had swum over 1 200 nautical miles southwards into the Roaring Forties and beyond the Antarctic Polar Front. Scientists and the public are eagerly following the pup's journey on the aquarium's Facebook page (Djunga 2020).

Pity the manatee: The African manatee (*Trichechus senegalensis*) is a unique marine mammal that eats both plants and animals. It occurs in 21 countries in West and Central Africa, from Senegal to Angola, living in rivers, estuaries and the inshore marine environment. They are related to the dugong (*Dugon dugon*), which lives off the east coast of Africa and elsewhere in the Indian Ocean. Lucy Keith Diagne (who founded the African Aquatic Conservation Fund with her husband, Tomas Diagne) has been studying African manatees since 2008. Using aerial and underwater drones, she has investigated manatee distribution, population genetics and feeding ecology, as well as the development of alternative livelihoods for manatee hunters from her base in Saly (Senegal). Her field research has revealed that manatees feed on a variety of organic material (including aquatic plants such as seagrasses and mangroves), the fruits and leaves of trees that fall into the water, and on snails and scaleless fish. Using a novel method comparing the carbon and nitrogen signatures in different foods with the contents of the ear bones of dead manatees, she determined the percentages of different foods consumed by manatees during their average lifetime (Lockwood 2020c).

Manatees have been hunted for centuries and it is estimated that there are now less than 10 000 left. They are listed as vulnerable on the Red List and are protected throughout their range. Although they occasionally come into conflict with fishers and may be killed in retaliation for damaging nets (and may even raid riverside farmers' crops in Cameroon), they tend to live in harmony with rural people. In Côte d'Ivoire and Senegal, fishers are known to share their catches with the gentle giants (Lockwood 2020c), which reach 450 kg in weight. Although manatees (and dugongs) are not among the most popular large mammals in Africa, they are important role-players in their ecosystems and need to be conserved with the same intensity as the rest of the continent's megafauna.

New marine protected areas in Africa: At the 5th IUCN World Parks Congress (WPC) held in Durban (South Africa) in September 2003, attended by over 3 000 delegates, many African countries were motivated to proclaim their intention to create new terrestrial and marine protected areas (MPAs). In particular, formal marine conservation took a leap forward in the highly biodiverse island country of Madagascar that made an initial commitment of 1 million ha of MPAs, which was reiterated at the WPC in Sydney in 2014 and has resulted in the establishment of a dozen substantial new MPAs with an area of over 1 million ha or about 11 per cent of the area of its continental shelf. The largest of these new MPAs (at almost 400 000 ha) is the Barren Islands Complex, an archipelago of coral islands within a large shallow marine region that is partially protected by offshore coral reefs. The major drivers of this outstanding achievement were the marine and coastal component of the national environment programme (1997–2002), the international MPAs movement and marine conservation NGOs that promoted community involvement in MPA management over the past 15 years (A. Cooke, personal communication, July 2020).

Innovations of significance have included the establishment of the Madagascar Protected Areas System that allows for the delegated management of protected areas by non-governmental entities, a new protected areas code (which integrates all six IUCN protected area categories), the establishment of the Fisheries/Environment Commission to identify MPAs and the provision for oversight of MPAs by the ministry responsible for fisheries. Complementary to these measures, the government of Madagascar is working with NGOs to promote the emergence of locally managed marine areas where the community-based management of MPAs is encouraged. Other African countries that have declared significant new MPAs since 2003 and 2014 include South Africa, Kenya, Tanzania, Mozambique, the Comoros, and the French island territories of Mayotte and Réunion.

In August 2019, South Africa declared 20 new or expanded MPAs, which ensures that 87 per cent of the country's different kinds of marine ecosystems are protected. The 20 areas include 17 new MPAs as well as expansions to three: the Aliwal Shoals MPA, the Bird Island MPA (to join the Addo Elephant National Park on land) and the St Lucia and Maputaland MPA (which has been consolidated into the larger iSimangaliso Wetland Park). This brings the total number of MPAs around South Africa to 41, with an additional large MPA in the country's Prince Edward Islands territory (Mann 2019). Mining and the exploration or exploitation of oil and gas are not permitted in any MPAs, and various zonations have been applied in the different MPAs for fishing. While some areas are zoned for 'no take', where no fishing is allowed, others are open to selected fishing methods.

Many scientists, policy-makers, research institutions and government departments were involved in the identification, delimitation, definition and proclamation of the MPAs, with Dr Kerry Sink (the principal scientist at the South African National Biodiversity Institute, who has dedicated many years to the achievement of this goal) playing a key role. MPAs can be equated to savings accounts where fish and other marine life are set aside to grow and 'reproduce', like compound interest. When the bounty spills over into adjacent areas, this 'interest' can be harvested; however, the sustainability of the system depends on having large MPAs that create a substantial 'principal' from which 'interest' can be extracted.

Mauritius oil spill: The catastrophic oil spill from the Japanese bulk carrier MV Wakashio, which ran aground on a coral reef near *Pointe d'Esny* off the south-east coast of Mauritius in July 2020, posed a severe threat to marine life in the pristine lagoons of Blue Bay and Mahebourg. It was one of the worst ecological crises ever to strike the small island country, with possible dire consequences for its economy, food security and health (Harris 2020). The Wakashio, which was carrying 4 000 tonnes of oil, was wrecked next to a nature reserve island (*Ile aux Aigrettes*), a few kilometres from a marine park (Blue Bay) and close to an internationally important wetland (*Pointe d'Esny* Ramsar site). The Mauritian authorities were poorly prepared for the tragedy, which highlighted the need for all African coastal countries to have an action plan that they can implement at short notice if an oil spill occurs (Harris 2020).

> There is no guaranteed safe way to extract, transport and store fossil fuel products. This oil leak is not a twist of fate, but the choice of our twisted addiction to fossil fuels. We must react by accelerating our withdrawal from fossil fuels. Once again, we see the risks in oil: aggravating the climate crisis, as well as devastating oceans and biodiversity and threatening local livelihoods around some of Africa's most precious lagoons.
>
> *Happy Khambule, Greenpeace Africa senior climate and energy campaign manager* (Harris 2020)

Flipflopi expedition: During its historic 500-km journey from the Kenyan island of Lamu to Zanzibar in Tanzania in early 2018, the Flipflopi (the world's first dhow made entirely from recycled plastic) raised awareness of the need to solve one of the world's biggest environmental problems – plastic pollution. The project was co-founded by Kenyan tour operator Ben Morison in 2016 and the dhow was built by master craftsman Ali Skanda and a team of volunteers using 10 000 tonnes of recycled plastic. The boat takes its name from the 30 000 recycled flip-flops used to decorate its multicoloured hull, mostly collected by flip-flop artist Benson Gitari.

The UNEP Clean Seas Campaign joined forces with Flipflopi for the voyage with the aim of inspiring people in Africa to become more aware of the dangers of plastic pollution. Workshops were conducted to give coastal communities a better understanding of the consequences of dumping plastic waste into the ocean and to show children how they could create useful objects out of used plastic bottles. During the Flipflopi's voyage, every port of call announced historic commitments to fight pollution, including a pledge to close the Kibarani landfill site in Mombasa that allowed toxic water to drain directly into the ocean. Kibarani is now being restored and planted with trees, while waste is dumped at a new site in a more environmentally responsible way. Furthermore, 22 hotels along the route committed to minimising their plastic waste by banning the use of plastic bottles and straws.

The UNEP Flipflopi expedition was part of Kenya's push to become a global pioneer in tackling plastic pollution. In August 2017, it introduced the world's toughest ban on plastic bags, with anyone producing, selling or using a plastic bag risking imprisonment of up to four years or a fine of US$40 000 (https://www.un.org/africarenewal/news/).

Emerging innovator: Monwabisi Sikweyiya has an unusual job: he is the shark spotter field manager at Muizenberg's famous surfing beach in Cape Town. Perched high up on the mountainside overlooking the beach, he and his assistants keep a close lookout for great white sharks, which can reach 1 000 kg and 4 m and pose a danger to surfers. In December 2017, he reported that no great whites had been spotted for six months but that they had seen orcas, which prey on sharks. Two fatal shark attacks took place at nearby Fish Hoek beach in 2004 and 2010, which led to the development of a unique 250-m-long shark net that is hauled across the beach every morning in summer. According to Sarah Waries, the project manager of Shark Spotters, the net is tapered to fit the contour of the sea floor and has a small mesh not to entangle any marine megafauna (Hagger 2017).

Emerging innovations: Durban University of Technology in South Africa has developed an alga-based fish feed that is a high-quality, low-cost alternative to expensive commercial aquaculture feeds such as fishmeal. As aquaculture represents the fastest-growing food production industry in the world, there is huge demand for inexpensive feed. Furthermore, as microalgal biomass contains all the essential nutrients for fish growth, it is a very suitable alternative to fishmeal. Nelson Mandela University in Port Elizabeth has developed an integrated microalgae cultivation system in conjunction with aquaponics and/or hydroponics in an environmentally friendly and sustainable process.

Discussion: The African continent is disproportionately affected by extreme climatic conditions, and the frequency of droughts and floods is increasing. Water stress is more than an environmental issue, as it also impacts politics, economics, human migration, health, agriculture and quality of life. Yet there are insufficient water professionals on the continent. This situation highlights the need for those water professionals who are active to integrate and coordinate their skills and resources, and collaborate in a structured way. The African Research Universities Alliance (Arua) has been established to 'facilitate collaboration, knowledge transfer, equipment sharing, pooling of resources and the development of mutually beneficial partnerships across Africa' (Odume & Slaughter 2018). Arua has proposed the formation of centres of excellence across Africa related to key thematic issues, including water. In 2017, Rhodes University in Makhanda (South Africa) was awarded the status of Water Conservation Centre of Excellence, led by the university's Institute for Water Research. Collaboration is anticipated with all Arua-affiliated countries, including Nigeria, Ghana, Tanzania, Kenya, Rwanda, Senegal, Uganda and Ethiopia.

5
Food and cooking

'Those who are at one regarding food are at one in life.'

Malawian proverb

Introduction: The feeding habits of our ancestors in Africa have been revealed at archaeological sites such as Border Cave in the Lebombo Mountains between Zululand (South Africa) and Eswatini (previously Swaziland). In 2016, researchers found evidence there that, like modern humans, early *Homo sapiens* ate roasted vegetables over 170 000 years ago (Wadley & Sievers 2020). The remains of cooked starchy underground stems (rhizomes) from the plant *Hypoxis angustifolia* (yellow stars) were found in the cave and it is thought that they were part of the staple diet of the cave dwellers. The rhizomes were roasted on hot ashes, which would have made them easier to digest by releasing glucose and breaking down fibre. For our large brains to function, we need about 100 g of carbohydrates per day; the *Hypoxis* rhizomes would probably have fulfilled this need in our ancestors. Furthermore, we know that the Border Cave dwellers ate meat and that they needed carbohydrates (and fat) in their diet to help them metabolise animal protein. Their balanced, healthy diet of cooked carbohydrate and protein – the 'real' paleo diet – would have increased their biological fitness and longevity (Wadley & Sievers 2020). Although they would not have appreciated the science behind their balanced diet, their trial-and-error method led them to the right dietary combinations.

Heirloom apples: Hortgro (the South African apple industry's research support body) and Tru-Cape (South Africa's largest exporter of apples and pears) brought back to the country the budwood of the first apple ever grown in southern Africa, the *Witte Wyn* ('white wine'), from the Netherlands. This apple variety was first reported in the diary of the Dutch governor at the Cape, Jan van Riebeeck, from the Company Garden in Cape Town on 17 April 1662. It was recently rediscovered in the Netherlands and brought back to South Africa, where it will be cultivated and sold as an heirloom variety. The *Witte Wyn* was described in the book *Fructologia* by Hermann Knoop in 1763 as a large, irregularly shaped apple with a smooth, red to yellow skin and a soft, very juicy flesh (Booysen 2018b).

Eating clay: In some parts of Africa, people have a craving for eating clayey soils (geophagia). This is most frequently practiced by women and children to relieve hunger and rectify nutritional deficiencies, or as folk medicine. Geophagic individuals tend to be very selective about the type of clay they consume, where it is obtained and its quality. In Ghana, kaolin clay (known as *ayilo* in Ga and *shere* in Twi) is eaten by pregnant women and people suffering from diarrhoea and nausea (Yuora 2019). The University of Venda has made it easier for discerning geophages by developing the Geophagic Chocolate Bar, which is clay based, safe for human consumption and contains many essential nutrients.

Favourite African dishes: The cultural and biotic diversity of Africa is reflected in the delicious and innovative dishes served as part of the daily fare in different cultures. Here are some favourites:

- *Red red* is a popular Ghanaian stew consisting of beans, red palm oil, tomato paste and flavourings of garlic, ginger, chili and onions, with red meat or fish added to the broth and served with fried plantains.
- *Daraba* is a Chadian stew made from chopped okra, mixed vegetables, tomatoes, stock cubes and peanut butter, served with rice, cassava or plantains.
- *Irio* is a hearty Kenyan dish of Kikuyu origin consisting of potatoes, peas and corn. It is especially popular when paired with grilled steak (*nyama na irio*).
- *Thieboudienne* is the national dish of Senegal, an aromatic combination of fish, vegetables and rice in tomato sauce flavoured with garlic and chili.
- *Domiati* is pickled white cheese from Egypt, made from water buffalo milk.
- *Lablabi* is a hot and spicy dish from Tunisia, essentially chickpea soup flavoured with garlic, cumin and harissa.
- *Wali wa kukaanga* is a traditional Kenyan dish made from boiled rice with turmeric, onions, oil, carrots, peas and corn, served with chicken.
- *Fufu*, Ghana's national dish, is a starchy side dish made from cassava and plantains as an accompaniment to meaty stews.
- *Kulwa* is a popular Eritrean dish consisting of lamb or beef chunks fried with *berbere* spices, ghee, tomatoes and onions, served with *injera* flatbread and paired with *tej* (a traditional honey wine).

- *Mchicha* is a creamy vegetarian meal (prepared with amaranth, grated coconut, peanut butter, tomatoes and onions) that is popular in Tanzania, usually served with samp and beans.
- *Attiéké* is a traditional Ivorian couscous dish consisting of fermented cassava roots with sliced onions, tomatoes, and grilled chicken or fried fish.
- *Ful medames* is the ancient national dish of Egypt, remnants of which have been found in 12th dynasty pharaonic tombs. It is a traditional breakfast consisting of simmered fava beans seasoned with lemon juice, olive oil, garlic and spices. It is often eaten to break the fast during Ramadan.

Spices and curries of Africa: Among the most sought-after food additives from Africa are spices, especially those from East Africa and the western Indian Ocean islands. Some spices have become more valuable than precious metals, for example in Madagascar vanilla is worth more, gram for gram, than silver. Zanzibar (part of Tanzania) is known worldwide as *the* spice island.

Like the best fusion foods, East African curry is a mix of brought and found ingredients, cooking methods and culinary traditions from Asia and Africa. Indian immigrants to East and southern Africa brought with them delicious, though hot, curry recipes. These recipes have been fused with traditional African dishes to produce a range of unique foods from Durban to Dar es Salaam. The Scoville Heat Unit, which measures the concentration of mouth-burning chilli capsaicinoids in curries, scores Durban curries as among the hottest in the world (Platter & Friedman 2019). This author once noticed that the hottest spice on sale at a Durban curry market was Mother-in-Law Hellfire!

The use of hot landrace chilli peppers from New Mexico and plenty of grated tomato accounts for the vibrant red colour of East African curry. Add ginger and turmeric from India, *infino* (wild cane herbs), *ibhece* (indigenous melons) and *amadumbe* (root vegetables) from KwaZulu-Natal, Mauritian *mazavaroo* (a spicy combination dominated by red chillies), and Mozambican *piri-piri* (African Bird's Eye chilli), and you have a uniquely African dish. Cape Town is also famous for its curries, but they are of Cape Malay origin and include prawn *ahkni* (an aromatic rice dish), *fluitjie* (chicken neck curry), *gheema* (beef) curry and *frikkadel* (meatball) curry (Sydow & Nordien 2018).

Kwa-Kwa chocolate: *Muquaqua* fruits, from the black monkey orange tree (*Strychnos madagascariensis*), are common in the Mpumalanga and Limpopo provinces of South Africa. Although they are rich in protein, magnesium, potassium and vitamin C, there are currently no products made from them on the market. However, the University of Venda has produced a nutritious chocolate bar from the *muquaqua* fruit by mixing dried fruit powder with Rice Crispies, caramel, peanuts and chocolate, and baking the bar at 120°C for five minutes. The same team has produced a rusk from wheat and *bambara* groundnuts, a cheap legume crop that is widely available in sub-Saharan Africa. Bambara groundnuts are rich in protein (19 per cent), carbohydrates (63 per cent) and fat (6.5 per cent, in the form of oil), and contain high-quality proteins with a good balance of essential amino acids (especially lysine and methionine).

Harvesting yams: Nigeria accounts for about 60 per cent of the world's yam production and 74 per cent of the West African harvest. Yam tubers are an important source of carbohydrates in the diet of African people, but their cultivation and harvesting have not always been carried out scientifically. As a result, the crucial decision on when to harvest the tuber is made using guesswork, which often results in them being harvested too early or too late. Bolani Akinwande, professor of food science at Ladoke Akintola University of Technology in Ogbomoso (Nigeria), has come up with a solution. By carefully measuring the sugar and starch content of the leaves and tubers of yams throughout their growth cycle, she found that the best time to harvest the tubers is when the leaves contain the least starch but the most carbohydrates. As the main nutritional value of yam is in the calories obtained from carbohydrates, the optimal time to harvest the tubers is when they contain the most carbohydrates – which is when the leaves turn yellow, an easy event in the plant's life cycle for the farmer to monitor. Harvesting the yams at the earliest possible time also allows farmers to plant a second crop before the end of the rainy season (Akinwande 2020).

Cooking fish and shellfish: Fish and shellfish are a staple component of the diet of many African people living along the marine coast or near rivers and lakes, and fish are preserved for future use and cooked in a variety of ways. The method of preservation depends on the species of fish (its size and oiliness), the availability of preservatives (especially salt), the requirements of and distance to the market, climate, and local customs and preferences. Fish and shellfish may be eaten fresh and fried; grilled; or incorporated into soups, stews or curries. However, most often they are sundried, salted and sundried, or smoked, and then eaten later.

Smoking not only preserves the flesh, but also imparts a pleasant taste depending on the wood used for smoking. In the Lake Mweru-Luapala fishery in Zambia and in the DRC, catfish (*bongwe*, *milonga* and *mbowa*), bulldog (*umukape*), butterfish (*lupata*), green bream (*pale*), tigerfish (*manda*) and vundu (*sampa*) are smoked, whereas smaller fishes such as *chisense*, juvenile bream (*kachenje*) and torpedo robber (*misibele*) are sundried (Gordon 2006).

In the Okavango Delta (Botswana), catfish are gutted, caked in wet mud and grilled over hot coals. In Zimbabwe and Zambia, *kapenta* (Kariba sardine) are flash fried in hot oil and eaten whole, while tilapia are gutted, scaled, diced and deep fried. In Uganda and Kenya, small cichlid fishes are sundried on reed mats and fried later. After the introduction of Nile perch into Lake Victoria, the nature of the fishery changed dramatically because these predators are too large to be sundried and their fillets have to be smoked, which places a heavy demand on wood (Bruton 2016).

Seafood delicacies: Many intertidal invertebrates are collected around the African coast for food, including mussels, scallops, clams, cockles, limpets, oysters, sea squirts, crabs and octopi. Edible species of sea cucumbers are harvested off Mozambique, Tanzania, Kenya and Madagascar, and exported to the Far East and Europe for the *bêche-de-mer* (dried sea cucumber) trade. Sea cucumber is also commonly served as a delicacy in restaurants with melon, dried scallop, shiitake mushrooms and Chinese cabbage. Octopus and squid are harvested in many African countries for 'First World' markets in the Far East and Europe, especially with the rising popularity of sushi, tapas and poke, and the desire for high-quality protein. Octopus (*pulpo* in Spanish, *tako* in Japanese) has long been a staple item in the diet of peoples in the Mediterranean and Far East, and is a valuable export product for African producers. However, overfishing has bedeviled many African seafood enterprises. Now warming, acidifying seas are causing further problems in the industry.

Octopus harvesting is a mainstay of the fishery economy of Mauritania, with most of their harvest being exported to Japan. Mauritian Minister of Fisheries and Maritime Economy Nany Ould Chrougha introduced strong measures to maintain the viability of the octopus fishery (Scigliano 2020). In 2000, Morocco was one of the largest octopus suppliers in the world, harvesting a staggering 99 400 tonnes; however, by 2004, overfishing resulted in the catch plummeting to just 19 200 tonnes. Now China (whose octopus catch increased from 4.6 tonnes in 2000 to over 120 000 tonnes in 2014), as well as Japan and Mexico, dominate world markets (Root 2019).

There are ambitious ventures in Mexico, Japan and Spain to grow octopi on farms (octoculture) but their complex life cycle and sensitivity to environmental fluctuations, as well as ethical issues, make this difficult. Octopus and squid are regarded as the most sensitive and intelligent invertebrates (Scigliano 2020) and Prof. Jennifer Jacquet from New York University has even argued that farming them would be unethical because 'a life in solitary confinement for a curious mind is ethically wrong' (Root 2019). Craig Forster's recent documentary film *My Octopus Teacher* (available on Netflix), which documents the remarkably intelligent behaviour of an octopus, is likely to support this cause. Whether or not these concerns are justified, African entrepreneurs need to investigate the aquaculture potential of octopi and other aquatic invertebrates.

Insect food: Insects are regarded as a highly nutritious food in many African countries. Mopane worms (the caterpillars of an emperor moth) are eaten in Botswana and Zimbabwe, grasshoppers in Uganda, termites and crickets in Kenya, and other species (such as beetles, bees, ants and cicadas) in other African countries. According to a recent study in Kenya, about 2.6 million of its population of about 50 million people face acute food insecurity and have required humanitarian aid since July 2019 (Mikia 2020). Kenya, therefore, has no choice but to explore alternative food sources.

James Muriithi began farming crickets in October 2018, after attending a training course at Jomo Kenyatta University of Science and Technology in Nairobi. His initial plan was to make cricket flour to feed his chickens; however, one day, he fried some crickets with onions and found that the crispy insects tasted delicious, with a nutty flavour and a hint of grassiness. He now actively promotes the commercial farming of crickets as food for humans. What is on his side is that research has revealed that crickets contain twice as much protein as, and more iron than, an equivalent amount of beef; they are also gluten free. Furthermore, they can be eaten fried or made into flour that is mixed with *mieliepap* (*ugali*, a porridge made from mealie meal), *chapati* (flat, unleavened bread) or pasta (Mikia 2020). Now crickets are internationally recognised as the perfect protein source. In 2019, the American company Exo launched a range of cricket protein bars that are high in essential amino acids, vitamin B12 and iron.

Eating swarming locusts is an obvious solution to the poverty problem in East Africa. It was widely practised in the past in Africa and Madagascar, as well as the Middle East, South Asia and North America. Many communities reversed their bad fortune by turning the plagues of locusts into human food or animal feed, including Muslims because grasshoppers and locusts are the only halaal insect food. However, this is no longer the case because locust outbreaks are managed using chemical insecticides (including organophosphates, carbamates and pyrethroidor) that make the locusts toxic and pose a severe health risk to humans. The FAO has therefore advised that swarming locusts should not be eaten (Van Itterbeeck 2020).

In Cape Town, the Woodstock-based start-up Gourmet Grub is redefining the way people think about insects as food and, specifically, as an alternative to dairy products. Leah Bessa and Jean Louwrens created a new product, Entomilk, from black soldier fly larvae, after first trying mealworms and crickets. Entomilk is a dairy alternative that can be used in a variety of products, from cheese to ice cream (Wentzel 2019).

Mrs Ball's Chutney: One of the most widely available African food additives is Mrs Ball's Chutney, which can be bought in many African cities as well as in the Middle East, Europe and North America. This chutney was first sold in South Africa with the unenterprising name 'Mrs Henry Adkins Senior, Colonial Chutney' and was made in Fort Jackson in the Cape Colony in the late 1800s. Mrs Adkins' daughter, Amelia Ball, inherited her mother's secret recipe and began production in her kitchen in Fish Hoek (Cape Town). She sought the advice of a Cape businessman, Fred Metter, regarding branding, bottling and marketing, and they were soon exporting their unique, eight-sided bottles of chutney to Europe and beyond. Today Mrs Ball's Chutney is produced by Lipton in Johannesburg, although copycat versions such as Amelia's Chutney (produced by Amelia's great-grandson, Desmond Ball) are sold at produce markets in Cape Town. The women's magazine *Sarie* recently let the cat out of the bottle by publishing the 'secret' recipe for the iconic chutney. Although Mrs Ball's Chutney is marketed as a nourishing food supplement, it contains an alarming 2.5 kg of white sugar per 18 bottles (Bruton 2017a)!

Ice cream spaghetti: We all love ice cream, but the Eish Kream outlet in Durban (South Africa) has repurposed ice cream presentation by serving squiggle-shaped ice cream 'spaghetti' in a take-away box in flavours that include bubblegum, café latte and Nutella. Rumour has it that in the 1950s a young chemist, Margaret Thatcher (the future UK prime minister), was part of the team at Dairy Queen Laboratories in the UK that developed methods to pump air into sweetened milk to create soft-serve ice cream. The aerated, minimally nutritious confection is popular throughout Africa and should be enjoyed despite its colonial origins!

Old-fashioned cheese: On the farm Latana in Oudtshoorn (South Africa), cheese-maker Dean Lategan uses traditional methods to make modern cheeses. His free-grazing cows are raised at the foot of the Swartberg Mountains and he cures his Latana sweet cheddar in a cave for a year. The cheese is covered with cloth and butter, and no additives, chemicals or colourants interfere with its natural curing. As the cave's humidity and air temperature are constant, the same high-quality product is produced year after year. Other innovative cheeses made in South Africa include Pepper Thom, which is made with paprika peppers; Kasselshoop white cheddar, which is made with stinging nettle, paprika peppers and calamata olives; and gouda, cheddar, feta, pecorino and halloumi cheeses, which are produced from Karoo goat's milk.

Eden peanut butter: Debbie Ncube inherited the secret recipe for a delicious peanut butter from her grandmother and established Eden All Natural Products in 2012 to make the delicious, preservative-free product on a large scale. Working part-time while studying accounting, she initially sold only 50 kg per month to friends and relatives, but now sells over 20 tonnes per month to the supermarkets

Pick n Pay and Spar as well as to Wellness Warehouse and Faithful to Nature shops throughout South Africa (Reid 2019).

There are many women who have made an impact in South Africa in the past and in the present. I believe that when you empower a woman, you empower a nation. Women see with their hearts and they see the best in communities. To the women out there, be the women you can be but make sure you do the best with your womanhood.

Debbie Ncube, founder of Eden All Natural Products (Reid 2019)

Mama Kena to the rescue: Kenalemang Kgoroeadira (aka Mama Kena) started Thojane Organic Farm in the North West province of South Africa in 2009, and runs it with the passion and wisdom of a real prophet. Her PhD thesis, which she completed at the University of South Africa, entitled 'Lifelong learning in the African indigenous perspective', sums up her life's work. She has close links with traditional healers and leaders (whom she regards as the custodians of her culture), as well as a solid understanding of the science of horticulture. Mama Kena encourages teachers to infuse their curriculum with indigenous knowledge and challenges the youth to look inward for affirmation before looking to the West.

It has always been my dream to go back to my roots, plough, interact with the soil like I used to as a girl, growing food, healthy organic food, not foods that are fed with chemicals for them to grow fast, losing all the nutrients … [Thojane is] a place where people can have self-determination and begin to live like Africans, where what you do or learn is for the benefit of the whole society.

One day this farm will supply all the hospitals, mines and even international markets. We will grow organic vegetables and herbs; there will be honey, mutton, indigenous chickens and eggs. You will be able to buy essential oils, herbal teas, bath soaps and creams with a Thojane label. One day.

Mama Kena, founder of Thojane Organic Farm (Kgoroeadira 2014)

Mama Kena is an expert on indigenous ways of farming and on the traditional uses of herbs and other plants. Her aim is to develop Thojane Organic Farm into a working example of how women throughout Africa can run small-scale sustainable farming units that provide locals with healthy food. She and her workers grow vegetables and herbs (including mint, yarrow, lemongrass and lavender) that are sold at local markets and to national retailers such as Food Lovers Market. They also donate produce to feeding schemes at local Phokeng schools. Her original 1-ha garden has grown into an award-winning programme with her being named the Best Female Subsistence Farmer in 2014 by the South African Department of Agriculture, Forestry and Fisheries.

Discomfort food: Members of the African diaspora are also making meaningful contributions to dietary innovations. Lagos-born chef and food writer Tunde Wey uses food as a tool for provocation and social transformation. Now based in New Orleans, his controversial methods have attracted both fans and detractors. In 2006, for his 'Blackness in America' dinner series, he invited white guests to engage with black diners on the impact of systemic racism in the USA over dishes of Nigerian *egusi* (ground melon seeds) stew. In 2018, he opened SAARTJ, a temporary food stall in New Orleans where black visitors were charged US$12 for a meal while white people had to pay US$30 – reflecting the city's racial wealth gap. In Pittsburgh, he matched immigrants and US citizens in a blind-date dinner series called 'Love Trumps Hate', inspired by his own status as a permanent US resident who was undocumented for 10 years. He describes his work as 'discomfort food' to illustrate how communal dining spaces can prompt meaningful action. In recognition of his work, *Time* magazine elected Wey and the Egyptian musician Dina el Wedidi as among the 100 most influential young people in the world in 2019 (Endolyn 2019).

African culinary excellence: In 2017, Chantel Dartnall of South Africa was named Chef of the Year for Africa and the Middle East and her restaurant Mosaic in Pretoria won the Luxury Restaurant of the Year accolade from *Travel & Hospitality* magazine. Dartnall uses only the best seasonal produce in her signature botanical cuisine, which takes its cue from traditional African dishes. The Karibu Restaurant on the V&A Waterfront in Cape Town, which serves only African cuisine, was voted Best Local Cuisine Restaurant in Africa by the Luxury Restaurants Association in 2016.

South African food abroad: Many travellers from Africa long for the taste and aroma of home cooking. I have tasted delicious *boerewors* (a type of sausage) rolls and *biltong* (dried, cured meat) in The Meat Company restaurant in Bahrain. Others have sampled *samoosas* (savoury triangular shaped pastries), *bunny chows* (a hollowed-out loaf of white bread filled with curry), *boerewors*, and samp and beans at the Lucky Tsotsi Shebeen and Bar in Sydney; genuine lamb *tjops* (chops) at Shaka Zulu in London; and *peri-peri* chicken flavoured with hot Mpumalanga Fire Sauce in the Africola restaurant in Adelaide. Just down the road from Buckingham Palace in London, the swish *Mzansi*-inspired bar offers venison burgers, grilled *boerewors*, springbok fillet, Cape Malay chicken curry and spring rolls filled with *bobotie* (baked curried meat topped with a milk and egg mixture). There is even a Braai Republic restaurant in Seoul that serves *bunny chow* and lamb shanks, and Pinotage (a South African eatery) in Beijing brings the taste of *Mzansi* (South Africa) to homesick South Africans and the rest of the world. In New York, you can order Cape Malay *bobotie* and oxtail stew at the Madiba Restaurant, lamb *potjie* (a stew slow cooked in a cast-iron pot over an open fire) and *melktert* (milk tart) at Peli Peli in Houston, or *biltong* and *mosbolletjies* (sweet-bun or bread traditionally made with wine) at Jan on the French Riveira.

Girth of a nation: As African people become more affluent, they tend to eat more processed food and less fresh fish, vegetables and fruit. One of the consequences of this trend is an increased incidence of obesity, with the concomitant health risks of heart disease and diabetes. This trend has also been recorded in countries like Bahrain where healthy diets of fresh fish, dates and vegetables have been replaced by fast-food diets dominated by processed food, combined with more sedentary lifestyles. Consequently, waistlines are expanding at an alarming rate in South Africa's black townships, triggering an explosion of hypertension, strokes and Type 2 diabetes. This trend is particularly marked among women, over half of whom are overweight or obese according to the 1998 South Africa Demographic and Health Survey (Lindow 2005). During the apartheid era in South Africa, activists chanted the slogan *phanzi* ('down with') at police during township protests. Now members of the Masiphakame Ngempilo Yethu ('Let us stand up for our health') Club in Khayelitsha outside Cape Town chant '*Phanzi blood pressure, phanzi! Phanzi diabetes, phanzi*!' in their war against a new kind of enemy – obesity (Lindow 2005).

Twiga Foods to the rescue: Peter Njonjo of Kenya started Twiga Foods in 2014 as a reaction to the inefficiencies in Africa's large but fragmented informal fruit and vegetable industry. The company uses a technology-enabled B2B (business to business) platform to source products from 17 000 farmers from 20 counties in Kenya, and deliver them directly to 2 500 vendors per day in Nairobi and its environs. Twiga Foods expanded into fast-moving consumer goods and Njonjo was planning to move into Nigeria and Francophone West Africa by late 2020. The company has been credited with raising farmers' yields while stabilising consumer prices. It was building a state-of-the-art distribution centre with cold rooms, conveyors and sorting equipment to support the entire supply chain (Ayene et al. 2020).

Intellectual property: Unbelievably, a Dutch company holds the patent for Ethiopia's most popular food, *injera*. According to the European Patent Office, the Dutchman Jans Roosjen 'invented' the flatbread made from teff flour in 2003, even though it is derived from a plant indigenous to Ethiopia that Ethiopians have been using for centuries to make spongy, fermented pancakes which they eat with their meals. Ethiopians – and chefs elsewhere such as Kassahun Gebrehana, the owner of the Little Addis Café in Maboneng (South Africa) – argue that *injera* is a traditional food in Ethiopia and they want their intellectual property back. According to Gebrehana, *injera* is so ingrained in Ethiopian culture that when he prays, he says 'Give us our daily *injera*' (Allison 2018).

The tiny teff grain, the smallest in the world, is gluten free and rich in nutrients. It is perfectly poised to take advantage of the global health food trend. Dutch researchers reckon that teff could become the next kale or quinoa, and have formed a company (which became Health and Performance Food International – HPFI) to explore options to grow and market it in Europe. After many negotiations with different government entities, the company reached a deal with Ethiopia to plant and distribute teff in Europe. In return, it undertook to send a generous slice of the profits back to Ethiopia. In 2003, Ethiopian officials sent 1 440 kg of teff seeds to the Netherlands and 91 Dutch agrarian entrepreneurs began growing it. But the project did not go accorrding to plan, as the demand for teff never materialised. The company soon went bankrupt and the Ethiopian government earned a pittance (€4 000) from the deal (Allison 2018).

However, HPFI had already applied for and been granted patents for the production and distribution of teff in Europe, and these patents did not lapse when the company went bankrupt. This was a problem for Ethiopia, as the patents were very broad, covering most forms of teff flour as well as all products resulting from mixing teff with liquids such as bread, pancakes, shortcake, cookies and – of course – *injera*. In early 2018, the Ethiopian Intellectual Property Office announced that it would use all legal means to reclaim its intellectual property (Allison 2018), but the battle still rages. This is an important lesson for African countries – do not give away the intellectual property in your natural resources.

Emerging innovator: Simphiwe Xulu is a butcher who specialises in crocodile meat in Hambanathi near Tongaat in KwaZulu-Natal (South Africa), selling about 30 kg of the reptile meat and fat per month. He says that about half his customers (mainly women) buy the meat for medicinal purposes to treat asthma and other chest problems, and recommends that it be boiled and served with a pinch of salt (Harper 2020b).

Emerging innovations: The potential of agroprocessing in Lesotho has not been fully exploited. Now the company Super-Foods (trading as Bohlokoa), in collaboration with the National University of Lesotho, is using modern techniques combined with traditional practices to help smallholder farmers develop sustainable sorghum farming enterprises and produce innovative products from sorghum such as muesli, nutty snacks, instant porridge and baking products. *Atchar* (a spicy condiment derived from Indian cuisine that is often eaten with curry) is popular in East and southern Africa. It is often made from unripe green mangoes and chillies, but a variety of fruits can be used. The University of Limpopo has developed a new *atchar* with a distinctive flavour made from the skin of the marula fruit, which is rich in essential nutrients.

Discussion: Despite all the advances in agriculture and horticulture in recent decades, feeding people is still a major challenge in Africa, especially following the socioeconomic fallout from the Covid-19 pandemic. In April 2020, the World Food Programme (WFP) estimated that food insecurity would affect 256 million people globally, which was nearly double the figure for 2019. 'Covid-19 is potentially catastrophic for millions who are already hanging on a thread', said the WFP's Chief Economist Arif Husain. 'We all need to come together to deal with this, because if we don't, the cost will be too high' (Sondo 2020b). We need to reach a situation where Africans consume what they produce and produce what they consume, and our agricultural and horticultural innovators will need to be on the frontline of this campaign.

The Covid-19 pandemic has restricted international trade in many food products and led to greater interest in eating home-grown produce, which has significant environmental benefits. Eating pomegranates and avocados from India, lentils from Canada, blueberries from the USA and goji berries from China results in an enormous carbon footprint. While going vegan has some environmental benefits, eating lamb chops from a local farm is better for the environment than buying avocados from India. Kenya, the world's sixth largest exporter of avocados, banned the export of the fruit in 2017 because the country's supply is at risk (Henderson 2018). This is yet another example of how the Covid-19 pandemic has forced us to look at old habits in new ways.

6

Drinks and beverages

'Only birds with long beaks can drink from deep pots.'

African proverb

Introduction: Tea is the most popular beverage in the world (beer is second) and Africa has become a major tea-growing continent in recent decades. Locally grown teas include not only the traditional green teas, but many craft condiments made from indigenous plants – of which *rooibos* ('red bush') tea is probably the most successful. *Rooibos* tea, which is derived from the indigenous South African plant *Aspalatus linearis*, has many beneficial effects; however, a recent study has revealed that the abundance of antioxidants in the tea can also reduce exercise-induced oxidative stress and cell damage during intense exercise. Oxidative stress is an imbalance between oxidants (also known as free radicals) and antioxidants, in favour of oxidants. Oxidants (which are unstable molecules that can damage genetic material, lipids and proteins) are derived from cells in the body and from the external environment, but can also be produced in response to strenuous exercise when our bodies use oxygen to produce energy.

To prevent free radical damage, our bodies have a natural defence system of antioxidants, but it is often not enough during strenuous physical activity. Drinking *rooibos* tea in a concentrated form – equivalent to six cups per day – helps the body to reset the balance. *Rooibos* can therefore reduce physical fatigue and improve performance. It could become the next must-have supplement for both elite athletes as well as anyone who wants to reduce the risk of cardiovascular diseases or diabetes (Anon 2018). African athletes, take note!

First alcoholic drink from Africa? Traces of alcohol do not preserve well in the archaeological record, nor do their containers made from skin and wood, so other evidence for the first fermentation of alcohol needs to be found. Neil Rusch at Wits University has investigated the possibility that fermented honey may have been used in Africa to make alcohol over 40 000 years ago, when honey was known to have been consumed in southern Africa. This is long before the previous earliest records in Israel (13 000 ybp) and China (27 000 ybp) (Rusch 2020). Rusch successfully repeated a fermentation method that had been told to the Swedish botanist Carl Thunberg by indigenous Khoi-San people in the 18th century involving mixing honey, water and *moerwortel* (*Glia prolifera*) to produce an alcoholic beverage. He then discovered scattered ethno-historical accounts on the use of honey fermentation, as well as rock paintings featuring bee-related themes, and examined a 40 000-year-old parcel of beeswax from Border Cave in Maputaland. These pieces of evidence led to his conclusion.

Rusch (2020) argues that the cognitive requirements needed to support an understanding of the chemical and technical processes of fermentation had already been shown by southern African people by the Middle Stone Age, including mixing and using ochre paint (100 000 ybp), making and hunting with bows and arrows (60 000 ybp), and using arrow poison (24 000 ybp). Fish poisons derived from indigenous plants have also been used in Africa for tens of thousands of years. Intentionally controlled fermentation fits comfortably within these 'techno-behaviours' but leaves no archaeological trace.

Research on indigenous drinks: Drinks produced from indigenous plants are a mainstay in the diet of many Africans, but they are often produced under suboptimal conditions with little quality control and few standardised production techniques. Werner Embashu from the University of Namibia in Windhoek has carried out pioneering research on the properties and production techniques of indigenous fermented drinks in Namibia, including *ontaku* (beer made from cereals such as pearl millet) (Embashu & Nantanga 2019) and *omalovu* (traditional opaque beer made from sorghum or pearl millet) (Embashu, Lileka & Nantanga 2019). Pearl millet (*Pennisetum glaucum*) is a staple component of the diet and an important source of nourishment for over 60 per cent of Namibian people. It is fermented to make *ontaku*, but also processed into flour and porridge. This indigenous cereal flourishes in hot, semi-arid areas where most other cereals do not survive and has the potential to become a staple food crop in regions of Africa that will be affected by climate change.

Currently the processing conditions for brewing *ontaku* are as diverse as the brewers. To improve the efficiency of *ontaku* production, Embashu recommends that starter cultures for the fermentation of pearl millet should be developed and that the quality and amount of the different ingredients, consistency of the pearl millet paste, temperature of the water added and duration of the conversion of the malt all need to be standardised to produce a high-quality product that is safe to drink.

Marula drinks: The marula tree (*Sclerocarya birrea*) is indigenous to Africa and its fruit is widely used to make beer and wine, which are tasty and highly nutritious. These drinks have strong antimicrobial properties and are high in tocotrienol and Omega 6 and 9 fats. This truly African beverage is traditionally made by women during the fruiting season and is quickly consumed at social events, as it does not have a long shelf life. Technologists at the University of Limpopo have produced a marula fruit wine with high longevity and strong sensory stability that will be available throughout the year. In addition, the company The Marula Guys (also in South Africa) produces marula-based natural oils, anti-ageing cream, facial and haircare products, as well as marula juice pulp, vinegar and activated carbon that can be used to purify gold and platinum, for water and air purification, and in aquaculture.

Amarula Cream, marketed as 'The Spirit of Africa', was ranked the sixth best liqueur in the world by 700 mixologists from 60 countries during the 2010 Soccer World Cup in South Africa. The tasty liqueur (a blend of marula fruit and cream with flavours of coffee, vanilla, chocolate, toasted nuts and citrus) is sold in 103 countries and is the second most popular cream drink of its kind in the world. The Amarula Trust has been established by the industry to promote elephant research and conservation. As marula trees are widely distributed in Africa and Madagascar, their full potential has yet to be realised.

Rooibos tea: South Africa's famous *rooibos* tea is derived from an indigenous plant that grows in the rugged Cedarberg region of the Western Cape. In 1904, the Russian immigrant Benjamin Ginsberg, whose ancestors were tea traders, realised that *rooibos* leaves had the potential to produce a world-class tea and began trading with local people who had been using health-promoting infusions from the plant for centuries. In 1930, P. le Fras Nortier (a well-known nature lover and medical doctor) developed the germination technique that made it possible for *rooibos* to be farmed commercially, and today it is sold worldwide. In terms of developing the full commercial potential of an indigenous plant, it is one of Africa's greatest success stories (Bruton 2017a).

Rooibos tea has so many health benefits that a device, slowRedT, has been invented to deliver regular doses of the drink to hospital patients. The device is an innovation of the CPUT that delivers extracts of *rooibos* as a dietary supplement over an extended period. It uses a controlled release minitablet to deliver the recommended daily requirement of six to eight cups of tea over a period of 24 hours if the treated individual consumes one capsule twice per day and eliminates the need to drink multiple cups of tea to enjoy its nutritional benefits.

Green teas: Technologists from the University of Limpopo have developed the herbal drink Harvest Green Tea from a blend of edible herbs and lemongrass (*Cymbopogon citratus*), the main ingredient. Harvest Green Tea is caffeine free; has no added preservatives; and contains calcium, copper, dietary fibre, fat, iron, magnesium, phosphorus, potassium, protein, and vitamins B1, B2, B6 and C. The Setšong Tea Crafters in Limpopo province make artisanal herbal tea infusions from wild-harvested indigenous plants. They work in collaboration with community elders and other indigenous knowledge practitioners to create a range of tea blends using the region's *diya* (red root) and *tepane* (black bush) leaves, which contain high levels of antioxidants and minerals that rejuvenate and balance the body. Their tea products have been consumed for over five generations by the Bapedi people in Sekhukhune. As a result, a great deal of their heritage is embedded in the brewing of their traditional beverage.

Ethiopian coffee: Coffee is unquestionably the most famous drink to come from Africa, with over 2.25 billion cups of coffee consumed each day worldwide. Ethiopia's main cash crops are coffee, oilseeds, cereals and cotton, with coffee accounting for over 74 per cent of all exports. The main coffee growing area is in the province of Kaffa, from which coffee probably got its name (although there are other claimants). Kaffa has ideal conditions for growing coffee: altitudes from 1 300 to 2 100 m, 1 500 to 2 500 mm rain per year, and slightly acidic topsoil (pH 4.5 to 5.5). From Kaffa, Arabica coffee spread to Yemen in the 14th century, where it acquired the Arabic name *qahweh* and became *kahveh* in Turkey, *untuk kopi* in Indonesia, *kopitiam* in Malaysia, *yi shi nóng suõ* in China, *caffe* in Italy, *Kaffee* in Germany, *café* in France and *koffie* in the Netherlands.

The most appealing of the many legends about the origins of coffee comes from the Ethiopian highlands, where a young goat-herder named Kaldi was surprised when he noticed that his lazy goats would become invigorated after eating the berries of a certain plant. He tried the berries himself and found them to be stimulating and invigorating, so he told a local monk about his discovery. The monk also tried the berries and found that after an arduous session of prayer, he remained wide awake. He passed the secret on to other monks and soon his brethren throughout Ethiopia were chewing the intoxicating berries and carrying out their devotions with bright and alert minds (Hancock et al. 1997). For several centuries after its discovery, coffee beans were chewed and not drunk. The berries were either eaten whole or crushed and mixed with ghee (clarified butter) – a practice that is still followed in Ethiopia. Later refinements included wine made from fermented coffee pulp, but it was not until the 13th century that the practice of brewing a hot drink from roasted beans was introduced. In Ethiopia, the whole process of pounding and then roasting the beans on an iron plate over a charcoal brazier, adding cloves and other spices, brewing in pots of hot water and serving in tiny china cups – all mingled with the heady aroma of incense – has taken on the trappings of a national ritual.

In addition to the commercially cultivated species from Ethiopia (*Coffea arabica*), there are many species of wild coffees in Africa and surrounding islands, some of them low in caffeine to suit modern, healthier diets. Over 50 species of wild *Coffea* are found in the rainforests of Madagascar that could be hybridised with *C. arabica* to introduce lower caffeine contents as well as better disease resistance in the commercially cultivated strain (Preston-Mafham 1991).

Congo coffee: Linda Mugaruka was the first coffee taster (or 'cupper') in the DRC, and does not plan to be the last. The eastern Congo was once one of the world's largest producers of coffee before war decimated the plantations, but Congo coffee is on the rise again. Linda is one of a handful of professional cuppers who help coffee growers to identify quality beans, improve their growing methods and attract international buyers who will pay a premium for high-quality processed beans. Magaruka (who works in a coffee-tasting laboratory in Bukavu) is training a new generation of cuppers and introducing Congolese women to the complexities, textures and aromas of their coffee terroir through weekly tastings. 'Together we will make sure that, when people hear the word "Congo", they will think of coffee, not war', she says. In addition to coffee tasters, new jobs are created for high-end chocolate- and tea-makers (Baker 2017).

Coffee in a Cone: Dayne Levinrad is the epitome of an entrepreneur. He started off mopping floors in a *Primi Piatti* ('first dishes') restaurant and now runs one of South Africa's most popular coffee shops, the Grind Coffee Company, in Melrose Arch (Johannesburg). For a while, he worked on creating unique ways to drink coffee while working as a barista in Los Angeles (USA), where the concept 'Coffee in a Cone' was born. Instead of launching it in Brazil as planned, he decided to bring it home to South Africa, where his scientific approach to creating high-quality coffee has been a huge hit.

> As a start-up, you constantly have to show your face, even when things are starting to move worldwide. I'm not complaining, because it's exceptionally rewarding. I guess when you start out with your own company, your right to a life is taken away for a bit. But whatever comes my way, I'm going to face it.
>
> *Dayne Levinrad, founder of the Grind Coffee Company and promoter of Coffee in a Cone (Fairlady 2016)*

Coffee in a Cone brings together ice cream, coffee and chocolate, topped with foaming milk, in a handy edible cone. Social media users went crazy over the invention and it soon became the most Instagrammed coffee in the world, with 16 million hits! When Jennifer Lopez posted a photo of Coffee in a Cone on her Facebook account, there were over 800 000 likes within minutes, with millions asking where they could get it (*Fairlady* 2016). From the beginning, Dayne has not had a day off, working 90-hour weeks and travelling internationally to launch his products but also maintaining a presence at the Grind Coffee Company.

Pinotage: Pinotage is the only grape variety that is unique to Africa. It was developed by Prof. Abraham Perold at Stellenbosch University in South Africa when he crossed Pinot Noir and Hermitage (Cinsaut) grape varieties in 1925. In 1961, Lanzerac was the first brand to use the name 'Pinotage' and they have since produced Lanzerac Pionier Pinotage 2013 as a tribute to this pioneering spirit. This bold and complex wine has a deep red colour and a full, velvety finish with taste notes of red berries, cinnamon, cloves and nutmeg. It has become South Africa's signature wine variety and is also grown in the USA, Australia and New Zealand (Bruton 2017a).

Craft beer: The recipe for beer is the oldest recipe in the world. The ever-popular tipple has been brewed in Egypt for over 3 000 years, where it was quaffed by both pharaohs and labourers, and regarded not only as a refreshment but (as early as 1600 BC) also as a medicinal remedy reputed to treat over 100 illnesses. Beer has remained a staple drink throughout Africa and is also a favourite among visitors to the continent. When US President Theodore Roosevelt visited Africa on a safari in 1909, he brought 500 gallons of beer with him to stay hydrated (Barnett 2006)! The first beer brewing plant in sub-Saharan Africa was established at Papenboom in Newlands (Cape Town) in 1713, and the largest beer malting plant in Africa is in Caledon in the Western Cape. Popular commercial beer brands in Africa include Club Shandy (Ghana), Kilimanjaro (Tanzania), Sibebe (Swaziland), Hansa (Namibia), Chibuku Shake Shake (Botswana), Tiemann Wild Dog Weiss (Zambia), Celtia (Tunisia), White Bull (South Sudan), Laurentina (Mozambique), Brau Zero (Libya) and Zambezi (Zimbabwe).

Craft beers have taken Africa by storm and many of the big commercial brewers are taking notice. Apiwe Nxusani-Mawela is the owner of South Africa's first black female majority-owned brewery, Brewsters Craft, in Johannesburg and is also a keen promoter of African traditional beer culture. In South Africa, many of the new craft beers have been named after animals that portray some aspect of their character. Master brewer John Morrow at the Nottingham Road Brewing Company in KwaZulu-Natal includes in his 'zoo' of beers Pye-eyed Possum Pilsner, Whistling Weasel Pale Ale, Tiddly Toad Premium Lager, Pickled Pig Porter, Swinging Samango Mango Ale, Tipsy Tiger India Pale Ale and Wobbly

Wombat Fruity Ale. The Old Main Brewery in Pietermaritzburg produces a Honey Badger Imperial Stout that is creamy white on top and dark underneath, like the real beast, as well as an Oyster Catcher India Pale Ale (Lawson 2017c). In South Africa, craft beers are made from imported and locally grown hops as well as from mangos, naartjies (tangerines), brambleberries and pineapples.

The Covid-19 pandemic has stimulated the development of at-home labs and small craft breweries, with innovation moving from corporate laboratories to private kitchens and basements. Digitisation has also made it easy to create collaborating communities, source information from experts and develop a mini-industry at minimal cost. This DIY approach has also spread into maker spaces, micro-electronics and material science.

Craft gins: In Africa, gin is traditionally produced from fermented sugar or sugarcane rather than grains (wheat or barley), which are twice as expensive, used in the northern hemisphere. According to expert gin distiller Simon von Witt of Cape Town, sugarcane produces a consistently neutral form of alcohol and when it is distilled, all the sugars are removed and converted to alcohol.

Waragi Gin has been produced in Uganda since 1965 and was originally sold in 100-ml plastic bags. *Waragi* is a generic word in Africa for moonshine, and the gin has a similar reputation as a do-it-yourself, underdistilled and sometimes dangerous drink. It is distilled from sugarcane (once Uganda's main export) but since the sugar industry collapsed in the late 1980s, it is also distilled from banana beer with juniper botanicals. Uganda Waragi Gin is now on sale in the USA (where it has a faithful following) and throughout East Africa, and was probably the first alcoholic spirit to be distilled commercially in Africa after independence.

The first craft gins produced in South Africa were distilled by Roger Jorgensen in 2007, long before the gin boom of 2013. The Cape Floral Kingdom around Cape Town is the smallest but most biodiverse plant kingdom in the world and can be 'experienced in a glass' by drinking Wilderer Fynbos Gin, which is distilled from sugarcane and flavoured with Cape fynbos botanicals, hibiscus, buchu, rose geranium, citrus peel, pine and spices. Another fynbos gin is Cape Town *Rooibos* Red Gin, whose flavourants include juniper berries, orange peel and cinnamon combined with the earthy, nutty flavour of *rooibos*. Other choices include D'Urban Scarlet Gin (which is infused with African botanicals and *cascara* [coffee cherries]) and Deep South Distillery's Ruby Gin (which gets its deep red hue from hibiscus flowers).

At Distillers & Union in Cape Town's creative hub Woodstock, Simon von Witt produces craft gins such as 5 Pence Cape Dry Gin (which has undertones of herbaceous citrus and an aftertaste provided by eight fynbos botanicals) and D&U Pink Gin (which is infused with Pinotage grape skins with hints of

Turkish delight and an aftertaste of rose geranium). In 2019, at least 280 brands of craft gins were produced in Cape Town alone and the explosion pushed gin consumption up from 518 000 9-litre cases in 2013 to almost 1.8 million in 2018, according to the drinks research company ISWR.

Duke craft gin has an interesting origin. First distilled in 2017, it evolved from an attempt to save the fragrant litchi orchards at Summerfields Rose Retreat & Spa in Hazyview, Mpumalanga (South Africa). At the time, litchi farming was in serious decline and the orchard was set to be replaced by a grove of macadamia trees, but André van Heerden had other ideas. After completing a gin course in Johannesburg, he immersed himself in the world of copper stills, litchi fermentation, juniper berries and triple distillation. His silky-smooth gin, named after the family's Vizsla dog, has a delicate balance between honeyed notes of rose, ginger and litchi grounded by earthier aromas of juniper berries and macadamia (www.dukegin.co.za).

After holidaying in a game reserve, biologist Paula Ansley asked her husband a strange question: 'Would it be possible to make gin from elephant dung?' After all, elephants selectively browse a range of botanicals, yet they absorb less than half of what they eat. The result is that much of their food (roots, leaves, fruit and bark) remains in their dung. The Ansleys reasoned that they should let the elephants do the hard work of collecting a range of natural ingredients that could be used to flavour and colour the gin, and they would do the distilling. After all, elephant dung has been used in East Africa for centuries to make tea and as a traditional medicine.

Although they had no experience in gin-making, they gave up academia and in 2018 started making Indlovu Gin 'infused with botanicals foraged by elephants' (*ndlovu* means 'elephant' in several African languages). The dung is dried, sanitised, rinsed and dried again, making it safe to consume. The dry product is then infused in the gin with other gin flavourants such as juniper and coriander. The resulting drink has a wooded, earthy taste; each bottle is marked with the GPS coordinates of where and when the dung was found so that one can compare a bottle from the Kruger National Park in winter to one from Botlierskop in summer. The marketing slogan for Ndlovu Gin is 'Made in Africa – designed by elephants' (www.ibhu.co.za).

Since the first batch was produced in November 2019, the company has produced over 6 000 litres of Indlovu Gin and exports it to Africa and Europe, with 15 per cent of all profits donated to the Africa Foundation for elephant conservation (Bailey 2020).

African rum: The first *rhum agricole* made from sugarcane juice (as opposed to *rhum industriale* made from molasses) to be produced in Africa was Tapanga

Rum, named after the Swahili word *panga* (the wide-bladed axe used for cutting sugarcane). At the Zululand Distilling Company run by Geoffrey Woollatt near Gingindlovu in Zululand (South Africa), the fermentation of their *rhum agricole* takes place in a majestic old copper still named Nandi, after the Zulu King Shaka's mother (Abbott 2019).

Alcohol-free beverages: As the trend towards more responsible drinking sweeps the world, more 'brewers' are catering for the needs of this mindful generation. Now, a Cape Town company has produced Abstinence, a blend of Cape spice botanicals added to cloves, angelica root, cassia bark, green cardamom seeds, whole peppercorns, buchu leaves and fresh orange juice, infused in alcohol for 24 hours. Water is then added to the infusion, which is distilled at an absolute pressure of 65 mbar, resulting in a boiling temperature of 37°C, which eliminates the alcohol. What is left is a non-alcoholic spirit to which nothing is added – no flavourants, colourants or sweeteners. Abstinence has a strong, herbaceous taste and is suitable as a mixer or a refreshing drink on its own. It was featured when Cape Town's Kirstenbosch Gardens hosted the Mindful Drinking Festival in October 2019 to showcase non-alcoholic beers, wines, bubblies, fruit juices and kombuchas (www.mindfuldrinking.co.za).

Lamugin, popularly known in Ghana as Hausa beer, is one of the oldest and most widely consumed non-alcoholic beverages in Muslim communities in Africa and is usually sold in informal markets in calabashes or bowls. Unlike traditional beer brewed with cereal grains (usually malted barley, wheat, maize, rice or cassava), the main ingredient of Hausa beer is pulp from the tamarind tree (*Tamarindus indica*), which produces bean-like pods containing seeds and fibrous pulp. As the fruit ripens, the juicy pulp that is initially sour tasting becomes sweeter. Lamugin is popular in *zongos* (stopovers for Muslim travellers) and is not only refreshing but also healthy, as it is known to treat diarrhoea, constipation and fever. Hajia Fati Shaibu, a typical vendor, has been brewing Hausa beer for 27 years and has sale points throughout Accra. She prepares the beer by soaking tamarind in water for an hour and then mixing it with chopped ginger, cloves, lemon juice and sugar. The beer has become so popular that it is now served in restaurants and at corporate events and weddings (Quansah 2019; personal observation 2019).

Healthy energy drinks: People who drink energy drinks are often frustrated by the unhealthy ingredients in these products. Now a South African company produces Beebad, a natural energy drink sweetened with honey and enriched with royal jelly, propolis, maca root extract, ginseng and vitamins that contains no refined sugar, gluten, taucine, artificial sweeteners, flavourants, colourants, preservatives or sodium, although it does contain caffeine (www.beebad.com).

Potable drinking water: Many African countries have experienced a steady decline in the quality of their drinking water, which creates an opportunity for innovators. San Aqua HCA in Cape Town is one of the companies that took advantage of the gap and designed and patented a water treatment technique called Hydrochemical Activation Treatment (HAT). Pamela Alborough, originally a science teacher, co-founded the company while caring for her young children at home. With her husband, Howard, providing the chemical engineering skills, she was able to combine her love of science, business and entrepreneurship and take the product to market. The HAT process involves applying an electric charge to water to sterilise it. It uses platinum group metals and other catalysts that eliminate the need to separate the anode and cathode, and therefore generates a stronger electrical current through the water. It treats water of varying quality cheaply and efficiently by coagulating harmful pollutants, producing a product that is safe to drink and use in cooking (Manyonga 2017a).

Discussion: In 1949, US scientists launched a new drug (cortisone) that could be used to treat a variety of ailments, from arthritis and allergies to lupus and skin conditions. The drug was initially made from diosgenin extracted exclusively from the Mexican wild yam (*Dioscorea villosa*), which triggered a global search for other sources. In the 1950s, botanists announced that diosgenin could also be extracted from elephant's foot (*Dioscorea sylvatica*), an indigenous southern African plant. This research benefited from earlier insights which revealed that African people had used the saponins in wild yams for centuries for medicinal purposes (Beinart 2020). Thereafter, the British pharmaceutical company Boots began to exploit elephant's foot on a massive scale, with over 6 000 tonnes being dug up and processed; however, during this period, indigenous people received no benefits from the industry. Soon the conservation authorities became concerned that the plant would be exploited to extinction and by 1960, all commercial collection permits had been terminated and Boots ceased production of diosgenin from *D. sylvatica* (Beinart 2020).

A dispute over the rights to another indigenous plant, the bitter ghaap (*Hoodia gordoni*), led to a landmark decision in 2003 that saw the South African Council for Science and Industrial Research (CSIR) sign a benefit agreement with the San Council for the exploitation of bitter ghaap, traditionally used as an appetite and thirst suppressant. The legal battle, which was led by the San Council on behalf of the San ethnic group in southern Africa, was eventually successful – and has influenced subsequent legislation on indigenous knowledge and benefit sharing (Maharaj 2011).

Although a Russian immigrant first realised the commercial potential of *rooibos* tea and took it to market, infusions from the *rooibos* plant have been used for centuries by the Khoi-San people to make refreshing drinks and medicinal remedies. Now the implementation in South Africa of a revolutionary access and benefit-sharing (ABS) agreement between the *rooibos* industry and the Khoi-Khoi and San communities will see indigenous people benefit from the commercial success of the tea. South Africa is a signatory to the Nagoya Protocol on Access to Genetic Resources, which requires industries that trade in indigenous biological resources such as *rooibos* to share benefits with the traditional knowledge holders in a fair and equitable way. The ABS agreement signed at the *!Kwa ttu* San Heritage Centre near Cape Town in January 2020 is the first of its kind, in terms of the Nagoya Protocol, in that it requires the entire *rooibos* industry – and not just specific companies – to honour the agreement. A benefit-sharing levy of 1.5 per cent of the farm gate price of *rooibos* has been applied and is expected to generate revenue of at least R9 million per year for the Khoi-Khoi and San communities (Anon 2020b).

This agreement has been hailed as a best-practice example that provides a robust framework for other bioprospecting access and benefit-sharing agreements in Africa and the rest of the world. It is also regarded as an important milestone in the history of global governance for the preservation of genetic biodiversity and associated indigenous knowledge. Other African countries that are signatories to the Nagoya Protocol include Algeria, Benin, Burkina Faso, Cape Verde, the Central African Republic (CAR), Chad, the Congo, the DRC, Djibouti, Egypt, Ethiopia, Gabon, Ghana, Guinea, Guinea-Bissau, Côte d'Ivoire, Kenya, Madagascar, Mali, Mauritania, Mozambique, Niger, Nigeria, Rwanda, Senegal, the Seychelles, Somalia, Sudan, Togo and Tunisia.

7

Health and medicine I: Covid-19

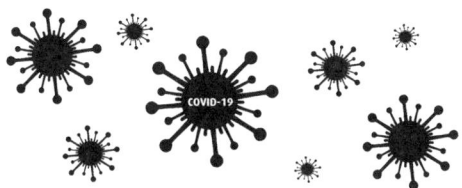

'Never before in history has innovation offered
promise of so much to so many in so short a time.'

Bill Gates, philanthropist and co-founder of Microsoft (Meier 2019)

Introduction: The biggest science news story of the past year has been the novel coronavirus (causing the disease Covid-19), which has tested Africa's ability to respond to yet another disease pandemic to the full. The pandemic started in Wuhan, a city of 11 million people in central China, and officials quickly identified the pathogen as a novel coronavirus related to the viruses that had caused the severe acute respiratory syndrome (Sars) outbreak in China in 2002/3 and the Middle East respiratory syndrome (Mers) outbreak in 2012 (Markotter 2020a). The way that the Chinese authorities handled the Sars crisis, by withholding crucial information for long periods, created a health fiasco and there were fears that the same would happen with Covid-19. Although there is evidence that the early warning signs of Covid-19 were suppressed, particularly between 14 and 20 January 2020, it is clear that the Chinese had developed a robust network for detecting novel viruses post-Sars and their scientists quickly isolated and sequenced the new virus. This allowed President Xi Jinping to release the viral sequences immediately so that the scientific community worldwide could start to develop diagnostics, vaccines and therapies.

As zoonotic viruses have a molecular 'clock' that allows scientists to estimate when they first jumped from wild animals to humans, scientists were able to conclude (from the limited genetic diversity found in the novel coronavirus) that the outbreak had been detected within weeks of it starting. The close similarity between Covid-19 and Sars also helped to speed up the process of identifying and characterising the pathogen. Initial estimates suggested that Covid-19 is 20 per cent less deadly than Sars but 23 times more deadly than flu, although it has the potential to spread far more rapidly than Sars.

Like other zoonotic diseases (transferred from wild animals to humans), such as Sars (probably transferred from bats to civets to humans) and Mers (transferred from dromedary camels), Covid-19 is hosted by wild animals such as civets, snakes or bats and then transferred to humans. The Huanan market in Wuhan – where meat from African animals such as snakes, monkeys, pangolins, porcupines and hedgehogs were on sale – was identified as the source of the outbreak. All these animals are being investigated as the possible source of the novel coronavirus (Campbell & Gunia 2020; Markotter 2020a). What has not been explained is why these zoonotic disease transfers have not taken place in the numerous bushmeat markets in Africa.

The procedures to adopt when a new infectious disease breaks out are not new: isolation and quarantine were carried out – with limited success – in ancient Athens when typhoid fever struck in 430 BC, during the Black Plague in Europe in the 1330s, during the 1519 and 1633 smallpox pandemics, and during the Spanish Flu pandemic of 1918. China responded to the Covid-19 crisis by suspending air travel in and out of Wuhan Tianhe International Airport, which offered direct flights to 15 countries, impacting the travels of over 15 million people during the Lunar New Year holiday. In an unprecedented step, the Chinese government later quarantined the entire city of Wuhan (Campbell & Gunia 2020; Park & Campbell 2020).

With the Covid-19 pandemic, African health officials have had the benefit of their experience in dealing with the Ebola Virus Disease (EVD) and Mers. They avoided the 'authoritarian trap' by quickly disclosing the extent of the disease outbreak and communicating information and advice to their citizens so that there could be a collaborative effort in combating it. They appealed to the international community to support their efforts and on 20 April 2020, the World Bank announced that it had committed US$160 billion to help developing countries (mostly in sub-Saharan Africa) respond to the pandemic, especially to support fragile healthcare systems and inadequate social safety nets but also to address high debt vulnerability (Dudhla 2020).

Precolonial reactions to epidemics: Shadreck Chirikure, a British Academy global professor at Oxford University and professor of archaeology at the University of Cape Town, conducts research on indigenous mining and metallurgy. He argues that remnants of precolonial technologies are an important part of our heritage and should be preserved because they may teach us lessons from the past. For example, his research at the early human settlement K2 (part of the Mapungubwe World Heritage Site in Limpopo Valley, South Africa) revealed that a sharp increase in infant mortality had prompted the sudden abandonment of the settlement in about 1 000 AD and migration to another site. Likewise, research in Ghana has revealed that some settlements in the Birim Valley were suddenly abandoned at the same time as the Black Death pandemic in Europe (Chirikure 2020).

Ancient African cultures were also in the habit of burning settlements where disease outbreaks had occurred and show evidence of deliberate 'social distancing' in the design of village layouts in Zimbabwe, Mozambique and elsewhere. While social coherence was the glue that held societies together, social distancing was used as an in-built mechanism to reduce the spread of diseases. Archaeologists' findings also reveal that it was taboo to touch or interfere with the remains of the dead lest they transmit disease. In the 17th and 18th centuries, the Shona people in present-day Zimbabwe isolated those suffering from infectious diseases such as leprosy in temporary huts. These behaviours were augmented by diversified diets of fruits, roots and other nutrient-rich foods that strengthened their immune systems (Chirikure 2020).

Predicting a new pandemic: For decades, researchers have investigated and public-health NGOs have predicted when another zoonotic disease epidemic would break out. In 2015, the WHO compiled a list of emerging zoonotic diseases 'likely to cause severe outbreaks in the near future'. Some of the most widely recognised and feared diseases started off in wild animal species, including salmonella, influenza, Lyme disease, tuberculosis (TB), anthrax, Aids, Sars, Mers, EVD and of course Covid-19 (Garrod 2020). The WHO states that zoonoses account not only for 60 per cent of diseases found in humans, but also represent 75 per cent of all emerging infectious diseases. It is likely that the risk is becoming more severe as habitat destruction and other environmental disruptions cause rural people to intensify their hunting and the bushmeat trade becomes increasingly global (Garrod 2020).

Possible African origins of the novel coronavirus: The Covid-19 outbreak in China in late 2019/early 2020 has had serious medical and economic consequences, and the search is on for the source of this deadly virus – possibly a bat (Hayes 2020). Bats are known to act as reservoirs of viruses, and horseshoe bats (genus *Rhinolophus*) have been found with viruses like Covid-19 (Markotter 2020b). There are 40 species of *Rhinolophus* bats in Africa but, so far, no viruses from them related to the cause of Covid-19 (or Sars) have been reported. Viral sequences distantly related to Sars have been identified in horseshoe bats collected in Rwanda and Uganda and free-tailed bats (*Chaerephon* species) in Kenya, but they do not at present pose a direct spillover risk and have not been shown to infect human cells. Although no outbreaks of coronaviruses linked to bats have thus far been reported in Africa, there is a need to be vigilant (Markotter 2020b).

Bats are excellent hosts for viruses, as their highly social lifestyle leads to the constant exchange of viral pathogens among them. But why do bats themselves not die from these infections? The answer may lie in a unique feature of these mammals – flight. The physiological requirements for flight have affected the bat's immune system. Flight causes bats to have elevated metabolic functions and raises their core body temperature above 38°C. This means that they are often in a state which for humans would be a fever, which may be a mechanism to help them survive viral infections (Grehan 2020). Coronaviruses and bats are locked in an evolutionary arms race in which the viruses are constantly evolving to evade the bat's immune system and the bats are evolving to withstand infection from coronaviruses. Considering that 20 per cent of all mammals are bats, it is not surprising that they are the primary mammalian carriers of diseases. Bats were also the source of the viruses causing EVD and rabies, Nipah and Hendra virus infections, Marburg virus disease and strains of influenza A (Hayes 2020).

Lessons from Ebola: Unlike the viruses that cause EVD, Sars and Mers that infected mainly poorer people, the novel coronavirus can and does infect all sectors of society. The first Covid-19 death recorded in Africa was an affluent man with a respiratory precondition: Rose Marie Compaoré, 62 years old, the vice-president of Burkina Faso's National Assembly. Another high-profile casualty of the disease was Jacques Yhombi-Opango, the president of Congo-Brazzaville from 1977 to 1979, who died in a Paris hospital in March 2020 at the age of 81 years. In April 2020, Guinea-Bissau's Prime Minister Nuno Gomes Na Biam and three of his ministers tested positive for Covid-19.

When the EVD outbreak erupted in West Africa in 2013, the world was not paying attention and by the time the virus was detected, it had been circulating for months. In South America, the Zika virus had been circulating for over a year before it was detected. In both cases, valuable time was lost in controlling the outbreaks, which gave the viruses time to mutate and become even more dangerous (Sabeti 2020). African countries are particularly vulnerable to these epidemics because the continent accounts for only 1 per cent of global health expenditure but it carries 23 per cent of the disease burden, according to the UN Economic Commission for Africa (Uneca). Furthermore, there is a chronic lack of medical facilities and staff, as the continent only has 1.8 beds per 1 000 people (Uneca 2019).

Prof. Janusz Paweska, the head of the National Institute for Communicable Diseases National Health Laboratory Service in Sandringham, Johannesburg (South Africa), found that several factors were important in controlling the EVD outbreak: quick development of diagnostic capability and experimental vaccines, research on the scene of the outbreak, securing community trust and engagement, follow-up care for survivors, political stability and support, and substantial international support (Paweska 2020). The same will apply to the Covid-19 pandemic. To prepare African countries for further disease epidemics, the Africa Centres for Disease Control and Prevention were established in Addis Ababa (Ethiopia). In addition, the Coalition for Epidemic Preparedness Innovations (a global group of public and private organisations dedicated to speeding up the development of new vaccines) was created and by late March 2020, had already issued three contracts to develop vaccines against Covid-19 (Piot 2020).

Sierra Leone suffered over 3 000 deaths during the deadly EVD outbreak from 2013 to 2016, but its health officials gained valuable experience in preparedness and the management of disease outbreaks. They realised the value of health education, raising awareness of the nature of the disease and how it is spread, and informing people about preventative measures (Kangbai 2020). Health officials in Sierra Leone embarked on a variety of community engagements, including workshops and radio and TV health programmes targeted mainly at community leaders and rural health caregivers because they had access to the broader community and played a major role in ending the EVD outbreak in 2016. To avoid a 'digital pandemic' of misinformation, South Africa has placed strong emphasis on transferring accurate information to people who are not literate or do not have internet access, as well as on involving traditional healers, empowering public-health officials and making health messages available in indigenous languages.

According to Dr Chikwe Ihekweazu (2020), Nigeria also learned from the EVD outbreak. The Nigeria Centre for Disease Control established the Coronavirus Preparedness Group that constantly reviews the risk of Covid-19 infections and improves the country's preparedness based on new findings. According to the commissioner for health in Lagos, Akin Abayomi, one of the key lessons learned from Ebola (which was named after a river in the DRC) was the need to build systems during 'peace time' that could be used during outbreaks, including rapid-response teams, efficient interstate communication systems, well-trained technical staff and properly equipped national reference laboratories that have primers specific for each new virus. Emphasis was also placed on developing emergency operations centres and issuing public-health advisories to all Nigerians (Kangbai 2020).

Kenya and Ghana learned from the EVD, Sars and Mers outbreaks, and put emergency measures in place. Ghana has repurposed EBV treatment centres and mobilised 35 000 military and police personnel to enforce isolation measures, as their health officials recognise that the novel coronavirus is more infectious than the Ebola virus because it can be transmitted in air droplets like the common cold. There are also reports that the novel coronavirus can be transmitted during the incubation period before the infected person develops a fever, which means that temperature screening has to be combined with rigid individual screening of high-risk people (Ziraba & Quashie 2020). By early June 2020, African countries reported far fewer deaths from Covid-19 than many European or American countries. Several reasons were put forward to explain this trend: the disease reached Europe and North America first from China, which allowed African countries to initiate enhanced surveillance early in the pandemic; the extent of air travel is much lower in Africa than in Europe and North America; and African countries were able to implement lockdown procedures relatively early in the pandemic cycle.

Measures to reduce the impact of Covid-19: Many experts believe that the policies adopted in Western countries to flatten the curve of Covid-19 infections (such as self-isolation, working from home, lockdown, and regular handwashing and social distancing) will be less effective in densely populated cities and informal settlements in Africa – Lagos has a human population density of 209 people per hectare compared to New York's 25 (Lawanson 2020). Social distancing is not practical on crowded public transport systems such as Lagos' Bus Rapid Transit (BRT) system and *danfos* (minibus taxis), which are used by 67 per cent of all commuters. Working from home is impractical for the 68 per cent of Lagosians, and hundreds of millions of other Africans, who earn a living from face-to-face informal trading in street markets and there are concerns that forced isolation will drive many of them below the poverty line. The Senegalese researcher Dr Massamba Gueye emphasises that lockdowns attack the social fabric of communities by banning religious gatherings and ceremonies which are the foundation of their society (Oudenhuijsen 2020). In response to these and other opinions, the Nigerian authorities decided to ease lockdown restrictions early on 4 May 2020 in Lagos and Abuja (Lawanson 2020).

Even regular handwashing is not practical for many urbanites in Africa, as it is estimated that only 56 per cent of the urban population of sub-Saharan Africa has access to piped water and many have to spend 30 minutes or longer sourcing it. How can you wash your hands regularly in Lagos when you must buy water by the bucket (Haas 2020; Lawanson 2020)? Scientists have also expressed concern about the indirect impacts of lockdowns.

But when we talk about saving lives, we should factor in the lives we are taking. The net number is what counts … The biggest public health risk in Africa is not

Covid-19 but the consequences of regional and global measures designed to reduce its effect on public health. The cost-benefit analysis of these measures yields a different result in Africa than in Europe, North America and large parts of Asia.

Prof. Alex Broadbent, director of the Institute for the Future of Knowledge, University of Johannesburg (Naidu 2020: para. 10, 14)

One of the biggest risk factors for serious or fatal Covid-19 cases is age. Piroth et al. (2021) estimate a fatality rate from Covid-19 of 0.32 per cent for people aged 60 years and younger and 6.4 per cent for those older than 60 years. The median age in Africa is 18 years, compared to 42 years in Europe; in terms of its human demography, Africa is by far the youngest continent. 'We must ask, then, whether African nations (including South Africa) have as much reason to fear Covid-19 as regions where so much of the population is older', Broadbent says (Ryan 2020). Perhaps we should be more concerned about the African people who lose their jobs, die of poverty or resort to violence to obtain food.

Each death resulting from the virus is a tragedy. But so is each death resulting from caged citizens and frustrated law enforcers, and so is each victim of home violence. And each malnourished child. And each newly unemployed South African.

Statement issued in April 2020 by the Democratic Alliance, South Africa
(Ryan 2020)

In March 2020, Mauritius implemented novel procedures to reduce congestion in supermarkets by allowing only 30 minutes for each shopping visit and allocating specific days for shopping depending on the initial letter of your surname: A to F shop on Monday and Thursday, G to N on Tuesday and Friday, and O to Z on Wednesday and Saturday. On 11 May 2020, Mauritius became the second African country (after Mauritania) to declare itself Covid-19 free at that time, although 10 people had died of the disease. The Ministry of Health and Wellness previously undertook a massive testing campaign in the small island nation, with 50 077 rapid antigen tests and 23 495 polymerase chain reaction (PCR) tests.

Fake cures: In April 2020, Madagascar's President Andry Rajoelina promoted homespun remedies such as drinking ginger and lemon tea and steam therapy for treating Covid-19, and then announced the launch of a herbal potion that he believed could cure the disease. The drug Covid-Organics (CVO), developed by the Malagasy Institute of Applied Research (Imra) in Antananarivo, contains extracts from sweet wormwood (*Artemisia annua*), a plant that has proven efficacy against malaria but no proven effect on Covid-19. Furthermore, Rajoelina issued a presidential decree that made it mandatory for CVO to be administered to all children returning to school in April 2020. He announced that CVO would be distributed 'free of charge to our most vulnerable compatriots and sold at very low prices to others. All profits will be donated to Imra to finance scientific research' (Shaban 2020). Rajoelina's announcement was met with scepticism internationally. In April 2020, the WHO announced that there was no proof that the CVO drink was effective against Covid-19 and insisted that it should undergo clinical trials before being sold. Unesco urged African people to heed accurate scientific advice and 'don't go viral'. In July 2020, the South African newspaper *Mail & Guardian* reported that hospitals in Madagascar had been overwhelmed by Covid-19 cases and that an analysis authorised by the Nigerian government did 'not show any proof that Covid-Organics can cure Covid-19'.

Tanzania's President John Magufuli framed Covid-19 as a war that should be waged using top-down tactics, and not a health and socioeconomic calamity requiring scientific consultation and community engagement. He downplayed the pandemic's threat, stating that faith was more important in healing the disease than science, publicly questioned the efficacy of Covid-19 tests and discontinued regular updates on Covid-19 deaths. He then sent a plane to Madagascar to collect that country's traditional remedy for the disease (Kwayu 2020). In March 2021, Magufuli died aged 61 years, after weeks of rumours that he had contracted Covid-19.

In Ghana, a Pentecostal pastor launched and sold Coronavirus Oil, telling his congregation that it was effective against Covid-19. Other quacks in Ghana claimed that sliced onion and their 'Covid-CURE' could cure not only Covid-19 but also cancer and Aids (Izugbara & Obiyan 2020). In most African countries, it is not illegal to claim that you have found a cure for a disease but it is illegal to sell a 'cure' as a 'medicine' if it has not been tested and approved. While some medicinal remedies derived from indigenous knowledge have been widely and successfully applied in Africa, there is no known cure for Covid-19 that is based on them. A heathy dose of humility and respect for science are key ingredients in dealing with the complex medical and socioeconomic impacts of the pandemic.

Although traditional Chinese uses of plants and animal parts came under intense scrutiny in 2020/1 as a result of the Covid-19 pandemic, we need to remember that one of the most important medical breakthroughs of the 20th century was thanks to Chinese traditional medicine, as described earlier. While searching for a treatment for malaria, the Chinese phyto-pharmacologist Tu Youyou discovered artemisinin, a drug derived from sweet wormwood used in traditional Chinese remedies, which subsequently has prevented millions of deaths from malaria. Her discovery earned her the Nobel Prize in Medicine in 2015 and won humanity important ground in the battle against one of history's deadliest diseases (Gates 2020). Tu described her team's findings, published in 1979, as 'a gift from traditional Chinese medicine to the world'. Artemisinin is being evaluated as a possible treatment for Covid-19.

Benefits of Covid-19: The extreme severity of the Covid-19 pandemic (over 3 000 people died in the first two months and 4.32 million had died by 31 July 2021) has ironically resulted in some consequences that are beneficial to wildlife and the natural environment. In early February 2020, the Chinese government issued a temporary ban on wildlife trade to curb the spread of the virus; on 24 February 2020, they banned the consumption and sale of bushmeat at wet markets. The temporary ban closed about 20 000 captive breeding enterprises involving 54 species (including snakes, ostriches, guinea fowls, peacocks, civets, porcupines, bats and monkeys) from Africa that are allowed to be traded domestically in China. The breeding centres operate through loopholes in Chinese domestic law that are arguably against the spirit of Cites (the Convention on International Trade in Endangered Species of Wild Fauna and Flora) (Evans 2020). However, the ban excludes species used in traditional medicine, such as sea cucumbers, perlemoen, seahorses and rhino horn. The international concern is that if bushmeat markets in China and elsewhere are not permanently banned, another even more harmful zoonotic virus with the high infection rate of Covid-19 but the mortality rate of the Spanish Flu (5 per cent) or Mers (35 per cent) could infect humans, with even more devastating impacts.

The Chinese government has promoted the idea that wildlife domestication benefits rural development and poverty alleviation, and a 2017 report by the Chinese Academy of Engineering valued their wildlife farming industry at £57 billion. Under their 'sustainable utilisation' policy, they allow civet farming – even though these wild cats have been identified as a carrier of Sars (Standaert 2020). The legal global wildlife trade is estimated to be worth about US$300 billion and the illegal trade between US$7 and US$23 billion annually. At least 5 500 different species are traded and wildlife breeding in China alone is estimated to be worth US$74 billion and involve 14 million people. Economists usually oppose bans because they tend to drive legal trade underground. Moreover, legal trade often has no ability to crowd out illegal activity because an illegal product is always cheaper to procure. Instead, in a corrupt world, it can create laundering opportunities for illegal supply chains. Improving regulations alone is also not the answer, as every improvement in regulation requires more law enforcement capacity, technology and manpower – which most countries do not have. Furthermore, the costs of policing bans are probably lower than the costs of policing improved regulations, and it is clear that a hybrid policy needs to be adopted. Whatever the solution, we cannot continue to use wildlife as if it is free capital, as ecological systems are at a breaking point (Harvey 2020).

The Covid-19 pandemic has revealed the full scale of the Chinese bushmeat industry to the world. It has finally made us realise that the illegal wildlife trade is not only an environmental issue but also a serious public-health issue, and there is now strong international pressure on China to permanently ban trade in species that are protected or endangered in and outside China. The temporary ban on bushmeat trading in China has had a direct impact on African wildlife conservation because the capture and export of wildlife, including endangered species such as the pangolin (a possible intermediary host of the coronavirus), has been stopped and will hopefully never resume with the same intensity. However, in 2003, at the height of the Sars epidemic, China also issued a temporary ban on wildlife trade; however, it was revoked just six months later, after international attention had shifted elsewhere, and the breeding and trading of wildlife soon resumed (Daly 2020). Let us hope that the extreme severity of the Covid-19 pandemic and the fact that assessments of the value of the bushmeat industry must now take into account not only biodiversity conservation issues but also biosecurity, animal rights, public-health and economic issues will result in the latest ban becoming permanent.

The second benefit was detected in early March 2020 when China reported the lowest levels of air pollution by CO_2, carbon monoxide and nitrogen dioxide in years in areas where economic and industrial activity had been impacted by travel bans. This soon became a worldwide trend. It is reminiscent of the significant reductions in ozone depletion and upper atmosphere pollution that occurred during the international air travel ban after the 9/11 attacks on the World Trade Centre in New York in September 2001. Conversely, the tens of millions of discarded face masks and disposable plastic gloves that have started to enter rivers and oceans pose a new environmental threat and may reverse significant gains in the reduction of plastic pollution in recent years.

Alternatives to bushmeat: While global bans on the bushmeat trade are being considered, an innovative alternative is being tried in Africa – the development of livelihoods that provide alternative sources of employment, income generation and protein at the community level, thereby addressing the problem at its source. These interventions reset past behaviours, rather than create new ones, and include activities such as beekeeping, fish farming, minilivestock rearing (including cane rats, rabbits and guinea pigs) and caterpillar farming. Wicander and Coad (2018) reviewed the development of alternative livelihood projects (many of them launched by NGOs with international donor support) in West and Central Africa, and found that they had been most successful in Ghana, the DRC, Cameroon, Liberia, the Republic of Congo and Burundi, with lesser success in Gabon. Van Velden et al. (2020) propose that micro-enterprises such as craft shops and beekeeping could help to solve the problem in Malawi. In many of the projects, the conditions for participation include an agreement by participants to reduce or cease hunting and selling bushmeat.

However, overall, there is no doubt that the Covid-19 epidemic is a catastrophe for the environment and for humankind. Budgets for biodiversity conservation have been cut drastically due to the loss of ecotourism revenues or the diversion of funds to health issues. Growing poverty and the loss of livelihoods are driving people to poach and there is widespread concern about deforestation and habitat loss, especially in Central Africa. A worldwide crisis in food insecurity and unemployment has also been precipitated. Notwithstanding this, the pandemic has benefitted small-scale vegetable and fruit producers who have been agile enough to change their business model and supply directly to consumers who are cooking most of their meals at home rather than buying them precooked. This has resulted in a matrix-like food supply chain that smaller businesses are better able to negotiate than big chain stores. It has also been a boon for local produce, as travel restrictions have reduced the amount of food that can be imported and exported.

There is also international concern that biodiversity loss, including the reduction of microbial diversity, is increasing the risk of humans contracting zoonotic diseases. Microbes (bacteria, viruses and fungi) are essential for maintaining ecosystem health as well as human health, as our digestive tracts, airways and skin harbour vast networks of microbes that perform vital life-supporting functions such as digestion and strengthening our immune systems. Exposure to a diverse range of microbes helps our bodies to mount an effective response against pathogens, but a lack of exposure (especially in sterile, urban environments) deprives us of this capacity. Germaphobia (the perception that all microbes are bad) exacerbates the situation, as it encourages people to sterilise their homes and stops children from playing outside in the dirt (one of the most biodiverse habitats). A key component of any post-pandemic recovery strategy should be encouraging people to spend more time outdoors and redesigning our cities to maximise interaction with green and blue spaces (Robinson 2020).

Taking its toll: The pandemic has taken its toll on some of Africa's most creative people. Two of West Africa's leading musicians, Aurius Mahele and Manu Dibango, died of Covid-19 in Paris in March 2020. Mahele (the king of Congolese *soukous* music – from the French '*secousse*', meaning shock or jerk) died at the age of 66 years and Dibango (the legendary Cameroonian saxophone and vibraphone superstar and funk, jazz and disco artist whose music had influenced Michael Jackson) died aged 86 years. Dibango was an artist whose creativity spanned the ages, as he brought traditional Cameroonian music back to life using modern instruments. The most famous albums of his band African Jazz include 'Soul Makassa' (1972) and 'Electric Africa' (1985). On 18 March 2020, Dibango posted a message to his fans informing them of his illness. His tone was upbeat – like his music – and he said that he expected to recover, but he died six days later. However, in July 2020, the *Mail & Guardian* reported that a 114-year-old Ethiopian monk, Tilahun Woldemichael, had recovered from Covid-19 after spending three weeks in hospital!

Impact of Gavi: Gavi is a globally collaborative vaccine alliance launched by Bill and Melinda Gates that vaccinates almost half of the world's children, which gives it tremendous power to negotiate vaccines at affordable prices. As a result of these market-shaping efforts, the cost of fully immunising a child with all 11 WHO-recommended childhood vaccines is US$28 in Gavi-supported countries compared to about US$1 100 in the USA. The pool of manufacturers producing prequalified Gavi-supported vaccines has grown from five in 2001 (with one in Africa) to 17 in 2017 (with 11 in Africa, Asia and Latin America).

> When Melinda and I started this work more than a decade ago, we were inspired by the conviction that 'all lives have equal value'. So, one of the first things we invested in was vaccines, which protect all children who receive them, no matter how rich or poor they may be. In short, vaccines work.
>
> *Bill Gates, co-chairperson of the Bill & Melinda Gates Foundation*
> (Gates & Gates 2021)

Gavi, which is active in 37 African countries, is distributing Covid-19 vaccines throughout the continent. According to Dr Marthe Essengue, the regional head of Gavi in Francophone Africa, 'All people in the world need to be protected by vaccines – not only those who can pay. This is about saving lives' (www.gavi.org/vaccineswork/covid19).

Africa's response to the pandemic: Dr Moses Alobo heads the African Academy of Science's response to the Covid-19 pandemic. Immediately after the pandemic started, the academy launched a survey among hundreds of African scientists to define research priorities across a range of disciplines so that it could provide guidance, networking opportunities and resources to fight the virus. In total, 845 respondents (79 per cent in Africa) from 56 countries (including 39 in Africa) responded to the survey. The respondents were policy-makers or from the biomedical sciences, clinical and epidemiological research, social sciences and management sciences (Alobo 2020). The priorities that were defined largely supported the WHO's Covid-19 Research Roadmap, although some additional points were raised such as mental health, food security, the indirect effects of the pandemic on maternal and newborn health, and the co-management of chronic conditions.

Bright Simons and mPedigree: Bright Simons – whom the *Financial Times* described as 'frighteningly clever' – is a Ghanaian social innovator, entrepreneur, writer and vice-president in charge of research at Imani (an African think-tank in Accra that applies free market solutions to social problems). He is also the founder and president of mPedigree, a social enterprise noted for its work in exposing the makers and distributors of counterfeit medicines and for creating the computer programme Goldkeys that enables the verification of medicinal, agricultural and other products. Manufacturers who sign on for the mPedigree scheme upload information on their medicine into a central registry and when consumers buy the product, they can query the pedigree information stored in the registry. Ghana, Kenya and Nigeria have announced their support for the mobile telephony anticounterfeiting system deployed by mPedigree, and other African countries are likely to follow suit.

In 2009, Simons was a TED Fellow. In 2012, the WEF recognised him as a Young Global Leader and the mPedigree network as a technology pioneer. In 2011, the Salzburg Global Seminar named him a fellow for expanding the mPedigree network from Africa into India and China. In 2013, the International Foundation for Africa Innovation gave Simons a lifetime achievement award for his contributions to innovation in Africa. In March 2016, *Fortune* magazine listed Simons among the World's 50 Greatest Leaders, placing him just ahead of Canadian Prime Minister Justin Trudeau; he was also ranked among the 100 Most Influential Young Africans in 2016 by Africa Youth Awards. In October 2016, he won the Innovator of the Year award at the All Africa Business Leaders Awards in West Africa; in 2018, he was inducted into the Power Brands LIFE Hall of Fame at the London International Forum for Equality.

Ventilator rush: When the novel coronavirus reached South Africa, there was no local manufacturing capacity for ventilators, which are used to supply oxygen to severely ill patients. This was concerning, as the demand for ventilators quickly surged and they were not readily available from global markets. In response, Minister of Trade and Industry Ebrahim Patel and Minister of Higher Education, Science and Innovation Blade Nzimande launched the National Ventilator Project to design and manufacture ventilators locally. Remarkably, they appointed one of South Africa's top scientific agencies, Sarao, to serve as project managers, given their world-class systems integration and systems engineering capabilities. Senior critical-care physicians advised the team to focus on the production of non-invasive continuous positive airway pressure (CPAP) ventilators, as they would have the greatest potential impact on saving lives. The CPAP ventilators developed in South Africa supply a pressurised mixture of ambient air and industrial oxygen to the patient through a mask or hood, and have been evaluated by engineers at Sarao to ensure compliance with international standards as well as usability in urban and rural hospitals. Within a few months, prototypes were designed, developed and tested. An initial order of 10 000 units was engineered by the CSIR, with the final assembly taking place at the Akacia Medical Facility outside Cape Town (Ebrahim 2020).

> Over the next month, we expect many thousands of non-invasive ventilators to be delivered to hospitals and medical facilities across the country, brought together through South African ingenuity and by South African hands … Within the space of four months, our country has gone from having no capacity to produce CPAP ventilators, to now having the first units coming off the production line.
>
> As we build our production capacity, we will support our neighbours across the African continent with these ventilators … The shortage of testing kits, ventilators, medical-grade masks and other PPEs has underlined the importance of Africa developing a strong innovation and manufacturing capability.
>
> *South African Minister of Trade and Industry Ebrahim Patel* (Ebrahim 2020)

Covid-19 antigens: Belinda Shaw, the founder and CEO of Cape BioPharms, is using biotechnology to manufacture Covid-19 proteins in plants. These proteins (called antigens) are used in rapid diagnostic test kits that detect antibodies against the virus in blood, indicating whether they have been exposed to the virus. If a person has Covid-19 antibodies in their blood, they will bind with the antigens in the test kit and trigger a positive response. According to Tamlyn Shaw, the director of Cape BioPharms, the protein antigens and antibodies that they are making could be used as vaccines and also in therapeutics. But the test kits and potential vaccines cannot be rolled out en masse as yet, as Cape BioPharm is struggling to secure a reliable supply of positive serum (blood from Covid-19 positive people) to test the kits, and they lack the advanced facility required to obtain a compliance certificate to make the vaccines and therapeutics (Shaw 2020).

> Africa will always be at the end of the queue when it comes to receiving vaccines and therapeutics. It's enough now. We have the capability, we have the expertise, we have incredibly clever people, to be able to ensure the security of

our own supply of pharmaceuticals. The business of having to import all this stuff should be, hopefully, almost a thing of the past.

Belinda Shaw, CEO of Cape BioPharm (Shaw 2020)

Africa's medical superstar: Her parents were both medical doctors and she swore that she would not become one, but she did. She vowed not to marry a medical doctor, but she did. Dr Matshidiso Moeti has spent her whole career in the public healthcare sector and is now, as the WHO's regional director for Africa, the spokesperson for the global organisation on the Covid-19 crisis. She was born in KwaThema in Gauteng (South Africa) and was a precociously intelligent child. At the age of just 10 years, she began high school at a boarding school in Eswatini (then Swaziland) and then moved with her family to Botswana, where her mother developed family planning services and her father became the smallpox commissioner for southern Africa. After obtaining degrees in medicine and public health in London, she joined Botswana's Health Ministry as a clinician and public-health specialist, and was later appointed to the UN's Children's Fund and then their HIV/Aids programme. She joined the WHO in 1999 and was appointed regional director for Africa in 2015, the first woman in this post.

Throughout the Covid-19 pandemic, Moeti has been the voice of Africa to the world in her weekly briefings from Brazzaville (Republic of Congo), where the regional headquarters of the WHO is located. Switching effortlessly between English and French, she has calmly detailed Africa's efforts to combat the crisis, a reassuring voice in a collective storm. But her main role has been to coordinate public-health responses in collaboration with leaders and ministries throughout the continent and to persuade them to make difficult decisions (Allison 2020c). She gained valuable experience in the fight against EBV and noticed some similarities with the Covid-19 crisis. 'To contain Ebola, cultural practices like washing the dead before burial had to change; for Covid-19, it is about persuading people to take physical distancing seriously', she says. But some aspects of the problem are different: 'When there was an Ebola outbreak, it was an African problem. This is a global problem. Some of the challenges relate to the fact that those countries from whom you would be expecting help are themselves struggling.' Furthermore, the draconian travel restrictions imposed by many African countries have meant that WHO experts and supplies cannot reach the places where they are most needed. She nevertheless remains positive about Africa's ability to fight the disease (Allison 2020c).

Artists against Covid-19: Ugandan pop star Bobi Wine has spent his career singing about social injustice; however, in 2017, he decided to take things further by running for – and winning – a seat in Uganda's parliament. Then the

37-year-old singer announced that he planned to run for the presidency in 2021, taking on President Museveni, who has ruled the country for the past 33 years. Although Wine has been jailed, beaten and charged with treason, he plans to run on a platform of reform (Ayene et al. 2020; Baker 2019b). In 2020, Wine led a star-studded galaxy of African musicians and politicians who composed and performed songs about the Covid-19 pandemic, including George Weah (the president of Liberia) and singers and songwriters from Senegal, Gabon, Nigeria, the DRC, South Africa and Ghana (Sosibo & Allison 2020). Wine's track was made with longtime collaborator Nubian Li and features a fusionist dancehall style that mixes African musical traditions with modern Jamaican 'riddims'. 'Sensitise the masses to sanitise, keep a social distance and quarantine', sings Wine, whose hit soon passed 1 million clicks on YouTube and earned international kudos (Makwa 2020)

Bobi Wine also launched a collective call to action, #DontGoViral, and invited content creators of every musical genre 'and creatives from all over the world' to share their work on #ShareInformation. He argues that in this time of crisis, humanity needs artists and cultural entrepreneurs to bring people together, to activate their collective intelligence and shared humanity, and to translate public-health information into everyday language that everyone can understand. He emphasises that while it is important to prevent Covid-19 from spreading, it is equally important to prevent misinformation and discrimination from going viral.

In Cape Town, The Kiffness (an electronic group founded by David Scott in 2011) embraced the challenges of the Covid-19 lockdown and parodied several well-known songs to raise awareness of the pandemic. The songs included 'Quaranqueen' (Dancing Queen: 'I've lost track of time and my life'), 'Yesterday' ('Yesterday, Covid-19 seemed so far away, now it looks as though it's here to stay') and 'Lockdown Rhapsody' (Bohemian Rhapsody: 'This is the real life, this isn't fantasy, caught in a lockdown, no escape from the quarantine'). The Nigerian singer-songwriter Asa (Bukola Elemide) composed and shared a song on reducing the risk of novel coronavirus infection.

Dominic Makwa of Makerere University in Kampala (Uganda) conducts research on the impact of pop music on society. He found that music has created awareness of the HIV/Aids epidemic and provides psycho-social support for stigmatised victims; he sees no reason why this should not be the case for Covid-19. Songs not only communicate information about a disease, but also shape popular opinion and sensitise people about how to prevent it. They can also be a mechanism for counselling due to the power of metaphor and their ability to turn despair into hope. Makwa (2020) has therefore recommended that the Ugandan government include Bobi Wine as well as Nubian Li, Bebe Cool and other musicians in their official campaign against Covid-19.

Sharon Refa, a hairdresser in Kibera (a poor suburb of Nairobi), has developed a spikey hairstyle called 'Coronavirus' that comprises a series of braided spikes radiating out from the head. Her innovation has rapidly become popular, as it is cheap, stylish (although a bit outlandish) and conveys a strong message. Sharon uses inexpensive cotton thread for the braids, so the hairstyle costs only 50 shillings compared to US$3 to US$5 for an average braided hairstyle. Most importantly, she says that the new fashion is a way in which children can convince their parents, and communicate to the public, that the Covid-19 crisis needs to be taken seriously because many adults in Kenya do not believe that the virus is real (Lynn 2020). Whoever thought that hairstyles could have educational value?

Song and dance can convey a variety of messages but none as strange as that of the dancing pallbearers of Ghana, founded by Benjamin Aidoo. The comedic grim reapers in black suits, sunglasses and leather shoes, who groove to a technobeat while carrying a coffin, have become the accidental face of the Covid-19 lockdown movement with their stark message 'Stay at home or dance with us'. They have made mourners grin through grief but also reflect on their own mortality. People worldwide have shared their video, which has garnered millions of clicks and an international fan base. Aidoo is planning to expand his venture worldwide (Paquette 2020).

Emerging innovations: *Afrika Umoja* (Swahili for 'Africa United') is an initiative created by Africa Oil & Power, HOTT3D and the African Energy Chamber to tackle the Covid-19 threat across Africa. Its aim is to build temporary medical infrastructure where medical professionals can work to combat Covid-19 in partnership with governments, the private sector and local entrepreneurs. The venture makes use of the design, construction and project management expertise of the founders to build temporary medical infrastructure and supply products within days/weeks at any location across the continent.

Engineers at the National Institute of Applied Science and Technology in Tunis (Tunisia) created an online platform that scans lung X-rays to determine whether a person is suffering from Covid-19. Thousands of X-rays have been fed into the system to enable it to recognise the novel coronavirus in the lungs. When an X-ray is uploaded onto the platform, it runs a test to detect signs of a possible infection and is 90 per cent effective in indicating the probability of infection (bbc.com/news/world-africa-53776027). Amid a shortage of ventilators in Covid-19 wards in Nigeria, 20-year-old engineering student Usman Dalhatu built a portable automatic ventilator to help people with respiratory problems.

Kenyan mobile money agent Danson Wanjohi built a wooden device (using a motor, rubber band and gears) that sanitises cash notes which are passed through a slot in the machine. South African techpreneurs Daniel Ndima and Dineo Lioma created a Covid-19 testing kit, qPCR, that produces results in just 65 minutes compared to three days for a normal test. The kit features technology used to measure DNA, but has to undergo regulatory approval before it can be rolled out (bbc.com/news/world-africa-53776027).

Discussion: The Covid-19 pandemic has had an impact on all aspects of African society and economics, and a multipronged approach is needed to alleviate its effects. As always, when a crisis strikes, there are entrepreneurs who take the opportunity and answer society's needs. Senzo Jiyane, a 32-year-old construction worker of Johannesburg runs a small factory that produces hand sanitiser. His sales, which have increased 13-fold since the pandemic started, mainly go to small businesses, hospitals and municipalities (Maeko 2020). Sadly, however, small-scale entrepreneurs and gig workers with insecure or casual employment are the most vulnerable to Covid-19, not only due to loss of income but also in terms of their exposure to the virus. They are typically on the frontline dealing with the public and, in the case of delivery drivers and rideshare hosts, are most likely to make direct contact with infected people. Furthermore, their lack of sick pay or medical aid support might lead many to continue working even if they are infected, rather than to self-isolate.

Julia Sunderland, a former director of the Bill & Melinda Gates Foundation's Strategic Investment Fund and now a co-founder and managing director of Biomatics Capital Partners in the USA, sums up the pandemic problem well:

> In the arms race with pathogens, there can be no final peace. The only question is whether we fight well or poorly. Fighting poorly means allowing pathogens to cause massive periodic disruptions and impose huge burdens in the form of lost economic productivity. Fighting well means investing appropriately in science and technology, funding the right people and infrastructure to optimise strategic preparedness, and assuming leadership over coordinated global responses.
>
> *Julia Sunderland, co-founder of Biomatics Capital Partners* (Sunderland 2020)

The focus on the Covid-19 pandemic may have compromised Nigeria's ability to deal with another health crisis – Lassa fever. This rat-borne disease was first diagnosed in two states in 1969 but spread to 23 states by 2019, with over 600 confirmed cases, 170 deaths, and fatality rates at a high 20 to 25 per cent. This is particularly worrying, as there is an effective treatment for the disease if it is detected and patients seek treatment early enough. Doyin Odubanjo of the Nigerian Academy of Science has made an appeal for Lassa fever to be recognised as an ongoing threat and for the capacity of the Nigerian national laboratory network to be further enhanced to improve the efficiency of diagnoses. Presently, only about 20 per cent of Lassa fever cases are diagnosed in time (Odubanjo 2020).

At the beginning of the Covid-19 pandemic in Zimbabwe, medical doctors went on strike to object to the way in which the soaring inflation rate had destroyed the buying rate of their salaries. In response, Zimbabwe's richest man, telecommunications billionaire Strive Masiyiwa, set up a US$6.25-million fund to pay doctors a subsistence allowance to encourage them to return to the frontline. Although the doctors accepted the offer, the funds would run out after six months and it was unclear what would happen after that (*The Millenial Source* 2020).

8

Health and medicine II: Other medical, hygiene and grooming innovations

HIV/AIDS

'Do not judge me by my successes, judge me by how many times I fell down and got back up again.'

Nelson Mandela, late president of South Africa
(excerpt from documentary *Mandela: The Interview*, 1994)

Introduction: Traditional and Western medicine exist side-by-side in Africa, but they have an uneasy relationship. Over 70 per cent of Kenyans still rely on traditional healers as their primary source of healthcare. Harrington (2018) estimates that in Kenya there is about one healer for every 950 patients, compared to one trained medical doctor for every 33 000 patients. Healers enjoy legitimacy and authority, serve as both herbalists and spiritualists, tend to be close at hand to attend to urgent needs, and are custodians of traditional knowledge and natural biodiversity. Many early independent African governments prioritised Western medicine and downplayed traditional biomedicine in their quest for modernisation. In 1969, Kenya's first president, Jomo Kenyatta, dismissed traditional healers as 'charlatans' while health officials called for them to be outlawed (Harrington 2018).

Notwithstanding these setbacks, traditional medicine has proved to be resilient in Kenya and elsewhere in Africa. It has withstood state neglect, hostility from medical professionals, biopiracy from foreign powers and the loss of biodiversity, yet it has adapted and extended its popularity on a rapidly urbanising continent. It deserves to be taken seriously. Today traditional healers enjoy improved status, with many countries putting policies in place that value and encourage their contributions – a shift that was partly initiated by the WHO in the mid-1970s when they promoted accessible primary healthcare by integrating traditional medicine into state systems, as had already been done in China, Vietnam and Korea (Harrington 2018).

Ongoing threat of HIV/Aids: HIV/Aids is still a serious global health problem and sub-Saharan Africa remains the most affected region. Worldwide, about 770 000 people died from Aids-related conditions in 2018, 160 000 of them in West and Central Africa. The standard treatment of the disease consists of a combination of at least three antiretroviral drugs, but inadequate nutrition may compromise its effectiveness. People with HIV have higher energy needs than non-sufferers, and there is a complex relationship between the disease and nutrition. Dr Temitope Bello, a postdoctoral fellow at the University of Johannesburg, has therefore developed and tested a nutritional education programme for adults living with HIV in Nigeria (Bello 2020).

She found that many Aids sufferers have unhealthy eating habits, unbalanced lifestyles and poor nutritional knowledge, which prompted her to develop educational aids on better nutrition and appropriate exercise and hygiene regimes. Patients were also advised on how they could vary their diet on a limited budget. The outcome of the programme was that the targeted participants functioned better physically, acquired better nutritional knowledge and eating habits, and had improved quality of life compared to a control group. The programme also demonstrated that people do not need more money to make better nutrition choices, but they can and do improve their well-being when they have more knowledge (Bello 2020) – a valuable lesson for all health practitioners.

Current commercially available rapid HIV tests are complex and require multiple steps that must be executed with precision. This can be time consuming for clinicians and could lead to errors during diagnosis; it also discourages self-administration. Incitech in South Africa has developed the Micropatch Rapid HIV Diagnosis Kit that removes this complexity by combining the numerous steps of existing rapid testing kits into a single, self-contained, easy-to-use device that produces accurate results at a comparable cost.

Tutu Tester: Archbishop Desmond Tutu, Nobel Prize winner and much-loved philanthropist, has made many contributions to community healthcare, but the Tutu Fleet launched by his HIV Foundation in 2008 probably has the greatest impact on the well-being of ordinary people. With their signature rainbow banners, the Tutu Tester, Tutu Teen Truck and Tutu Kwik Testers bring essential healthcare services directly to communities in peri-urban Cape Town, where levels of HIV are high but levels of treatment are low. About one-fifth of South Africans aged 15 to 49 years have HIV and the communities that are most vulnerable to the disease live in remote and densely populated areas (Reid 2020).

While the Tutu Tester specialises in HIV testing, it also offers a wealth of other essential services for free (including blood pressure measurements, family planning and TB testing) and has already seen over 50 000 patients. As TB is the main cause of death for HIV-positive people, it is essential to stop both. The Tutu Tester sees a completely different demographic from traditional healthcare facilities, with about half of its clients being male and a quarter under the age of 25 years. Phillip Smith, the mobile services manager for the Tutu Fleet, notes: 'Our research shows that mobiles diagnose HIV at an earlier stage when compared with traditional facilities.' The Tutu Teen Truck facilitates visits by teenagers in an adolescent-friendly environment and ensures that they can receive help outside school hours (Reid 2020).

According to the WHO, a key factor that influences whether adolescents seek healthcare is if they could get into trouble with their parents or guardians. Furthermore, strong societal expectations may be placed on young people to avoid premarital sex, which can deter them from seeking help for sexual health problems. Dorothy Zakariya, a clinical nurse on the Tutu Teen Truck since 2015, commented, 'Our clients need an environment where they feel safe to share and that is what we are trying to provide for them'. The staff often hear heart-wrenching stories of rape, relationship violence and peer pressure – which make the judgment-free environment of the Tutu Teen Truck important. The Tutu Fleet is an essential service that reaches isolated communities and helps to keep young women in work and school while being able to access contraception (Reid 2020).

Other disease outbreaks: As if Sars, Mers, EVD and now Covid-19 do not put enough strain on public-health facilities, some African countries like Kenya are still plagued by outbreaks of Rift Valley fever, malaria, dengue fever and *chikungunya* (a Makonde term meaning 'to become contorted' or 'that which bends up') fever – all of which are transmitted by mosquitoes. Malaria is caused by a protozoan, whereas the others are caused by viruses. According to Eunice Owino, a medical entomologist at the University of Nairobi, heavy rains in Kenya are a major contributor to outbreaks of these diseases because stagnant water is an ideal breeding ground for mosquitos. They deposit their eggs in dry soil, where they can lie dormant for years until they are moistened by persistent rains, which causes them to hatch. She advises that one of the most effective ways to control these diseases is by restricting the breeding of the mosquitoes by spraying stagnant pools and draining water from garbage and discarded containers (Owino 2020). A new technique developed at Duke University in collaboration with Nasa in the USA makes it possible to use satellite imagery to monitor rainfall patterns and predict where and when mosquito outbreaks will occur before they happen.

Technology and women's healthcare: Internet access, automation and other tools of the digital age have helped to remove the barriers that previously kept women from participating in the health economy, but access remains a problem in rural and low-income peri-urban areas (especially in Africa). Gabrielle Lobban, the founder of the health and wellness platform Zumbudda, is optimistic about the future of technology on the continent. She believes that developments driven by mobile technology are useful for solving local problems, as one can only understand the challenges of a particular community if one has either lived in or engaged with that community. Because of this, locally focused programmes have the potential to make massive impacts (*Top Women* 2020).

> Despite a history of low-tech capability, the past decade has seen Africa leapfrog over these challenges to become a highly connected region. Instead of progressing from written communication to computer-driven communication, Africans have become citizens of a mobile technology ecosystem. From fintech to healthtech, Africans are solving local challenges and, in turn, galvanising others into designing new innovations that are uniquely African.
>
> *Gabrielle Lobban, founder of Zumbudda* (*Top Women* 2020: 76)

A good example is Taungana Africa, a programme that provides rural African girls with the opportunity to access and explore science, technology, engineering and mathematics (STEM) fields. The founder of Taungana, Sandra Tererai, is passionate about financial independence for women through STEM. Once a year, her organisation invites girls from rural communities in Zambia, Zimbabwe and South Africa to attend a one-week, multi-industry immersion and entrepreneurship programme. These young women have all been exposed to the daily challenges of water, electricity and sanitation first-hand and, through Taungana, have been given the opportunity to acquire the skills and develop the ideas needed to create sustainable solutions.

According to Lobban (*Top Women* 2020), healthcare in Africa is set to benefit from technological developments. The continent has some unique health challenges, including frequent disease outbreaks, high incidences of maternal deaths and elevated risks of TB and HIV infection, as well as problems caused by migration, poor rural infrastructure and natural disasters (*Top Women* 2020). However, recent developments in remote care and telemedicine have catapulted the health technology sector into new territory. Also, new technologies such as apps, video and text consulting platforms make it possible for community healthcare practitioners to receive training online and reduce the need to travel to cities, which also reduces the burden on city healthcare centres. In South

Africa, text-based information services like MomConnect and NurseConnect have had a positive impact on communities by bridging the knowledge gap in maternal healthcare.

> Traditionally, these communities have managed their own healthcare. But as they have become less self-reliant and more dependent on Western medicine, there has been a need to educate and provide more access to centres of knowledge and care, which are predominantly located in larger towns and cities.
>
> *Gabrielle Lobban, founder of Zumbudda (Top Women 2020: 76)*

As with countless other issues, technological innovations that help women will uplift the broader community and filter equal opportunities down the value chain, which will have a significant impact on economic transformation. What is key is that women themselves must become involved in the technology sector so that female-centric innovations are developed that fully embrace and take advantage of the opportunities that technology presents to women. 'Women are both the problem and the solution. The best way we can empower each other is by supporting one another and ensuring better outcomes for all', Lobban says (*Top Women* 2020).

From shepherd to biotechnologist: Tebello Nyokong was born into a poor family in Lesotho in 1951. As a young girl, she was sent to live with her grandparents in the mountains and learned about science by observing wildlife while she looked after the sheep. She would spend one day at school and the next day as a shepherd, and her main ambition at the time was to own a pair of shoes! When she started school, she was steered away from science but, with two years to go, she changed direction and (showing remarkable perseverance) completed the three-year course in two years.

After completing her university studies in Lesotho and Canada, she received a Fulbright Fellowship to carry out postdoctoral studies at the University of Notre Dame du Lac in the USA. She then returned to southern Africa and joined the staff of Rhodes University in Makhanda in 1992, where she is now a distinguished professor. Her fields of research include nanotechnology and photodynamic therapy, with the latter paving the way for safer cancer detection and treatment without the debilitating side effects of chemotherapy. In 2007, she was rated one of the top three publishing scientists in South Africa and she was awarded the Order of Mapungubwe (Bronze), the South African Chemical Institute Gold Medal, the L'Oreal-Unesco award for Women in Science and the National Research Foundation's Lifetime Achievement Award. She was named one of the top 10 most influential women in science and technology in Africa by IT News Africa and was elected a fellow of the Royal Society of South Africa.

Smart lockers for medicines: South Africa has the world's largest antiretroviral therapy programme for patients living with HIV and there has been a steady increase in the number of patients with non-communicable diseases (NCDs) who require chronic therapy. A patient's experience tends to be one of long waiting times, typically over three hours, which may exacerbate health problems and lead to loss of income. Technovera in South Africa has developed Pelebox Lockers, a smart locker system that allows patients to collect their repeat chronic medication in under 22 seconds. Working with the Aurum Institute, City of Tshwane, City of eKurhuleni and the national Department of Health, they have leveraged the power of the IoT through internet-enabled smart locker devices in public healthcare facilities that remotely track medication collection and treatment compliance. Prepacked medications for a patient are loaded into an internet-enabled smart locker and the system then sends an SMS with a one-time-pin (OTP) to the patients. The patients visit the collection unit, authenticate themselves using the OTP together with their cellphone number, and the cubicle with their medication pops open. The system keeps track of all collections and can be integrated into a patient's records management system. Technovera has already achieved over 10 000 patient collections and wants to create opportunities to reach more sites and patients.

Polymer heart valve: Murray Legg is an entrepreneurial thinker with a track record of growing innovative technology businesses that make a difference. With Prof. David Wheatley (a surgeon), he co-founded SA Cardiosynthetics, a company that develops polymer heart valve replacements for emerging market patients. They have also turned their attention to developing products that will help treat rheumatic fever, a disease that affects hundreds of thousands of people each year but goes largely untreated because there are no suitable products available.

3D printing of prosthetics: Nneile Nkholise, a young entrepreneur from South Africa, started working with the iMED Tech Group in 2015 while studying for her master's degree in mechanical engineering at the Central University of Technology in Bloemfontein. iMED provides innovative medical solutions, including the 3D printing of prosthetics for cancer patients and burn victims, that impact healthcare throughout the continent. iMED has also produced a range of 3D-printed external breast prostheses in a range of skin tones for mastectomy patients and has received a patent for its breast prostheses retention bar. By starting iMED Tech, Nkholise wanted to show that women have the power to run businesses with the potential to become global conglomerates within the medical technology sector. She has won the SAB Foundation Social Innovation

Award and the WEF in Davos has recognised her as one of the most influential African innovators. She was selected as one of 100 top entrepreneurs at the 2016 Global Entrepreneurship Summit.

Microbe that fights malaria: Dr Jeremy Herren and his co-workers at the ICIPE in Nairobi, and collaborators in the UK, have discovered a microbe that prevents mosquitoes from being infected with malaria. They are investigating whether they can release mosquitoes infected with the microbe into the wild, or use its spores to suppress the disease, in order to reduce the spread of malaria. The agent that they use is the malaria-blocking parasitic fungus *Microsporidia MB*, which was discovered in mosquitoes on the shores of Lake Victoria in Kenya and appears to occur naturally in about 5 per cent of wild mosquitoes (Herren et al. 2020).

The way in which the fungus blocks malaria infections is still being investigated. One possibility is that *Microsporidia MB* primes the mosquito's immune system so that it is better able to fight off infections; another is that the presence of the fungus could have an effect on the mosquito's metabolism that makes it inhospitable to the malaria parasite. The researchers estimate that at least 40 per cent of mosquitoes in a region would need to be infected with *Microsporidia MB* to make a significant dent in the spread of malaria. The concept of disease control using microbes is not unprecedented, as a type of bacterium called *Wolbachia* reduced the spread of dengue fever in real-world trials (Herren et al. 2020).

Malaria diagnosis and prevention: On 25 April 2020 (World Malaria Day), the WHO announced that malaria caused about 400 000 deaths per year, of which 90 per cent occured in Africa. It predicted that deaths from malaria were likely to double in Africa in 2020. There was also a risk that the ongoing threat of malaria would be neglected in the rush to quell the Covid-19 pandemic.

After Ghanaian software engineer Brian Gitta was misdiagnosed with malaria for the third time, he decided to do something about it. He was aware that malaria is one of the leading causes of death in Ghana and that the best way to combat it is through rapid and accurate diagnosis, treatment and containment. His experience was that malaria blood tests were time consuming, required trained laboratory technicians and were sometimes inaccurate. For the past six years, the 27-year-old computer science graduate from Uganda's Makerere University has worked with scientists, doctors and software engineers to develop a simple method to test for malaria without a blood sample, microscope or trained technician.

After much frustrating toil, they developed the matiscope (from the Swahili word *matibabu*, which means 'treatment'), a portable, shoebox-sized device that works with a smartphone. Patients with malaria-like symptoms (including high fever, chills and headaches) place a finger in the device's cradle, which uses magnets and a beam of red light to detect changes in blood cells caused by the malaria parasite. The readings are analysed by the smartphone, diagnosed and, if positive, can be uploaded to a nationwide grid so that national healthcare authorities can monitor for outbreaks. The device is reusable and provides results within two minutes!

According to Dr Jimmy Opigo, the manager of Uganda's National Malaria Control Programme, the matiscope is a game changer because it provides quick and accurate diagnoses that allow for appropriate medical prescriptions to be given on the spot, thus avoiding overmedication that can lead to treatment resistance. The real-time outbreak-monitoring capacity of the device could pave the way for targeted prevention campaigns. 'This system could be a key part of ending malaria worldwide,' Opigo says, 'I am proud it is coming from one of our own' (Baker 2019c). However, before the matiscope can be introduced, data on the accuracy of the device (especially true and false positive and negative rates) need to be evaluated.

The development of the device was not easy but, despite early setbacks, it is in its fifth iteration and is in clinical trials in Ghana and Angola. Early indications are excellent, as the results are on par with those using the blood-sample-and-microscope method. Opigo expects to take the device to the market within three years. 'If I had known how difficult this would be, I'm not sure I would have started down this path,' he says, 'It's OK to fail, as long as you keep pushing through to your idea' (Baker 2019c).

One-pill malaria treatment: Prof. Kelly Shibale did not have an easy start to his academic career. After attending school in Zambia, he was unsuccessful in his application to study accounting at a university and was forced to reconsider his options. He remembered his fascination with chemistry at school and decided to study this subject instead. His passion eventually enabled him to complete a PhD in organic chemistry at the University of Cambridge in England. After doing research in the UK and at the Cripps Institute in the USA, he moved to the University of Cape Town in South Africa, where he leads a team that discovered a malaria drug development candidate with the potential to be used as part of a single-dose cure for all strains of the human malaria parasite. From a library of over 40 000 drug leads, he and his team identified one compound that showed potential and has since passed several human trials. This is the first time that an African-led drug discovery project has taken a compound from screening to human trials (K. Shibale, personal communication, June 2020).

Asthma Grid: Moses Kebalepile (a doctoral student in the Faculty of Health Sciences of the University of Pretoria) won first place at the International Pitchfest of the Swiss–South African Venture Leaders Programme (SSAVP) in Zurich for his innovative early warning system, Asthma Grid, that predicts an imminent asthma attack. The SSAVP is administered by the Innovation Skills Development section of the Technology Innovation Agency, an entity of the DSI. The SSAVP is a framework for providing motivation, entrepreneurial know-how and support to scientists from Switzerland and South Africa. The Asthma Grid is a wearable device that determines the likelihood of an attack based on measurements of environmental pollutants and aerial allergens via a nanosensor. The device uses this information and a mathematical algorithm to predict an asthma attack, which enables asthma sufferers and their caregivers to be prepared. It has been welcomed in South Africa, which has one of the highest asthma-related death rates in the world, mainly affecting people 5 to 35 years old. Kebalepile believes that his invention could revolutionise the self-management of asthma and reduce the economic burden of managing the affliction, particularly in emerging economies in Africa (Hlabangane 2017).

> Winning first place at the international Pitchfest validates the potential [that the] Asthma Grid has in addressing the global need for a patient-centric solution to asthma treatment. It also opens doors for international collaborations and partnership, which is good for commercialisation.
>
> *Moses Kebalepile, inventor of Asthma Grid* (Hlabangane 2017)

Threat of snake bites: About 50 000 people in sub-Saharan Africa die from snake bites every year, and another 400 000 survive with amputated limbs and other permanent disabilities, yet few resources are available to combat this (Nicolon 2020). Major factors of the crisis include an acute shortage of antivenoms combined with the high cost of treatment and distrust of Western medicine. Furthermore, as snake bite is prevalent among the poor, policy-makers have tended to ignore it. To draw attention to this crisis, the WHO added snakebite poisoning to the roster of 'neglected tropical diseases' (which includes rabies, dengue fever, leprosy and trachoma) in 2017.

According to Baldé Cellou, a biologist at the Institute for Applied Biological Research in Kindia (Guinea), elevating snake bite to this level of concern 'will hopefully serve as a shock to health ministers' in Africa. Cellou, a world-renowned lecturer on snake bites, began studying the topic 25 years ago when a 12-year-old girl who had been bitten died in his arms. He stopped studying insects and turned his attention to how best to treat snake bites (Nicolon 2020). In his search for snakebite treatments, Cellou has experimented with indigenous remedies and with *Fav-Afrique* (an antivenom made by the French company

Sanofi), which is effective against the bites of about 10 snakes. In 2013, the Mexican company Inosan Biopharma (with help from Cellou and other experts) perfected Inoserp Pan-Africa, an antivenom that neutralises the toxins of at least 18 snake species – more than any other antivenom available in Africa. Its broad coverage of species means that it can be used if it is unclear which snake bit the victim, and it has few side effects.

At Cellou's institute, snakebite mortality plummeted from 18 per cent in the 1990s to 1.3 per cent in 2019, thanks largely to Inoserp Pan-Africa. But there is a severe shortage of the antivenom, as less than 5 per cent of the 2 million vials needed every year in sub-Saharan Africa are produced. Now companies and philanthropists are providing free Inoserp and medical training to health centres in Guinea, Kenya, Sierra Leone and the Republic of the Congo, and the James Ashe Antivenom Trust buys antivenom for hospitals in Kenya's Kilifi County so that patients can receive free treatment (Nicolon 2020).

Undernourished children: Undernutrition (when children do not ingest enough nutrient-rich food for normal growth) is a worldwide problem but it is particularly prevalent in Africa, where it causes childhood stunting. In children younger than five years, this can have a long-term effect on physical and cognitive development and educational performance, and eventually on economic productivity in adulthood. Stunting also affects women's ability to give birth to normal-weight children. Globally, stunting in children declined from 32 per cent to 21.9 per cent between 2000 and 2018. However, research by Blessing Akombi (2020) has revealed that the prevalence of stunting among young children in Nigeria, which has Africa's largest economy, has remained unchanged since 2013 at a shocking 3 per cent.

Child stunting is most common in the north-west of Nigeria (57 per cent) and least common in the south-east (18 per cent). This trend is attributed to the status of women in these regions, with the rates of teenage pregnancy being extremely high and educational levels of women very low in the north-west. Akombi (2020) recommends the following interventions: improve women's nutrition to reduce the incidence of low birth size, improve household hygiene to reduce infections such as diarrhoea, and promote breastfeeding and other appropriate child-feeding practices. At the community level, cash transfer schemes are needed to enable uneducated mothers of low economic status to obtain information about the causes of childhood stunting and buy nutritious food to remedy it. Research conducted in Ghana and Tanzania has also shown that improving maternal education and household income are key strategies in reducing child stunting (Akombi 2020).

Female genital mutilation (FGM): FGM is the ritual cutting of some or all of the external female genitalia. Unicef estimated in 2016 that 200 million women living in 30 countries, including 27 African states, had undergone the procedure. The African countries where it takes place extend east to west from Somalia to Senegal and north to south from Egypt to Tanzania. The highest concentrations in the 15 to 49 years age group are in Somalia (98 per cent), Guinea (97 per cent), Djibouti (93 per cent), Egypt (91 per cent) and Sierra Leone (90 per cent). The practice is rooted in gender inequality; attempts to control women's sexuality; and ideas about purity, modesty and beauty. It is usually initiated and carried out by women, who see it as a source of honour and fear that failing to have their daughters and granddaughters 'cut' will expose them to social exclusion, despite the proven health risks. FGM's origins in north-eastern Africa are pre-Islamic, but the practice became associated with Islam because of the religion's focus on female chastity and seclusion. There is no mention of it in the Quran (or the Bible).

Many women have called for the banning of FGM. Egyptian physician and feminist Nawal El Saadawi criticised the practice in her book *Women and sex* (1972), but the book was banned in Egypt and she lost her job as the director-general of public health when it was eventually released there. In 1980, she followed it up with the chapter 'The Circumcision of Girls' in her book *The hidden face of Eve: Women in the Arab world*, in which she described her own clitoridectomy when she was six years old.

I did not know what they had cut off from my body, and I did not try to find out. I just wept, and called out to my mother for help. But the worst shock of all was when I looked around and found her standing by my side.

Nawal El Saadawi, physician, author and critic of FGM (El Saadawi 1980: 8)

The Inter-African Committee on Traditional Practices Affecting the Health of Women and Children (which was founded in 1984 in Dakar, Senegal) called for an end to the practice, as did the UN World Conference on Human Rights in Vienna in 1993. The conference listed FGM as a form of violence against women, marking it as a human-rights violation rather than a medical issue. Throughout the 1990s and 2000s, governments in Africa (and the Middle East) passed legislation banning or restricting FGM. In 2003, the AU ratified the Maputo Protocol on the rights of women, which supports the elimination of FGM. By 2015, laws restricting FGM had been passed in 23 of the 27 African countries where it is concentrated, although several fell short of a ban and the legislation is not strictly enforced.

Mental health in the Central African Republic: Flora Pasquereau is the only practicing clinical psychologist in the CAR, which has a population of 4.6 million people. Since the civil war in the CAR began in 2012, it has left thousands of people dead, hundreds of thousands homeless and millions traumatised. The country desperately needs counselling, but how can one person provide it? 'Breathing exercises, and dancing,' says Pasquereau. Given the scale of mental health problems in the country, she has had to spread her expertise as wide as possible by offering large-group therapy sessions at outdoor venues in Bangui that include dancing, singing, deep breathing, stretching, yoga, meditation and listening (Baxter & Allison 2020).

> Even if you take one minute ... for one minute you forget about everything, so when things that are difficult happen later on, you can go back to this kind of therapy and have your relief. A miracle solution does not exist. Those little things like yoga or meditation can help people. The first thing I try to show them is how to forgive themselves. Today you are a victim. Tomorrow you are not. You work on their self-esteem because they have no self-esteem. You focus on the positives. But you must be realistic, because you don't know what is happening next.
>
> *Flora Pasquereau, clinical psychologist in the CAR* (Baxter & Allison 2020)

Pasquereau grew up in Bangui, but her family left the CAR when she was 15 years old. She studied in France and then moved to Canada to start a private practice, but her thoughts were never far from home. She decided that she wanted to be part of the solution and moved back to Bangui in 2017 and set up an NGO, *Obouni* (which means 'no matter what, we will succeed' in Mbaka). She has trained a dozen clinicians and together they run sessions on coping mechanisms for groups of people who have been traumatised by sexual abuse, torture and/ or losing a close relative or friend. They try to reframe people's experiences by emphasising resilience, stressing how they have already overcome the kinds of emotional challenges that would have broken others. In its first three years, *Obouni* held nearly 3 000 consultations and opened a safe house for women who had suffered gender-based violence. Pasquereau has also succeeded in putting mental healthcare on the national agenda for the first time, and has had positive feedback from the government (Baxter & Allison 2020).

Community healthcare app: Staff at the Centre for Community Technologies at Nelson Mandela University in Port Elizabeth (South Africa) have developed an integrated mobile app called *Ncediso* ('help' in isiXhosa) to upskill community healthcare workers (including nurses and clinic practitioners) in areas where basic healthcare, first-aid skills and formal training are scarce. The app facilitates the early detection of various disabilities and diseases among children, and provides guidance on first aid, child nutrition, chronic disease management, and the treatment of infectious and non-infectious diseases and other medical conditions. The objective is to make healthcare and medical information more accessible to those with no or limited medical training. The app has undergone strict user testing and is available to users through Google Play Store (www. innovationbridgeportal.info).

LifeBank: Temie Giwa-Tubosun of Nigeria believes that a business must be successful if it is to do good, and her medical supplies delivery start-up LifeBank lives up to this maxim. LifeBank is a healthcare technology and logistics company based in Lagos that facilitates the timely delivery of blood to hospitals and patients. Giwa-Tubosun initially established an NGO – the One Percent Project – in 2012, whose goal was to increase voluntary blood donations across Nigeria. In December 2015, the NGO was converted into a commercial enterprise, LifeBank, which delivered 9 000 pints of blood by motorcycle in its first 24 months and made US$90 000 in revenue. After receiving a grant of US$250 000 from the Jack Ma Foundation, Giwa-Tubosun was able to refine her business model and delivers blood as well as medical oxygen, vaccines and antivenoms to over 170 hospitals. LifeBank also runs regular blood donation drives in collaboration with the state government's blood transfusion services. In 2018, she announced plans to add drones to her delivery service and to expand to Abuja, Kaduna and other parts of northern Nigeria where maternal mortality rates are the highest in the country (Egbejule 2020). Giwa-Tubosun was named one of the BBC's 100 Women in 2014 and has been featured as a TEDx speaker. Mark Zuckerberg highlighted her as one of the most inspiring entrepreneurs he met during his visit to Nigeria in 2016.

> Becoming an entrepreneur was never something I aspired to. In fact, I didn't have any particular interest in technology either. I started LifeBank out of sheer necessity because the problem of people dying from lack of access to blood in Nigeria was worsening every day and no-one was focusing on solving it. I had a good job in health management, which I really loved and didn't want to give up, but I kept seeing people around me dying unnecessarily. For years, I tried to find different ways to solve this problem while remaining comfortable in my job, but these solutions didn't have the level of impact needed.
>
> At some point, I realised I needed to quit my job and focus on solving this problem if I really wanted to make a difference. The catalyst was when someone very close to me lost their dad who bled to death in a small town in eastern Nigeria. When she told me her story, I knew I just couldn't look away any longer. I believed the most sustainable way to solve this problem would be to set up a venture rather than a non-profit. And I also knew that I would

need technology to overcome many of the inefficiencies relating to cost and infrastructure. That's how I ended up setting up a health-tech company. It wasn't something I had initially planned in terms of my career path.

Temie Giwa-Tubosun, founder of LifeBank (Africa.com 2018: para. 5)

Health insurance made easy: Lilian Makoi is a Tanzanian fintech (financial technology) entrepreneur who founded Jamii, a mobile micro-insurance start-up that facilitates access to affordable health insurance for low-income communities. After running two start-ups (providing a diet food delivery service and then a housekeeping and cleaning service), Makoi joined the telecommunications sector, which allowed her to witness the impact that mobile technology made on people's lives and gave her exposure to the rapidly growing mobile payments space. She discovered that over 70 per cent of Tanzania's population struggled with healthcare financing.

It's been so exciting to see the impact that technology can make and I am now very passionate about using technology to solve problems for low-income communities.

Lilian Makoi, founder of Jamii (Africa.com 2019: para. 5)

Makoi investigated the healthcare financing industry and concluded that the major challenge was high insurance administration costs. After further research, she came up with a solution for all the administrative tasks to be carried out via cellphone, thus reducing the overall cost. She was then able to build a mobile platform that made health insurance cashless and paperless, which enabled her insurance partner to launch a 1-dollar health insurance product that was affordable for low-income families. Makoi has won numerous awards, including being named one of the most innovative women in technology in Africa by the WEF in 2016.

Black Like Me: During the apartheid years in South Africa, Herman Mashaba refused to be a servant or a gardener. Instead, he honed his skills as a salesperson, first as a teenager selling dope and then as a commercial salesperson. One day in 1982, he saw a billboard advertising products to relax the hair of black people. He pondered the business opportunity, then decided to take the plunge. He joined Superkurl, which makes hair relaxers. Three years later – together with his wife, Connie; a colleague from Superkurl, Joseph Molwantwa; and a pharmacist, Johan Kriel, he launched Black Like Me, which makes a range of hair and skincare products for black people. After overcoming many obstacles, the company was soon producing and selling a wide range of hair relaxers, shampoos and sprays. In 1997, Black Like Me amalgamated with Colgate-Palmolive, with Mashaba retaining a 25 per cent share; however, it did not work out, so he bought back the company. In addition, he is the executive chairperson of Lephatsi Investments (which operates in the mining, construction and logistics sectors) and served as the mayor of Johannesburg (Mashaba & Morris 2012).

Oil of Olay: While working for the Industrial Development Corporation in South Africa, ex-Unilever chemist Graham Wulff helped to build factories where grease was recovered from wool. At the time, wool grease was the only lubricant available for the heavy steel roller mills at the giant steel-maker Iscor. He met Herman Beier at Congella Woolwashing in Durban and they decided to start a company, Adams National Industries, that produced lanolin from wool grease. Lanolin is a wax secreted by sheep that keeps their wool and skin clean and protects them from the weather. Wulff reckoned that it could do the same for humans. In 1952, he set up a laboratory in his home to concoct a lanolin-based cream that would penetrate rapidly, restore and retain moisture, protect the skin and disappear in 10 minutes, leaving a supple, matt finish – a tough ask. One morning while shaving, he juggled in his mind the letters in 'lanolin' and came up with 'Oil of Olay'. By 1959, they had introduced their products to the UK (as Oil of Ulay), continental Europe (Oli of Olaz) and Australia (Oil of Ulan). In 1985, Oil of Olay was acquired by Procter & Gamble and, in 2013, it became their 13th billion-dollar brand (Bruton 2017a).

Natural hair restorers: Valerie Buhlmann-Strydom suffered from an illness that caused her to lose her hair, but used this as an opportunity to develop a potion that restores hair using natural ingredients. Her product, Herb-Hair (which contains lavender, rose petal, lemon concentrate and natural herbs but no oils or hormones) has been so successful that it is sold commercially in southern Africa and Europe. Valerie has also developed Herb-Wound (for skin wounds and diseases) and Herb-Coat (hair restorer for pets) (V. Buhlmann-Strydom, personal communication, May 2020).

Botswana's mompreneur: Motherhood is a full-time job and running a business while being a mother is a task for superwomen. Mother-of-three Tuduetso Tebape of Botswana is the superwoman behind the natural beauty product brand Nubian Seed. Since 2014, her business has grown from a cottage industry selling shea butter and African black soap into a thriving leader in the 'green beauty' industry. 'Nubian Seed' is a metaphor for Africa's rich heritage and refers to the historic state of Nubia on the Nile River that was the birthplace of one of the continent's earliest and most powerful kingdoms. Today, 'Nubian' implies anything related to Black culture. 'Seed' refers to the origins of the company's product range, including natural and organic butters, carrier oils, essential oils, clays and soaps – all made from plants and soils indigenous to Africa. In addition to being commercially successful, Tuduetso has another goal: to impart knowledge on the power of indigenous natural ingredients so that customers can experience their complete healing properties (Tebape 2019).

> There is no blanket solution for the path to success. I don't believe that at all. So, with that in mind, I don't believe that the only way to success is through entrepreneurship ... I haven't thrived within the boundaries of a formal employment situation so this is why I've chosen this path ... What I do with this business, as cliché as it sounds, is an expression of passion.
>
> *Tuduetso Tebape, founder of Nubian Seed* (Tebape 2019)

Dust Bunny: In Ghana, many people die each year from infectious diseases such as diarrhoea and lower respiratory infections. As lack of clean water and poor sanitation are the main causes, good hygiene is the best way to break the chain of infection. Dr Emmanuel Tsekleves, working with Ghanian parasitologist Prof. Daniel Boakye, has estimated that over 15 000 people (including 2 250 children younger than five years) die each year in Ghana from poor hygiene and has launched the international research project Dust Bunny to design novel solutions to reduce infections in the home. They have worked closely with householders to design new cleaning practices on the basis that people are more likely to adopt new ways of doing things if they have been actively involved in their design. Their key message is that citizens can do as much as scientists in the fight against drug-resistant bacteria. The Dust Bunny project is being piloted in Ghana and will be rolled out in other African countries (Tsekleves 2020).

Drones and flying doctors: Every day, hundreds of people in Africa die for lack of medical supplies, from vaccines to snakebite antivenom. In 2014, Keller Rinaudo decided to do something about it and developed a drone-enabled delivery system for vital medicines in Rwanda. Through his California-based start-up Zipline, he has developed battery-powered, fixed-wing zips that have covered more than 1.2 million miles and made over 23 000 emergency deliveries (Baker 2019b). In another successful venture, Ola Orekrunin founded Flying Doctors Nigeria (an air ambulance service) before she was 30 years old. It is an ambitious and high-risk project and she deals with life-and-death situations every day, but she has persevered and is providing a vital service.

Afrobotanicals: Ntombenhle Khathwane is a born entrepreneur, but that does not mean that she has always been successful. After failed ventures making asphalt and then selling stationery, she went back to the drawing board. In 2010, she entered and won the business plan competition Pitch & Polish, which earned her a trip to the USA where she was advised by industry experts on product development. Today her haircare products company Afrobotanicals has a full range of products that is on the shelves of national retailers, and she is exploring the idea of manufacturing shampoos and body washes for the hotel industry.

Emerging innovators: In Burundi, Gérard Niyondiko (a former high-school chemistry teacher in Burkina Faso) and his partner, Moctar Démbélé, are battling one of Africa's deadliest diseases with the simplest of weapons: a bar of soap. Their Faso soap is made of age-old ingredients such as shea butter, lemongrass, African marigold and other locally derived repellents to fend off mosquitoes and reduce the risk of people contracting malaria as well as *chikungunya* ('to become contorted'), yellow fever or dengue. The soap offers up to five hours of protection and is comparable in price to normal soaps. In April 2013, they became the first non-American winners of the University of California's prestigious Global Social Venture Competition for social-minded entrepreneurs (Liem 2014). In 2016, a crowdfunding campaign raised over €70 000 to help them take their product to market. At the age of 11 years, Sandra Mwiihangele from Namibia won the Expo for Young Scientists competition for a top-quality matte lipstick that she had developed. Today, she is a trained cosmetic chemist and her company Kiyomisandz Beauty Products won Namibian innovation grants. She will soon export her products throughout Africa and beyond.

Discussion: The vast and highly profitable medical establishment is under threat from medical devices such as Tricorder (partly inspired by a similar device in *Star Trek*) and other devices mentioned in this book that enable people to check their own health using their smartphone. Using a Tricorder, you can do a retinal scan, take a blood sample and analyse 54 biomarkers from your breath that will help you to diagnose most common diseases. IBM's Watson can instantly diagnose diseases better than human doctors and will eventually be able to predict and even pre-empt an individual's future health problems. The Honor Play 4 Pro android smartphone can already measure human body temperature using an infrared sensor. Many people with access to the internet already perform ad hoc diagnoses and treat themselves at home using information systems such as Google and Wikipedia. Despite these high-tech advances, it is widely believed that family and clinic doctors and nurses – supported by low- and mid-tech initiatives such as LifeBank, Ncediso, the Tutu Tester, Pelebox Lockers and SolarTurtle products – will continue to play a leading role in healthcare in rural and peri-urban areas of Africa for many years to come.

9

Sustainable living

'The future is not what it used to be.'

Clem Sunter, South African futurist (Sunter 1987: 1)

Introduction: Cleantech is any process, product or service that reduces negative environmental impacts through significant energy savings (the sustainable use of resources or actions that offer environmental protection). Cleantech includes recycling, renewable energy, information technology, green transportation, electric motors, green chemistry, low-energy lighting, grey water re-use and many more. Although clean technologies are typically competitive with – if not superior to – their conventional counterparts and offer significant additional benefits, their introduction has been suppressed by vested interests, high initial capital costs and 'innovation inertia' (i.e. initial reluctance to adopt new technology). Cleantech is not only relevant in developed economies but also has the potential to significantly improve the quality of life of people in developing economies in Africa. Cleantech went boom in the early 2000s and then bust by the end of the decade, but it is now making a comeback with more savvy management and a deeper understanding of the severity of the problems that have to be addressed. Many of the leading innovators in cleantech are from Africa.

Water wars: Conflicts in the 21st century will be fought over water, not land. The Grand Ethiopian Renaissance Dam on the Nile River is a case in point. Its recent completion, five years behind schedule, is important for Ethiopia because the 720 MW of hydroelectricity that it will produce will help to keep the Addis–Djibouti railway line and many other projects functioning. However, downstream countries such as Sudan and Egypt are concerned that they may be deprived of the lifegiving waters of the Nile during droughts. The conflict has been mediated by the USA and South Africa, but a final resolution has not been reached. Ethiopia's Minister of Water, Irrigation and Energy Seleshi Bekele has drawn his line in the sand: 'If there is a single word in the document to be signed that could compromise Ethiopia's right to use the water, Ethiopia will not sign it' (Kiruga 2020).

Water purification and desalination: A locally designed solar-powered water purification system that uses an activated carbon filter made from waste macadamia nut shells was launched by Kusini Water in Cape Town in 2018. The filter removes 99.99 per cent of disease-causing bacteria and parasites from water using a mechanical nanofilter that physically separates out particles. It also uses an activated carbon block made from locally sourced macadamia nut shells to absorb chemicals like chlorine and pesticides. The Kusini Water filtration system produces 40 times more water than reverse osmosis (the current best practice) and uses about half the energy. Brine discharge is done through the existing sewerage system, where it is diluted to the extent that it has no negative environmental impacts.

Kusini Water filtration systems are monitored and controlled using a smart water meter that is IoT enabled so that each site can be operated and maintained remotely. According to Kusini Water's founder Murendeni Mafumo, their first water filtration system was launched in Shayandima, Limpopo province, in 2016 and their new site at the V&A Waterfront in Cape Town produces enough desalinated freshwater for 4 800 households per day (www.kusiniwater.co.za).

Gambia's Queen of Recycling: Isatou Ceesay is a Gambian activist and social entrepreneur popularly known as the 'Queen of Recycling' because she initiated a successful recycling movement in The Gambia called One Plastic Bag. Ceesay has been empowering women for over 17 years, teaching them how to turn plastic waste into an income and to play a more influential role in society. With over 2 000 members in 40 groups, and projects with the EU and the UNDP, Isatou's NGO – Women Initiative The Gambia – is making a difference. She has also launched a programme to make cooking briquettes from dried grass, mango leaves, coconut fibre and paper that provide a cheaper and less toxic source of household fuel than charcoal briquettes, which contribute to deforestation (Riché 2019).

We still have a long way to go in order to educate the population on the relationship between our health and the environment. This is a key issue, as 75% of the population in The Gambia does not have access to proper education.

Women have a key role to play in this endeavour. Throughout the world, women carry an incredible responsibility; they are by nature the engine of human development. I love them so much. Their commitment and their strength are unrivalled. We have fallen behind in our development in Africa by not including them.

Isatou Ceesay, Gambian environmental activist (Riché 2019)

McNuts Briquettes: Ken Robertson is a politician who is prepared to roll up his sleeves and get his hands dirty in service to his community. He and his wife Brigitte noticed that macadamia nut farming produces a huge amount of empty nut shells that are discarded as waste. They decided to turn the shells into charcoal that could be used in fires for cooking and heating. After several years of research and development, they came up with the right recipe and made their first deliveries of McNuts Briquettes to local stores and communities in Limpopo province (South Africa), where they have been very well received. The briquettes burn for a long time and produce excellent coals with no nasty chemical smells.

Green bricks: Recycling polyethylene terephthalate (PET) plastic bottles has become a top priority, as it is estimated that the production of each 1-litre bottle consumes 7 litres of water and 200 g of crude oil and produces 30 g of waste that will take up to 700 years to decompose. Used bottles often end up in the oceans, where they partially decompose into plastic beads (microplastics) that are consumed or filtered out by marine animals and then enter the food chain.

One way to re-use 2-litre PET bottles is to make ecobricks, which are plastic bottles densely filled with non-recyclable waste (not organic waste). Packed together, the stuffed bottles form building bricks that are robust and affordable and can be used with plaster or cement to make walls, benches, raised vegetable beds and even entire buildings. In 2000, the German architect Andreas Froese used sand-filled PET bottles to make community buildings in South America; in 2003, Alvaro Molina built a primary school in Nicaragua using ecobricks. Since then, ecobricking has spread to Africa and they are in widespread use as a cheap building material in Zambia, Zimbabwe, Botswana, Uganda and South Africa. Technologists at the University of Zululand have developed another kind of green brick made from waste materials that are abundant in KwaZulu-Natal (such as sugarcane bagasse ash), which they combine with cementitious binders (such as fly ash and lime) and compress into a building brick that can be used to make informal dwellings.

Green toilets: The Water Technologies Demonstration Programme in South Africa has developed Arumloo, the world's greenest toilet, which only requires 2 litres per flush – one-third of the normal volume. The toilet was inspired by the elegant form of the arum lily and has a redesigned pan with an altered trap to reduce water usage. It is available in ceramic or cheaper plastic options.

Flynn Goodwin (an experienced commercial plumber) established the company PlanetSaver to develop Lilydome, a waterless urinal waste system that can be retrofitted into most urinal bowls, which reduces water usage and creates a more hygienic and odour-free environment in restrooms. Lilydome replaces the conventional urinal trap seal with a disposable one that contains a specially designed valve which traps odours without the need for additional water supply. Goodwin estimates that a single Lilydome will save between 130 000 and 200 000 litres of water per year, enough to provide one person with 2 litres of drinking water per day for 226 years (Manyonga 2017b)!

Dyllon Randall of the University of Cape Town has developed a novel toilet hub, SaniHive, which converts urine and faeces into fertiliser and compost. It uses no electricity or water, and differs from similar systems in that it separates the urine and faeces within the toilet. SaniHive, which can be used in both informal townships and upmarket suburbs, won the Global Sustainability Prize at Unleash 2018 (a global innovation festival held in Singapore).

Recycling car tyres: Although used rubber tyres have many purposes, most of them clutter up landfill sites or despoil the natural environment. Through the Cubic 38 project, Tshwane University of Technology in Pretoria has perfected recycling tyres using the process of pyrolysis (i.e. the thermal decomposition of organic materials such as rubber at elevated temperatures in an inert atmosphere, which changes their chemical composition but produces no toxic fumes). The by-products that are created include oil, high tensile steel and waste tyre char, with the latter being used to make products for cleaning car tyres and polishing black leather shoes.

Ceiling in a can: Scientists at Mangosuthu University of Technology in Durban have created Ceiling in a Can (CiC), a patented innovation that allows poor people to improve their homes by making an instant ceiling. The innovation comprises two 5-litre cans containing different polymers, the base and the activator. The two polymers are mixed and painted onto layers of newspapers laid on the floor to form a stiff foam board. Once it is dry, the board can be lifted and attached to the rafters to create a low-cost, easy-to-install ceiling. In addition, the ceiling has fire retarding qualities, excellent insulating properties and is waterproof. It is also lightweight and requires no supporting frame, and costs less than half the price of competitive products. The CiC is ideal for people living in low-cost housing or shacks in informal settlements.

The last straw: Single-use plastic drinking straws are on their way out because they take nearly a century to biodegrade, but the search for an alternative has been frustrating. Cardboard and paper straws go soggy, and consumers frequently forget to carry their kits of re-usable glass or bamboo straws with them. Bonnie Bio Products in South Africa has developed straws made from natural corn and vegetable ingredients that are as sturdy as plastic straws but also biodegradable and contain no toxins. Bonnie Bio Products also produces biodegradable refuse and shopping bags and clingwrap (www. bonniebio.co.za).

Another useful straw innovation that could become a lifesaver in rural Africa is the LifeStraw Personal Water Filter, a unique portable water filtration system developed in Switzerland by Torben Vestergaard Frandsen that enables users to drink water directly from rivers, lakes and dams or to store water and drink it later. The filtration system removes 99.5 per cent of all bacteria (including *Salmonella*, *Shigella*, *Enterococcus* and *Staphylococcus*) and other pathogens and weighs only 57 g. Its longlasting membrane microfilter can filter up to 4 000 litres of water, enough for an individual for over five years. However, LifeStraw filters do not remove dissolved salts and should therefore not be used to drink undiluted urine or sea water.

Solar cooking: Billy Hadlow was motivated by the large number of shack fires in South African townships to develop a safe cooker that would replace paraffin and charcoal stoves and open fires – all of which pose a severe fire risk. He launched Khaya Power, a clean-power company that has developed the Khaya Tri Cooker, a solar-powered appliance that can be used for cooking, baking and heating. The Khaya Cooker is a fan-assisted gasifier that uses a two-stage combustion process that ensures full and efficient use of the fuel. The first stage heats the biomass (usually wood waste) through pyrolysis, which releases the volatile gases in the fuel. The second stage injects air jets into these gases and burns them, creating more heat and eliminating harmful emissions (Manyonga 2017c).

The gasifier has a double insulation layer so that the body of the stove stays cool while the fire chamber reaches temperatures of over 700°C. The Khaya Cooker costs less than a paraffin stove; is easy to load; can be used indoors, as it produces no smoke or harmful gases; and can be run off a battery, mains electricity or solar power. Khaya Power is also developing a franchise model to roll out Energy Shops throughout Africa that will provide clean technology for households not connected to the electricity grid.

We aim to power rural and township homes in Africa, as well as charge low-powered devices like cellphones with portable solar-powered batteries called 'Juz Boxes' that can be leased from local franchises, thereby creating employment and sustainable income in local communities. I believe that the technologies used in the past to build large power stations and supply power via regional grids will be surpassed and replaced by locally distributed power plants and supply companies.

Billy Hadlow, founder of Khaya Power (Manyonga 2017c)

DryBath: At school, Ludwig Marishane was a natural tinkerer, full of ideas that he tried to take to market. When his father gave him a chemistry set at the age of 10 years, he nearly killed himself by swallowing copper sulphate – but that did not deter him from trying to develop diesel biofuel from algae. He hit on his 'big idea' while chatting to a friend in Mototema Village in Limpopo province. He was aware that millions of people do not have access to running water and decided to invent a 'waterless bath'. He partnered with Dr Hennie du Plessis (a chemical engineer who had invented an easy-to-open sachet) to develop, make and sell the germicidal moisturising gel DryBath, which you can rub on your skin to clean yourself without using water. DryBath comprises a blend of naturally antiseptic essential oils, bioflavourants and tawas (potassium alum) that clean the skin and kill odour-forming germs. Through his company Headboy Industries, he has continued to refine the DryBath product, which is now widely on sale. Marishane is South Africa's youngest patent holder and was elected one of Google's 12 Brightest Young Minds in the World. He was also the first African to be declared the Global Student Entrepreneur of the Year (Bruton 2017a; Maggs 2018).

Emerging innovations: Lithabe Pasika from South Africa has developed a solar window blind that not only shades a room but also uses the sun's rays to light up the house, thus saving electricity. Former surfer Grant Vanderwagen has developed the H_2O Catcher, a system that harvests water from fog and mist. The mesh-netting device is strung up on existing structures such as telephone poles and can harvest up to 12 000 litres per day. Farmers who have plenty of space on their land for the H_2O Catcher could harvest enough water to cover most of their irrigation needs.

During Cape Town's water crisis in 2018, Abel Namuramba (the founder of Isiphiwo Fusion, whose motto is 'Never let a good crisis go to waste') developed a series of water-saving devices, including affordable flow restrictors and showerheads, that reduce water consumption by up to 30 per cent. (*Isiphiwo* means 'gift' in Zulu.) Johannesburg-based computer scientist Tshepo Nkopane, the co-founder of EnviroCentral, has developed software that measures water consumption using a portable device. He was inspired to develop his idea when he read about the excessive cost of water leakage in South Africa.

Emerging innovators: Luvuyo Ndiki has patented the first 3D-printed biodegradable cups in Africa through his company Red Cup Village. Ndiki remembers that while growing up on his family's farm, one of their precious cows died after eating plastic. This set him on a mission to change the way that people think about plastic pollution. His Ecocup is made using polylactide filament and bioactive thermoplastic aliphatic polyester derived from renewable resources such as sugarcane and cornstarch that breaks down in compost. Ndiki was also the organiser of the first Cape Town Biodegradable Festival, held in February 2020, aimed at promoting the global move towards more sustainable living and responsible tourism.

Discussion: Global warming, sea-level rise, the increased incidence and wider amplitude of spikes in extreme weather events, and shifting bioclimatic zones will all have a disproportionate impact on Africa (especially southern Africa and the Sahel). Forced human migrations, viral epidemics, conflicts over water, and poverty and civil unrest are all likely consequences because global climate change is far more than an environmental event; it is also a socioeconomic tragedy. The Covid-19 pandemic has shown that humans can act cooperatively in the face of a severe international crisis. It will hopefully teach us to recognise that climate change is an equally threatening crisis that also requires a unified global response. Interestingly, in a survey of 80 000 people in 40 countries, researchers at Oxford University found that people from two African countries showed the highest levels of concern (Kenya 90 per cent and South Africa 87 per cent) for the impacts of climate change (Andi & Painter 2020). But standing up to climate change denialists will require more than scientific evidence – it will also demand political will and leadership, and what late South African President Nelson Mandela referred to as a 'stubborn sense of fairness' (Mandela 1994).

10

Energy and alternative energy

'Innovation distinguishes between a leader and a follower.'

Steve Jobs, American technology visionary and co-founder of Apple
(Jobs 2005)

Introduction: Water technologists are looking at ancient technologies to meet our modern energy needs. The kinetic energy of running water has been used for millennia to turn water wheels and do work, such as grind corn. This idea led to the development of giant hydropower plants on large reservoirs, of which there are many successful examples in Africa. But because we are running out of large, dammable rivers, technologists are looking at creating miniaturised versions of the ancient water wheels to generate energy. Improved laminar flow turbines make it possible to harvest vast amounts of energy from existing elevated water infrastructure such as water towers and small urban reservoirs. For example, the water supply entity for Bloemfontein in South Africa, BloemWater, receives all its electricity from a nano-hydropower unit, making it the first water utility in Africa to operate off grid (Naidoo 2018).

Renewable energy is renewed naturally and includes solar energy (photovoltaic generation and thermal heating), windpower (windmills and wind turbines), hydropower (still the cheapest, if available), wavepower (very expensive), bio-energy (limited application) and geothermal power (in Africa, only viable in East Africa). When using electricity, the best way to reduce energy usage is, of course, to use electricity as little as possible by turning off unnecessary lights and appliances, turning the geyser down to 55°C and installing alternative sources of energy. In South Africa, the state power generator Eskom calls this form of saving 'negawatts'.

Several African countries (particularly Egypt, South Africa, Kenya, Namibia and Ghana) already have substantial arrays of photovoltaic panels. In South Africa, the energy that they produce is 40 per cent cheaper than that generated by fossil- or nuclear-fuelled power stations. The main drawback is that solar power can only be generated when the sun is shining and must be stored. Africa, with its abundant solar energy resources, is ideally placed to develop this form of renewable energy and several multinational projects are developing this potential. The World Bank's Scaling Solar project and the Deutsche Bank's GET FiT initiative have made major strides in this regard. In Uganda, GET FiT has mobilised US$450 million in investments to increase energy production by about 20 per cent, which guarantees electricity supply for at least 200 000 households. Two solar plants in Senegal, each of which has a capacity of 30 MW, have been built through the Scaling Solar project that supplies the cheapest available electricity in the country. The enormous Noor solar power plant near the Moroccan town of Ouarzazate was developed with the Spanish consortium consisting of TSK, Accione and Sener (Haag 2020).

There is, however, a downside to these developments. Lamine Ndlaye (a pioneer in renewable energy in Africa) has pointed out that Scaling Solar will sell electricity at a price of less than four eurocents per kWh (kilowatt hour), which may kill the sector because no other player will be able to sell at that price (Haag 2020). Boniface Mabanza, an academic in the DRC, believes that the current regulatory environment benefits foreign investors rather than leading to substantial transformation of the alternative energy sector in Africa. He proposes that African countries should reject aid programmes that promote private foreign investment until they are properly regulated (Haag 2020).

Within days of South African President Cyril Ramaphosa's call in 2018 for more power to promote economic growth, the Norwegian company Scatec Solar (Africa's largest independent solar power producer) announced that it was ready to offer a combined capacity of 258 MW to boost the country's electricity supply. Scatec Solar recently completed its fourth solar power plant in South Africa near Upington in the Northern Cape and has its eyes set on other African countries with abundant solar resources. The costs of generating solar energy have declined so sharply in recent years that it is now the cheapest electricity available for countries with enough solar resources that are looking at new power generation. Australia, which also has abundant sunshine, added 2 130 MW of power to its national grid from solar power plants in 2019 (Davie 2019) and Africa should follow suit.

Graça Machel, the former education minister of Mozambique and widow of Nelson Mandela, argues that Africa could be the launchpad for a green-energy revolution that inspires the world if it refuses to accept further foreign financing of fossil-fuel projects on the continent. She points out, for example, that Japan's three largest banks (Mizuho, Mitsubishi and Sumitomo) have been some of the biggest lenders for coal development across the globe, including Africa. Furthermore, the UK has spent as much underwriting fossil-fuel projects overseas this decade as it has in international climate funding. She rightly argues that if we are to limit post-industrial global temperature increases to 1.5°C, we should not be investing in new fossil-fuel infrastructure. In Kenya, campaigners against the country's first coal-fired power station succeeded in getting an environmental licence revoked, which halted the project (Machel 2019). Machel has reason to be worried, as cyclones Idai and Kenneth devastated parts of Mozambique (as well as Zimbabwe and Malawi) in early 2019, with the city of Beira 'all but wiped from the face of the Earth'. But there is also good news. In Ghana, the World Bank is helping to fund solar mini-grids that will supply electricity to about 10 000 people on islands in Lake Volta, and a utility-scale solar power plant is being built in rural Mozambique (Machel 2019).

SolarTurtle: SolarTurtle, the Cape Town-based renewable energy start-up, was founded in 2016 by social entrepreneur and engineer James van der Walt (now the chief technology officer [CTO]), CEO Lungelwa Tyali, Chief Security Officer Charlene Barnes and Sales Director Ursula Julius. Tyali was previously a vice-president at Ericsson Telecom, responsible for expanding the company into sub-Saharan Africa, but she became disillusioned with corporate life and returned to her home village with the dream of making it on her own terms. The goal of SolarTurtle is to design and manufacture a series of scaled, mobile renewable-energy kiosks for use in rural and off-grid areas. Their first product was the SolarTurtle Hub, which is built around 6- or 12-m shipping containers and can be used to provide power and connectivity to any rural micro-utility or to set up an internet hotspot, rural bank, clinic, classroom or office. The SolarTurtle Mini is a lightweight fibreglass solar-energy kiosk that offers secure trading space for small-scale entrepreneurs and vendors, and the SolarTurtle Trolley is a hand-pushed, mobile charging station that is ideal for entrepreneurs wanting to offer charging and internet services to their customers.

> The SolarTurtle feeds just like a turtle. In the morning, when it is safe, the panels unfold from their secure location to feed from the rays of the sun. In the evening, when it is unsafe, the panels fold away into the hard shell of the container.
>
> *Lungelwa Tyali, CEO of SolarTurtle* (Malinga 2018)

SolarTurtle's latest range of products (launched online in August 2020) includes the portable SolarCase, a suitcase-sized, solar-powered unit; SparkBike, a solar-powered trailer pulled by a bicycle; and SolarCart, a larger solar-powered trailer pulled by a vehicle. All the units are ideally designed to supply electricity and connectivity to small business enterprises, farmers, clinics, schools and disaster relief efforts. Through their NPO, the SolarTurtle Foundation, the multi-award-winning company (which recently won the Nation Builder Social Innovation Challenge run by Stellenbosch University's Innovation Lab) also plans to use the scalability of renewable energy to start micro-energy franchise businesses focused on empowering women in rural communities.

> Our vision is to change the way people do business in remote parts of Africa and beyond. We offer a universal solution to business development in off-grid locations: from the smallest energy kiosk to the largest mining solar installation. We want to train Turtlepreneurs (the people who run the SolarTurtles) to run a successful energy business and ensure that the community stays brightly lit. This is done via an innovative pay-as-you-go model, where each entrepreneur pays for rental of the kiosk, in addition to any electricity they use, which they mark up and sell.
>
> *James van der Walt, founder and CTO of SolarTurtle* (Malinga 2018)

Dangote refinery: The mindset of Africa's richest businessman, Nigerian Aliko Dangote, is epitomised by the spirit of *harambee*. He has no plans to rest on his laurels and has set his sights on dominating several market sectors. In the cement market, where he first made his fortune, he plans to expand capacity on the continent by 29 per cent. In November 2019, he announced plans to build a US$2-billion fertiliser factory in Togo; in 2021, he will likely oversee the completion of Africa's largest oil refinery, despite numerous delays (Payne & George 2018). The Dangote refinery will process about 650 000 barrels of crude oil per day and will also start producing refined petroleum products within months of its completion. This will reverse the decades-old system in Nigeria of exporting crude oil and having to import refined oil-based products, and will free up billions of dollars of Nigerian currency.

Small-scale hydroelectric generator: Hydrotechnology typically requires the building of permanent infrastructure (e.g. weirs and dams), which are not only expensive but have a negative impact on river ecology. In a bid to provide reliable access to green electricity without harming the natural environment, technologists at the Walter Sisulu University in Mthatha (South Africa) are developing the river hydroelectric generator that uses a paddle wheel to generate electricity from rivers. Although it is aimed at providing cheap and reliable power in rural settings, it has the potential to augment power supplies in urban and peri-urban environments.

Solar energy in Egypt: Ahmed Zahran of Egypt is one of the most influential solar power trailblazers in North Africa. After gaining experience with Shell International and Tri Ocean Energy in Tunisia, the UK and Egypt, he formed the company KarmSolar in Cairo to shift Egypt away from its dependence on centralised power to more sustainable options (especially solar). KarmSolar was the first private solar company in Egypt to obtain licences to generate, sell and distribute electricity to consumers and to operate a feed-in tariff station that sells power to the national grid. The French group EDF announced in late 2019 that it would invest up to US$25 million in Zahran's start-up, making it a leading supplier of solar power in Egypt – a country that has set a target of generating 42 per cent of its electricity from renewable resources by 2035 (Ayene et al. 2020). Zahran is also the co-founder of Nahdet El Mahrousa, a social change incubator in Cairo that trains and empowers social entrepreneurs.

Wind turbines: Technologists at Nelson Mandela University in Port Elizabeth (South Africa) have developed Twerly, a renewable energy-powered streetlight that provides off-grid lighting. It is powered by a combination of a vertical axis wind turbine, a solar panel and a rechargeable battery, and can be used in residential areas because it is very quiet. There is considerable potential for offshore wind turbines to be installed around the coast of Africa, especially off West Africa, as 80 per cent of all uninterrupted wind blows offshore (Gourvenec 2020). Recent projects off the coast of Scotland (Hywind) and Portugal (Windfloat) show that it is possible to build floating, 6-MW wind turbines; however, construction and maintenance costs are high.

Ivuthakahle **low-smoke fuel:** In South Africa, more than 1 million households still use coal or wood as their main form of household energy and are paying the price for it through poor air quality, ill health, and accidental death from carbon monoxide poisoning and fires. In addition, about 40 per cent of all visible air pollution in South Africa is caused by household fires. Creative technologists at Nelson Mandela University have developed a low-emission solid fuel as a substitute for coal and wood that is just as cheap or even cheaper. The low-smoke fuel comprises a combination of microalgae and discarded coal fines that are processed through pyrolysis, which brings about a change in their chemical composition. *Ivuthakahle* has excellent combustion properties and burns with almost no particulate matter emissions. Two new solid-fuel stoves have also been developed.

Dung Beetle Project: A collaborative project between the NPO Alliance Earth, South African inventor Pierre Pretorius and American artist Nathan Honey has given rise to an educational sculpture that turns single-use plastic into usable fuel. Called the Dung Beetle Project, it is inspired by the insect's ability to clean up waste and transform it into something useful. Pretorius initially developed a system to produce fuel from macadamia nut shells, but then turned his attention to turning plastic waste into fuel. In his new system, shredded plastic is gasified in an oxygen-free chamber, the gases are collected, and the plastic residue is recirculated and burned again. After running through cooling ribs, the gases are condensed into liquid fuel. The system produces low-emission diesel, petrol or syngas (synthetic gas) without any harmful emissions.

The prototype (built by Pretorius, Alliance Earth's director Jeffrey Barbee and 25 volunteers from around the world) comprises an enormous dung beetle sculpture made from scrap metal with a reactor inside. The structure is attached to a heavy-duty trailer and has a stage that unfolds to form a teaching and training platform. The building plans have been made available free on open-source software so that anyone can build it using locally available materials. Pretorius' intention is to travel with the Dung Beetle Project around South Africa to encourage rural communities to make their own simple gasification units in order to recycle plastic, make fuel and create jobs. The project will also educate rural people about the perils of plastic pollution (Samson 2018).

Biogas production: Gordon Ayres (the CEO of Agama Biogas) believes that Africa has hardly scratched the surface of the enormous potential of biogas as a sustainable energy source, waste management system and producer of fertiliser. Agama Biogas has been installing prefabricated biogas digesters throughout South Africa for a decade, including 50 government- and USAID-sponsored systems in rural schools, with excellent results.

Biogas is produced in digesters through the decomposition of organic material such as food waste and plant matter. The organic material is eaten and digested by bacteria to produce biogas, a mixture of CO_2 and methane, which can be used as a fuel to generate electricity, to power farm equipment, for lighting, in gas cookers for cooking and even in vehicles. In Germany, 10 per cent of the power feeding the national grid is derived from processed sewerage and wheat silage; in the USA, biogas digesters (mainly producing fertiliser) are widely used on dairy farms. As a low-cost, low-technology solution to alternative energy production and waste management in Africa, biogas has the potential to contribute significantly to the development of a sustainable, closed-loop economy (Ayres 2020).

> The capturing of methane through the use of biogas technology has an immensely important role to play in rural energisation, poverty alleviation and development … If we succeed, biogas digesters will be widely recognized as a simple, cost-effective solution that will eventually become the norm for households or businesses.
>
> *Gordon Ayres, CEO of Agama Biogas* (Ayres 2020)

Water hyacinth (*Eichhornia crassipes*) is an aquatic weed that has one of the fastest growth rates of any plant but this property has resulted in it becoming a major problem in waterbodies, where it spreads rapidly over the water surface and causes problems with navigation, recreation, irrigation and hydropower generation. The elimination of water hyacinth from waterways is almost impossible because they produce hardy seeds that remain viable for up to 20 years; the complete removal of the plant is not wise because it plays a role in reducing water pollution. The control and sustainable use of the plant are therefore the best management option, as this approach means that an unlimited supply of organic matter can be used to generate bio-energy. The Agricultural Research Council in South Africa has developed an anaerobic digestion process that turns harvested hyacinth into biogas. Furthermore, the liquid by-product of the digestion process will be sold as a soil ameliorant and the solid by-product as a pelleted animal feed.

Scientists at Walter Sisulu University are investigating the efficacy of producing biogas from the high-temperature anaerobic degradation of dairy wastes and scientists at Nelson Mandela University are developing a product called Coalgae through a microalgae-based technology that selectively removes mineral contaminants from waste coal fines. The resulting Coalgae is a biomass-rich fuel that can be used instead of coal.

In Bronkhorstspruit (South Africa), Bio2Watt has developed a biogas facility that produces 4.5 MW of electricity and has been supplying green energy to BMW's Rosslyn plant since 2015. Organic waste is obtained from the food industry and cattle feedlots, processed in digesters to make biogas and then fed into gas engines to produce electricity. President Cyril Ramaphosa's decision in 2019 to allow municipalities to buy power from independent power producers (IPPs) has created new opportunities for green power entrepreneurs. According to Sean Thomas (Bio2Watt's MD), 'this could be a great opportunity for both municipalities in good standing and IPPs to diversify their energy mix; it will also help to decentralise power generation' (*Mail & Guardian* [online] 2020b).

Methane from Lake Kivu: Lake Kivu stores huge amounts of CO_2 and methane gas, which Rwanda is extracting to produce electricity. The CO_2 originates from two active volcanoes on the lake's northern shore, Nyiragongo and Nyamuragir; whereas the methane derives from the degradation of organic matter on the lake's surface and the conversion of CO_2 into methane in its depths. Such a large accumulation of methane in a lake is unique and results because the gases are trapped beneath a layer of warm, salty, CO_2-rich water. It is estimated that 45 km³ of methane is economically viable for extraction, which would generate over 500 MW of electricity over 50 years.

The extraction of methane is beneficial for two reasons. First, the dissolved gases pose a risk to the local population because they could potentially erupt from the lake. In Cameroon, a gas burst at Lake Nyos in 1986 caused the death of 1 700 people. Second, the methane can be used to produce electricity, which is crucial for Rwanda's development because only 30 per cent of Rwandans have access to electricity through the national grid. The first phase of the KivuWatt power plant, run by Contour Global, already generates 26 MW of electricity and will reach 100 MW in the second phase. The Rwandan government plans to achieve an electrification rate of 100 per cent by 2024 (Tofield-Pasche 2020).

Discussion: Renewable energy resources hold great promise in Africa, but only if the energy generated from them can be efficiently stored. The concept of lithium-ion (Li-ion) batteries was first developed at the CSIR in South Africa, but the product was not taken to market at the time. Today, almost all Li-ion batteries are produced in the USA, Poland, South Korea, Japan and China, although there is one university-based centre – the Energy Storage Innovation Lab (ESIL) at the University of the Western Cape in Cape Town – that is producing the batteries on a pilot scale in Africa and laying the foundation for industrial-scale production.

Li-ion batteries are rechargeable and are widely used in power tools, toys, electric bikes, laptops and cellphones. They can also be used as a buffer between power generation and consumption by being charged when power is available from, for example, a wind turbine, solar panel or wave generator, and then releasing the power when it is not. If Li-ion batteries could be made in Africa on an appropriate scale, they would become cheaper and power users would be able to rely more on renewable energy sources, which would have many financial and environmental benefits for the continent.

Manganese is an important component of Li-ion batteries and South Africa has about 80 per cent of the world's known reserves of the mineral. There is therefore a huge opportunity to develop a Li-ion manufacturing capacity in Africa. However, according to Bernard Bladergroen (the head of the ESIL), this would only be commercially viable if millions of excellent quality cells were produced per day in a large-scale facility at a cost of US$300 per kWh (Bladergroen 2018). Economy of scale would be crucial to achieve these costs, but it would be a game changer for energy generation in Africa.

11

Town planning and urban living

'Change: Organising tomorrow, today.'

Jay Naidoo, South African politician and businessman (Naidoo 2017: title)

Introduction: Ancient cities were built in various parts of Africa at about the same time as those in Persia, India and China. There is also evidence that many infrastructural innovations were introduced in early African cities. Addis Ababa (Ethiopia) is young by African standards, having been founded about 135 years ago by Emperor Menilek II and Queen Taytu. The settlement was originally called *Filwoha* ('hot spring'), as it was built on thermal springs, but it was later renamed Addis Ababa ('new flower') by Queen Taytu on account of the flowering mimosa trees that are common there. The emperor built a palace in the city in 1892, and it was rebuilt in 1894 after a fire. An interesting innovation in the palace was a fountain fed by water piped from aquifers in the surrounding hills. The emperor was also involved, with support from the French, in the development of a railway line from Djibouti to Addis Ababa using heavy iron sleepers to stabilise the line on the mobile desert sands. These customised sleepers are still known as 'menileks' in the French iron industry (Hancock et al. 1997).

Menilek's palace was the site of huge, sumptuous banquets. In 1897, he commissioned the construction of an enormous *adarash* (reception hall) that had a three-gabled roof and could accommodate over 7 000 people. It also had 16 chandeliers of electric lights suspended from the roof (a first in Africa) and there were over 50 additional buildings in the palace compound, including one of the first technology hubs in Africa that provided facilities for the advancement of metallurgy, agriculture, horticulture, pharmacology, building, jewellery, weaving, weaponry, saddlery, carpentry, carpet-making and mead brewing. The compound also had a mint, dispensary, vegetable garden and zoo. Addis Ababa also boasted many technological and other innovations during Menilek II's reign (which ended in 1913), including macadamised roads (in 1903); a sophisticated telegraph station where payments were made using salt or bullets (1905); the first bank (1906), hotel (1907), modern school (1908), sewing machine factory (1909) and hospital (1910) in that part of Africa; and a printing press and hydroelectric plant (1911) (Hancock et al. 1997).

Recreating an African city: Karim Sadr (professor of geography, archaeology and environmental studies at Wits University) and his students have used LiDAR to rediscover and map a large Tswana 'lost city' located 60 km south of Johannesburg on the Suikerbosrand hills. LiDAR is a method of measuring distances by illuminating a target with laser light and then measuring the reflection with a sensor; differences in laser return times and wave lengths are used to create digital 3D representations of the target. The city, provisionally named SKBR until a suitable Tswana name can be adopted, measures 10 km × 2 km, was occupied from the 15th century until the late 1800s and includes over 800 homesteads. SKBR consists of stone-walled dwellings and cattle drives, large cattle kraals and strange artificial mounds comprising masses of ash from cattle dung fires mixed with livestock bones and broken pottery vessels. These mounds are the remains of feasts and symbolise the generosity and wealth of the nearby homestead. There are also many short stone towers, possibly graves or grain silos (Sadr 2018).

Antipandemic architecture: Some of the design features of historical architecture have provided useful insights into how Covid-19 and other pandemics can teach us to be mindful when designing buildings. In historical buildings, it was common for water points to be freely available outside the front entrance or inside near the front door, but today they are tucked away deep in the building in private places. Early architects were sensitive to the transmissive danger of unclean gatherings, and in Cairo, Timbuktu and Mapungubwe, fountains or taps were commonly available outside buildings. The location of outdoor cleansing rooms to support the Islamic practice of *wuḍū* (ritual cleaning before prayers) also had public-health benefits. We now know that the novel coronavirus attaches well to shiny surfaces most often associated with cleanliness like glass, stainless steel and plastics, but rougher surfaces such as fabrics, wood, ceramics and even copper (which are often used in African architecture) appear to break up its RNA chains (Le Roux 2020).

Courtyards, lapas, verandas and porches are a common feature of older buildings in Africa. They provide residents with access to fresh air and sunlight without having to leave the building, whereas modern apartments and shacks do not offer this amenity. In places where there is sufficient space, medicinal gardens are an important feature of traditional African dwellings. They provide not only medicines and grooming products, but also release vapourised antimicrobial oils and, in the case of herbal hedges, can filter out transmissive agents. In South Africa, flu symptoms are alleviated using bushes of *umhlonyama* (*Artemesia afra*) and *imphepho* (*Helichrysum* species), and ubiquitous aloes provide a supply of gels to rehydrate hands that have dried after frequent washing. African architecture needs to reclaim its connection with public health (Le Roux 2020).

Traditional homes: Although modern building methods have been widely adopted in Africa, many traditional techniques continue to be used. In the Ethiopian highlands, the Dorze ethnic group build beautiful beehive-shaped houses up to 12 m tall, with the smooth curve of the continuous roof and walls thatched with *ensete* (false banana leaves) stretched over a complex bamboo frame, broken only by a small door. As a result of the high, vaulted ceiling, the spacious sleeping and living quarters (which have floors of compacted earth) are cool and dark. Each house has its own stockade garden and is surrounded by other, smaller structures of the same design that are used as cattle byres, kitchens, guest cottages or workshops. Unlike many traditional dwellings made from natural materials, Dorze houses are durable structures that last 40 years or longer. Furthermore, if the lower part of the wall is attacked by termites or starts to rot, the whole house can simply be lifted up and carried on poles to a new location, there to be trimmed and re-established, slightly shorter than before (Hancock et al. 1997).

Hemp houses in Morocco: In the remote High Central Rif of northern Morocco, the local tribespeople (Riffians) have lived for centuries in relative isolation despite attempts by Morocco and even Spain to displace them. Now they face a looming ecological disaster. Deforestation resulting from tree felling to provide fuel for heating and cooking, and to clear land for cannabis cultivation, has decimated forests and led to soil erosion. A further problem is that there is insufficient wood to build new houses. However, a novel solution is being investigated by the German architect and hemp pioneer Monika Brümmer (the founder and owner of Granada-based Cannabric). She noticed that one of the by-products of cannabis cultivation is cannabis straw, used traditionally for making buildings in the Rif (Brümmer 2019).

Brümmer hopes, in collaboration with the local authorities, to move the Rif's cannabis-based economy away from using naturally occurring marijuana strains that are processed into *kif* (the potent form of cannabis made into hashish) towards more positive goals. While *kif* production is illegal in Morocco, it does generate enough income to support over 80 000 Riffian families, but they live under constant threat of prosecution and receive a pittance from the criminals who market the hashish internationally at a huge profit. While Moroccan officials are considering laws to legitimise cannabis production for medicinal and industrial purposes, Brümmer has other plans. She has found that over 8 000 tonnes of cannabis straw go to waste in the Rif region each year and plans to build houses, schools and other public buildings with this material. In 2019, together with students from the National School of Architecture, she designed and built a unique, spherical solar-powered house from locally sourced hemp, plant bioresins and other non-synthetic materials (https://www.cannabis-mag. com/a-revolutionary-eco-building-in-Morocco-combines-hemp-and-solar-energy/).

She also plans to use the waste material to make briquettes and pellets for biofuel, which will further reduce the harvesting of trees. With a population of over 450 000 people, the High Central Rif is the most densely populated region of Morocco and will benefit from the many jobs that are created by this innovative venture (Brümmer 2019).

Sustainable architecture in Niger: When Niger-born architect Mariam Kamara was presented with the 2018 Rolex Mentor and Protégé Arts Initiative award, she was given the opportunity to work with British-Ghanaian architect David Adjaye on her dream of designing an arts centre for Niger's capital city Niamey. Kamara, who founded the sustainable design practice Atelier Masomi in 2014, consulted widely on the design of the building, particularly with the young people of Niger. They told her that they wanted a place to read and participate in the arts, as well as reflect on their identities and the future (Maditla 2020). Her design of the Niamey Cultural Centre includes an auditorium, art gallery and the city's first public library. Like her previous projects in the city, it will be built using compressed-earth bricks, a breathable material that is responsive to Niger's desert climate and a strong reflection of local vernacular architecture. The design also encompasses four 47-m curved towers that will cool the building using natural ventilation and create shaded public courtyards and passages (Maditla 2020).

Kamara left Niger to attend university in the USA in 1997 and originally qualified as a computer scientist. In her 30s, she decided to follow her passion for architecture and retrained at the University of Washington. Although she now splits her time between Niger and Providence (Rhode Island, USA), it is her projects in Niger, blending contemporary and traditional influences, that have earned her the most accolades (including the prestigious Prince Claus Award in 2019). A market in the remote village of Dandaji that she designed and includes colourful recycled metal canopies surrounding a sacred tree has become a vital space for vendors. Hikma, the religious complex that she designed for Dandaj in 2018, saw her transforming a derelict Hausa mosque into a library with the help of the adobe masons who had originally built it. She created a new, larger mosque alongside it from materials sourced within a 5-km radius. Kamara said the limitations of working in a country with so few resources have forced her to hone her approach, as has the climate change crisis.

> There is something very useful about that scarcity. It clarifies your thinking. It really makes it about space rather than bells and whistles ... It becomes more about what the space is doing and how it is bringing people together. That for me is what architecture is about ... It's impossible to ignore all the climate-related issues. We need to figure out what progress looks like when it's not fed by exploitation, and recognize our mistakes in terms of construction, material and resource depletion. Now Asia, South America and Africa can be trailblazers. For me, that means there is an opportunity to make a real contribution.
>
> *Mariam Kamara, award-winning Niger-born architect* (Maditla 2020)

First garden city: The first 'garden city' in Africa was Pinelands in Cape Town which was designed along the lines of Ebenezer Howard's (1898) idea that a city should be planned around a central park, public buildings and a commercial centre, with adequate roads, cycle tracks and pedestrian lanes, and that every effort should be made to preserve naturally occurring trees and other plants. The layout of Pinelands was completed in 1921 and the first houses were built in 1922 – all with thatched roofs. Care was taken that each house was different from its neighbours, with the original Cotswold houses proclaimed national monuments in 1983. Today, Pinelands remains a spacious and attractive garden city that is proud of its natural and built heritage (P. Beck, Genealogical Society of South Africa, personal communication, 2019).

Modern cities: Today cities cover less than 2 per cent of the Earth's surface, yet they are home to more than 50 per cent of the world's population and generate about 80 per cent of global GDP. But they also consume two-thirds of the world's energy and generate nearly 70 per cent of worldwide greenhouse gas emissions. So, although climate change is a global problem, many of the day-to-day causes and effects will have to be dealt with at city level. In 2019, the non-profit research group CDP produced an A-list of the world's cities that had made the most progress in both setting out their climate adaptation plans and in publicly reporting the extent to which they had reduced their greenhouse gas emissions. CDP provides a global platform for over 800 cities to measure, manage and disclose their environmental data each year in order to manage emissions, build resilience, protect themselves from climate change impacts, and create better places for people to live and work.

On the 2019 A-list, which included 105 cities (up from 43 in the 2018 listing), the most compliant cities were Palo Alto (USA), Stockholm (Sweden), Toronto (Canada), Manchester (UK) and Iskandar (Malaysia). Only two African cities, Durban and Cape Town in South Africa, were on the A-list (Hodges, Lombrana & Pogkas 2020). In 2019, the eThekwini Metropolitan Municipality (City of Durban) cemented its position among the world's climate leaders by joining the A-list for the first time and being the first African city to complete a Paris Climate Change Agreement-aligned Climate Action Plan (CAP) in collaboration with the C40 Leadership Group. In their CAP, the city sets out ambitious but achievable emission reduction targets of 40 per cent by 2030 and 80 per cent by 2050; it details actions to improve its resilience against climate change and its ability to support vulnerable populations in the transition to a low-carbon economy.

The eThekwini CAP addresses several key areas from energy, buildings, transport, solid waste and air pollution to adaptation (water, health, biodiversity, food and sea-level rise) and cross-cutting issues impacting vulnerable communities. eThekwini's reduction targets anticipate a decline in emissions per person from 5.95 tCO2e (tonnes of CO_2 equivalent) in 2015 to 0.8 tCO2e in 2050 (the equivalent of a reduction from 2 536 to 341 litres of gasoline/petrol consumed per person), with the highest emitting sectors (electricity and transport) being prioritised (Hodges et al. 2020).

In eThekwini, electricity consumption is the primary source of greenhouse gas emissions. To reduce emissions, the city is focusing on decarbonising its grid and promoting renewable energy use. In South Africa, which generates 94 per cent of its electricity from coal, the Durban Climate Change Strategy aims to achieve a target of 40 per cent of electricity consumption being met by renewables by 2030. In addition, the CAP aims to ensure that by 2050, 70 per cent of private electricity demand will be provided by self-generated renewable energy. As eThekwini's transport systems account for 30 per cent of the city's 20.8 million tonnes of annual greenhouse emissions, it is planned that at least 50 per cent of its citizens will use public transport by 2030. To achieve this, the city is expanding its network of electric buses and aims to reduce private car trips by 50 per cent by 2050. EThekwini has also set up an Agro-Ecology Unit and aims to achieve a 50 per cent increase in locally produced food and to reduce the volume of good quality leftover food waste by 80 per cent by 2030. EThekwini's climate action plan is a bold commitment to addressing the drivers and effects of climate change, and demonstrates the crucial role that African cities can play in facing up to the challenge of implementing the Paris Climate Agreement (www.cdp.net).

Lagos smart city: On its own, Nigeria's economic capital Lagos would rank as Africa's seventh-largest economy. Its population swelled to over 20 million in 2019 (making it Africa's largest city) and is projected to double by 2050 (Egbejule 2020). Known for both its entrepreneurial dynamism and its traffic 'go-slows', Lagos is a megalopolis on the move in slow motion. In December 2019, Babajide Sanwo-Olu (the governor of Lagos state) unveiled the Lagos Innovation Master Plan and announced a Naira 250-million technology fund for research and development in the state. The master plan seeks to make Lagos a smart (and fast) city and provide a foundation for the dynamic technology sector in Yaba to grow.

But Sanwo-Olu also has his detractors who have noted that, like his predecessors, he evicted residents of an informal township (the Tarkwa Bay beach community) in favour of developing an upmarket residential area, and banned the ubiquitous *okadas* (motorcycles) and *kekes* (motor tricycles) from the city centre despite the fact that they help to ease the problems created by serious traffic congestion and a dysfunctional mass transit system. This decision created a social media storm and local people joke that it takes less time to fly from Lagos to London than it takes to cross from one side of the city to the other. Various apps have been developed to facilitate usage of the *okadas* and *kekes* but policy-makers view them as inappropriate in a city the size and stature of Lagos, a classic example of the gap between haves and have-nots. Lagos is also plagued by dysfunctional sanitation, waste management and water drainage systems, and lacks an interstate railway connecting it to Ibadan 129 km away (Egbejule 2020).

A partial solution has been reached with the completion of the BRT system connecting Lagos and Victoria Island. The BRT operates 16 hours per day and its 22 buses in their dedicated lanes transport more than 200 000 passengers daily. The project (which was funded and supervised by the World Bank) cuts journey time by 25 minutes, reduces greenhouse gas emissions by 20 per cent and carries out 25 per cent of all trips with just 4 per cent of the vehicles (Adeshokan 2020).

Steven Jennings (the CEO of Rendeavour and an African new-city builder who cut his teeth on Kenya's Tatu City project), has commented that 'almost without exception, sub-Saharan African cities are broken' (Norbrook 2020a) with regard to traffic congestion, poor infrastructure, lack of planning, poorly developed building-approval processes, land ownership rights and logistics problems. Rendeavour is engaged in developing Alaro City in the Lekki Free Zone in Lagos in an effort to create more chaos-free urban environments. According to Yomi Ademola (Rendeavour's head of West Africa) the lessons learned in Kenya were invaluable, especially the sequencing and rollout of infrastructure and the creation of complete, livable and workable enclaves not disrupted by subsequent construction. Another new-city development in Lagos is Eko Atlantic City, which is being built on reclaimed land. Competition between the two new-city developments is keen, but Jennings points out: 'Their reclamation costs are around $1 000 per square metre. Our site levelling costs at Alaro are about $1 per square metre. So, they have this massive cost structure they have to try to recover from their clients' (Norbrook 2020a). In October 2019, the global ride-hailing company Uber Technologies launched a pilot taxi boat service in Lagos and their WaXi water taxis are now plying the waterways of the commercial hub to ease traffic congestion.

Urban renewal in Addis Ababa: Biruk Terrefe is a PhD candidate in the Department of International Development at Oxford University who has analysed the social impact of urban megaprojects in Addis Ababa. He sees urban spaces as sites of 'continuous renewal' that new administrations use to stamp their mark on cities. He has found that urban investments across Africa are at an all-time high, but that they do not always benefit everyone. In Addis Ababa, a new urban aesthetic is emerging that targets urban elites, the Ethiopian diaspora and international tourists. Since coming to power in 2018, Prime Minister Abiy Ahmed's administration has initiated several urban megaprojects that support this new aesthetic, including the 36-ha real-estate project LaGare; a 56-km riverside renewal scheme called Beautifying Sheger; hiking trails, viewpoints and recreational spaces on Entoto Mountain; and a new science and technology centre. These projects will give Africa's political capital a facelift and generate revenue through higher land values and urban tourism, but Ahmed's focus on luxury real-estate projects and urban tourism represent a clear ideological break from the past and his plans are reminiscent of the flashy, hypermodern skyscrapers in Dubai and Abu Dhabi.

LaGare and Beautifying Sheger represent a new era of urban development in Ethiopia and raise fundamental questions about spatial justice. They cater to a growing urban elite and are in stark contrast to the pro-poor focus of the previous administration. The new megaprojects will make the city less accessible and affordable, and exacerbate its major transport, sanitation and housing shortages. Dancing water fountains will decorate the main plaza of Beautifying Sheger at a time when most people have no reliable access to water. As one of the fastest urbanising nations in the world, Ethiopia needs to ensure that its urban investments provide adequate infrastructural and social services to its most vulnerable citizens. The creation of elite enclaves and a singular focus on urban tourism, while important, are unlikely to trickle down to the majority of the city's residents (Terrefe 2020).

Africa's tallest buildings: Africa is not known for tall buildings. For millennia, the Great Pyramid of Giza in Egypt (built in the 3rd century BCE) was the tallest building on the continent. This ancient structure originally reached a height of

146.5 m, but has eroded down to 138.8 m. It was only surpassed in 1300, when the Lincoln Cathedral (159.7 m) was completed in England. Since then, Africa has languished far behind in the tall building stakes. Architects and property developers argue that it does not make sense to build skyscrapers in Africa because land is relatively cheap and cities typically tend to grow outwards rather than upwards, with the result that African cities are characterised by low-rise buildings and urban sprawl. In Harare and Maputo, for example, more than 30 per cent of the land within 5 km of the central business district remains unbuilt yet massive informal townships have developed around their peripheries. Another reason is that land ownership in African cities is often disputed and no-one can risk making large investments on land that might later be taken away from them (Allison 2019).

For 45 years, the tallest modern building on the continent was the Carlton Centre in Johannesburg at 222.5 m. Now the Leonardo, also in Johannesburg but in the upmarket Sandton district, has become 'Top of Africa' at 234 m. Other tall buildings in Africa include the Britam Tower in Nairobi (built in 2017, at 200.1 m), Ponte City in Johannesburg (1975, 185 m), the UAP Old Mutual Tower in Nairobi (2016, 163 m), Necom House in Lagos (1979, 160.3 m) and the Ports Authority Headquarters in Dar es Salaam (2016, 157 m). Kenya plans to take top ranking in 2023, with the new Pinnacle Tower (320 m) rising above the streets of Nairobi, as the construction of the ambitious Palm Exotica coastal resort in Watama (370 m) has been delayed due to opposition by environmentalists (Allison 2019). But there is trouble on the horizon for the Pinnacle Tower, as the high court in Nairobi has issued arrest warrants to two Dubai-based real-estate speculators backing the project as a result of a dispute over who owns the land on which it is being built. Africa's old problem of land ownership has come back to haunt them.

Egypt also has lofty ambitions with its Iconic Tower in the new iCity complex in Al Qahera Al Gadida, between Cairo and the Red Sea. In addition to the tower, the development will have 60 residential buildings and 98 villas, twin houses and stand-alone units. According to Housing Minister Asem al-Gazzar, the project – which will be developed on a land area of about 700 km² (12 times the size of Manhattan Island) – incorporates pioneering sustainability technology and will be the first in Egypt to be built under a public–private partnership. The Iconic Tower, inspired by the ancient obelisk nearby, is under construction by the China State Construction Engineering Corporation and will reach a height of 390 m when it is completed in 2023.

The iCity project is an important landmark in China's Belt and Road Initiative aimed at promoting win-win partnerships between participating states through trade partnerships and infrastructure projects. The cutting-edge iCity, which is laid out in a novel set of interconnecting 'power rings', has a budget of US$58 billion and is one of several megaprojects that have been launched to create jobs and drive investment in Egypt. As very tall buildings are often a symbol of wealth and power, it is not surprising that the most superlative examples are those in the oil-rich countries of the Middle East. The Burj Khalifa in Dubai (830 m) is 2½ times taller than any building in Africa, and they will all be dwarfed by the stupendous Jeddah Tower (1 008 m) if it is completed in Saudi Arabia. There was steady progress on this ambitious project until January 2018 when building owner JEC halted structural concrete work, with the tower about one-third completed, due to labour issues.

Shelter Afrique: Zimbabwean Andrew Chimphondah (a chartered accountant with extensive experience in real-estate financing) is the CEO of Shelter Afrique, the only pan-African financial institution that exclusively supports the development of the housing and real-estate sector on the continent. Shelter Afrique, which was launched in 1984, is owned by 44 African member governments (Class A shareholders) and the African Development Bank and Africa-Reinsurance (Class B shareholders), and has completed over 17 500 projects and issued loan approvals worth over US$1.147 billion since its inception.

The profit-making, multinational finance company (also known as the Company for Habitat and Housing in Africa) is headquartered in Nairobi, with regional offices in Abidjan and Abuja. Their partners include the Entrepreneurs Development Bank, the French Development Agency, UN Habitat for a Better Urban Future and the European Investment Bank. An example of their projects is the Eden Beach Resort & Spa in Mombasa, Kenya. Their raison d'etre is that, left untouched, Africa's informal settlements will exacerbate the continent's most pressing problems. They recognise that housing millions of people who are without adequate shelter in quality, affordable homes is one of humanity's greatest challenges but also one of its greatest opportunities – and they seek to be in the forefront of this massive endeavour.

Design with Nature: Although American landscape architect Ian McHarg launched the Design with Nature concept over 40 years ago, it has many modern adherents. Yolandi Schoeman of South Africa launched Baoberry in 2014 to provide innovative solutions and services in the field of ecological engineering. Like other cleantech companies, Baoberry endeavours to provide solutions that make both ecological and economic sense. One of their recent projects was the development of a cost-effective grey water re-use system for domestic and corporate applications that will not only provide cleaner water, but also lead to job creation and better water and food security (Manyonga 2017d).

Jean-Charles Tall is a Dakar-based green architect who co-founded the *Collége Universitaire d'Architecture de Dakar* (University College of Architecture of Dakar) when the Senegalese government decided to close the only official architectural school in the country. He has a deep concern for the development of bioclimatic architecture that is embedded in African traditional knowledge, and has showcased the diverse architectural styles and urban development patterns in Senegal in his African Mobilities project. For example, Tall supports the use of the bulrush (*Typha australis*, which grows abundantly along the Senegal River) for thermal insulation in homes and the traditional cooling and ventilation systems used in Jolofi houses in southern Senegal. The Jolofi houses use earth-and-thatch roofs, recycled rain water, natural cross-ventilation, and water-filled clay pots to humidify and cool the air (Matsipa 2020).

> I found that even in the villages everybody is abandoning the spirit of traditional architecture. There is a fight for power that has been won by the Europeans. They impose their way of understanding the world and try to erase completely our traditional ways of understanding our world … In all the cultures in the world, the architecture that has been produced by the people themselves takes into account the climate, the situation, the society. I am teaching the way that the climate enhances the houses.
>
> *Jean-Charles Tall, leading Senegalese green architect* (Matsipa 2020)

Vere Shaba, a South African mechanical engineering graduate, established Shaba & Ramplin Green Building Solutions in 2016 to promote an integrated green design approach to engineering. Tired of the assumption that, as a mechanical engineer, she could not also be a green building consultant, Shaba launched her own consulting firm that specialises in green building engineering and interior services. She is now a multi-award-winning entrepreneur who was selected as one of the 10 Top Women in Engineering for the Standard Bank Topco Top 100 Women in Business and Government, the *Mail & Guardian*'s 200 Young South Africans for Environment and the Kingdom of the Netherlands' Inspiring 50 Women in South Africa in STEM, and also as a finalist in the Gauteng Women Excellence Awards for Young Achievers.

Community gardens: In 2013, when Xolisa Bangani approached a school in his hometown of Khayelitsha near Cape Town for permission to create a community garden, he was met with some scepticism. His plan was to turn the unused grounds of the school, which had become rubbish dumps, into a flourishing garden. Although he was just 20 years old, the Isikhokelo Primary School granted him a three-month temporary concession. After three months of hard work clearing the land and planting seeds, the school was so impressed that they granted him a 20-year concession. Now, seven years later, the Ikhaya Garden not only provides fresh organic vegetables to the school's kitchen, but is also changing young people's perceptions about gardening and drawing tourists and revenue to the area.

> I grew up here and I always wanted to be a part of the change. I wanted to change the negative representation of Khayelitsha that the media tends to focus on, and I wanted to make gardening cool … Back in 2013, as we started preparing the land for cultivation, the kids from the school wanted to find out what we were doing. When we told them that we were going to grow food, they were amazed. They thought that seeds came from the shops, not the earth … Growing up in Khayelitsha, we only ever saw butterflies on TV. I wanted to bring more greenery into the township.
>
> *Xolisa Bangani, founder of Ikhaya Garden*
> (*https://ikhayagarden.wixsite.com/ikhayagarden*)

With a flourishing food garden, regular food markets, community discussions on optimising food production and using organic waste, and events at which visiting experts share their knowledge, Xolisa has unquestionably achieved his goal – and many others are following his example (https://ikhayagarden.wixsite.com/ikhayagarden).

Rooftop gardens: Nhlanhla Mpati is the agropreneur in charge of an ambitious rooftop garden project led by the innovation incubator Wouldn't it be Cool (WIBC) in collaboration with the public–private Inner City Partnership of Johannesburg. Working atop the soaring Chamber of Mines building in the central business district, he uses hydroponics to grow vegetables in special water solutions without the need for soil or large open spaces. Furthermore, the plants grow faster and use 80 per cent less water than in conventional farming, and rooftop gardening cuts down on 'food miles' (the distance food needs to be transported from producer to consumer), thus reducing carbon emissions.

'You're creating a perpetual cycle of sustainability. The farm's sustainable, the project is sustainable – until we run out of buildings', says WIBC's founder Brenda Martens (Lazareva 2017), who plans to establish another 100 inner-city rooftop gardens in Johannesburg. Moroka Mokgoko, who works for the urban eco-farming enterprise Rooftop Roots, predicts that rooftop gardening will help to alleviate poverty and provide jobs for urban youth (Lazareva 2017). It is an idea worth trying in other African cities.

Reel gardening: When 16-year-old Claire Reed of Johannesburg was asked by her father to plant a vegetable garden, she was appalled at how much seed and fertiliser she had to buy and how much wastage there was. She experimented with various ways of planting economically and submitted a project to the Expo for Young Scientists in 2002, where she won the gold medal. Her high-school idea has since blossomed into a successful commercial company, Reel Gardening, which has developed award-winning products that simplify home gardening. Her start-up, which was launched in 2010, sells reels of paper strips that are prepackaged with seeds (of vegetables, herbs and companion-planting flowers) and organic fertiliser – all correctly spaced so that they can easily be planted and maintained. The advantage of using strip planting is that birds cannot eat the seeds and you only need to plant as many as your space allows. The biodegradable strips use 80 per cent less water than traditional gardening, as they only need water at the exact location of each seed, and were immediately successful. Reel Gardening also sells patented planting pots, decorated with *shweshwe* (a printed dyed cotton fabric used for traditional southern African clothing), that are properly aerated and have an internal drip tray.

Hot Spot geyser: Electric geysers consume a substantial proportion of a typical household's electricity, but switching them off when you do not need hot water is not a practical solution. Sandiswa Qayi established Amahlathi Eco Tech Africa (AET Africa), with her partners Michael Romer and Tamsanqa Gxowa, to address this and other cleantech challenges for society. Sandiswa is not just a pretty face fronting for her male partners, but a real-life inventor who has found a way to reduce a geyser's energy consumption by using a simple but effective contraption that heats water only when necessary. Her Hot Spot device is a geyser sleeve that can be retrofitted over any standard geyser element. It works by pushing hot water from the bottom to the top of the geyser using a thermosiphon that delivers 50 litres of water within 30 minutes at 50°C. By October 2017, her company had retrofitted 7 million geysers (Manyonga 2017e).

AET Africa is based in East London's Science and Technology Park in the Eastern Cape (South Africa). The goal of the company, which was established in 2012, is to identify and solve energy challenges. In addition to developing the Hot Spot geyser sleeve, AET Africa manufactures jet pulse filter bags and is working with Hydrogen South Africa (HySA) and the University of Cape Town on the development of fuel cells that use heat exchange technology. Due to their innovative approach, they have received funding for research and development from the Technology Innovation Agency, The Innovation Hub, Grassroots Innovation, the Small Enterprise Development Agency, and the Department of Science and Innovation in South Africa. AET Africa is also a key player in promoting science and innovation as a vehicle to foster socioeconomic transformation, and is planning to offer its products and services throughout Africa in future (Manyonga 2017e).

Loo Cap: The Water Research Commission in South Africa has developed the Loo Cap, a handwash basin that fits onto an existing toilet cistern and replaces the lid. The Loo Cap is fitted with a handwash tap and a connection to the main water supply, and can be retrofitted onto any cistern. Furthermore, grey water from the handwash basin is used to fill the cistern, which saves 800 to 1 200 ml of water per usage. The device is ideal for use in schools, malls and hospitals, as well as in homes.

Importance of waste pickers: Every large African city has an army of men and women weaving their way through the traffic pushing flatbed carts piled high with plastic, which they sell to recycling depots to support their families. These unsung heroes (the waste pickers) play a vital role in the recycling process and, at the same time, reduce the amount of plastic entering the oceans.

Professional big-wave surfer Frank Solomon has developed a deep love for the marine environment and has organised numerous conservation projects, including beach clean-ups and the creation of the Sentinel Ocean Alliance (which works with underprivileged children in Hout Bay, South Africa, teaching them how to surf, protect the oceans and become lifeguards). Recently he teamed up with adventure film-maker Arthur Neumeier to create *Street Surfers*, a documentary about the waste pickers who surf the streets of big cities, sometimes hundreds of kilometres from the ocean. Two waste pickers in Johannesburg, Mokete Mokete and Thabo Mouti, start their day at 03h00 and work non-stop until dusk. They have to collect 100 kg of plastic to earn just R300, which is their only source of income. Although the community of waste pickers is responsible for up to 90 per cent of the plastic waste that reaches recycling depots in South Africa, they receive little recognition for their vital work. Now *Street Surfers*, which won Best International Short Film at the London Surf Film Festival, has highlighted their plight and funds have been donated to the waste pickers by viewers worldwide through the website www.backabuddy.co.za/street-surfers (Potgieter 2020).

The global trend towards privatising waste management has come into conflict with the livelihood of waste pickers. Waste that was previously available free to poor people has been appropriated by private corporations that generate profits but deprive waste pickers of their livelihoods. In Johannesburg, the Genesis Landfill was privatised and waste pickers were violently evicted. In Egypt and Ghana, waste pickers have been excluded from contributing to municipal waste management services. The incineration of waste also threatens waste pickers who tend to recycle their material. The first large waste incinerator in Africa was built in Ethiopia in 2018 with Chinese money and Danish technology even though research had shown that recycling is environmentally preferable to incineration. The Global Alliance of Waste Pickers has been formed with agencies in 28 countries. Its aim is to include waste pickers in decision making, improve their working conditions, develop their capacity and achieve recognition for their work (Demaria & Todt 2020). In South Africa, after initially being ignored, waste pickers were accorded the status of 'essential services' in April 2020, which allowed them to continue plying their trade during the strict Covid-19 lockdown.

Waste management in Equatorial Guinea: Equatorial Guinea is an unusual African country. It is the only nation on the continent that has Spanish as its official language and it is the smallest African country to be a member of the UN. Although it is one of the richest countries in Africa due to its oil revenues, this wealth is not evenly distributed and most of its people live in poverty. The president has been in power since 1979 and his son is the vice-president. The capital Malabo is located on an offshore island, although a new capital (Oyala) is being built on the mainland. Notwithstanding all its challenges, Equatorial Guinea is famous for the quality of its waste management services – a field in which it sets the standard for the rest of Africa. Several companies, including Golden Swan and a subsidiary of the Indian multinational Shree Hari Group, have developed state-of-the-art waste management services for the disposal or recycling of oil sludge; waste chemicals; paints; biomedical waste; waste metal; glass, wood and plastic; e-waste; and used filters and batteries.

SweepSouth: Aisha Pandor, whose PhD in genetics resulted in her winning the 2011 South African Woman in Science Award, decided after only two years of formal employment to venture into entrepreneurship. Her big breakthrough came when she made a prizewinning pitch at the SiMODiSA Startup SA technology conference. In 2012, she launched the app SweepSouth, an all-in-one, no-fuss service that recruits vetted domestic workers. The concept behind the SweepSouth app is like that of Uber and Airbnb, and came about as a result of her struggle to find a domestic worker. The company recently celebrated its fifth anniversary, having created employment opportunities for over 15 000 previously unemployed or underemployed workers in South Africa, and is growing its online shop that sells home products. SweepSouth is set to expand into gardening, plumbing and electrical services.

When setting out in business, particularly if you're going to be striking it out on your own as an entrepreneur, think big. Focus on unique challenges and solutions that have the potential to change people's lives … When you have a big vision, you can't expect to reach your outcomes on day one. Success is not a destination, and it is natural for goal posts to shift. There is no such thing as overnight success, and every stage of development comes with some frustration.

Aisha Pandor, co-founder of SweepSouth (*Independent On-line* 2019)

Emerging innovators: Agnes Ngoyi, a women's rights advocate who is Uganda's deputy national coordinator for the prevention of human trafficking, was forced to flee her home during the 20-year insurgency in northern Uganda by the Lord's Resistance Army (LRA). The war ended in 2005 when the LRA was ousted and fled to South Sudan. In 2013, Ngoyi returned to her home to help other victims of war. Working with Paska Akello (a single mother who had been kidnapped by the LRA when she was pregnant) and 15 other women in Paicho county, she launched a movement that saw women adopt a role that they had never carried out before – building houses. Further west, in Pida Village, a group of 40 women (many of whom had been abducted by the LRA, raped and then shunned by their families for returning with 'rebel babies') set up their own community in 2011. They called it *Waroco kwowa* ('Let us rebuild our lives' in Acholi) and set about not only building houses but also teaching tailoring, craft-making, farming and business skills to a new generation of teenage mothers. Some men applied to join the *Waroco kwowa* community but only five, who offered skills that the women lacked, were accepted. In Paicho, the builders' group admitted one man because he could read and write, but he was fired when he stole 100 000 shillings. 'We don't need men because they can't be trusted', said Akello (Lazareva 2018).

Discussion: By 2050, 75 per cent of the world's population will live in cities, so the design of these conurbations will have a major impact on the quality of life of most people. City planners will have to take into account climate change and the threat of future pandemics. However, the question is: Will a newfound respect for science and a fear of future shocks lead us to finally wake up, or will the desire to return to normal overshadow the threats that still face us? The timeframe for effective climate change mitigation action has always been tight, but the Covid-19 pandemic (the most significant disruption yet to the post-war fossil-fuel order) has shrunk it further. According to the IMF, the global economy contracted by more than 5 per cent in 2020 and the challenge is so big that it has created a once-in-a-lifetime opportunity to change direction.

Rwanda, which has a GDP of about US$9 billion, has stepped up to the plate by adopting an US$11-billion plan over several years to reduce emissions and adapt to climate change, including a push for buses, cars and motorcycles to go electric. Further afield, in Paris, Mayor Anne Hidalgo is determined to use the Covid-19 lockdown as an opportunity to create a greener city and has quietly converted 50 km of roads into bike lanes (*corona pistes*) to add to the 1 400 km of bike lanes that already exist.

There is no doubt that technology will be an integral component of future cities as millennials look for environments where they can co-create a better future using the coordinated power of digital technologies and the IoT. Smart cities will embrace alternative energy and energy conservation measures by providing energy-saving heat pumps; light-emitting diode (LED) lighting; electric car recharging stations; mass parking for e-cycles and bicycles; solar panels; alternatives to water-wasting air conditioners; automatic, dynamic self-shading systems; green roofs and walls planted with xerophytes; heat-reflecting paint; and even, in coastal areas, on-site desalination plants.

Many new developments already offer precinct-wide Wi-Fi, fibre-optic cabling to the home and office, and 24/7 digital security surveillance, and are installing sophisticated environmental monitoring sensors that automatically alert the authorities to pollution and other problems. There is also a strong trend towards mixed-use precincts where people live, work and play in the same place, and towards green spaces and pedestrianised roads that have the feel of traditional villages but also have every high-tech amenities. Developers are also investing in water treatment plants and calcamite sewerage systems, and even in fish culture projects, bat hotels, owl houses and beehives.

For too long humans have relied on human-built, 'grey' infrastructure to solve water supply problems in cities. Now increasing attention is being paid, from Unesco to local water supply entities, to the use of green technologies derived from indigenous knowledge and biomimicry that imitate the ways in which nature has stored and supplied water for aeons. These methods include soil moisture retention and the use of naturally occurring soil moisturisers and mulchers, relying on wetlands and floodplains to capture and store water, and the use of vegetated green roofs and walls.

12

Education and the youth

'Education is the most powerful weapon
you can use to change the world.'

Nelson Mandela, late president of South Africa
(speech at launch of Mindset Network 2003)

Introduction: Africa has a proud tradition of education. The University of Al-Qarawiyyin in Fez (Morocco) is recognised as the oldest degree-granting institution in the world. It evolved from a mosque and *madrasa* (educational institution associated with a mosque) established by a woman, Fatima al-Fihri, in 859 CE. Fatima inherited a large amount of money from her father, a rich merchant, and used it to build the mosque and *madrasa*. From the 10th century onwards, the Great Mosque of Al-Qarawiyyin became a famous religious and teaching institute with 'teaching chairs' in various disciplines. Other early tertiary learning institutions in Africa include Al-Azhar University in Cairo (established in 975 CE), which also offered postgraduate degrees, and the Sankore Mosque and *Madrasa* in Timbuktu (989 CE). Fourah Bay College in Freetown (Sierra Leone), which was established in 1827, is the oldest institution of higher learning in modern sub-Saharan Africa after the collapse of the *madrasa* in Timbuktu. The University of Cape Town was founded on 1 October 1829.

Several African countries have stayed at the forefront of educational practice. For example, Kenya's first president in the independence era, Jomo Kenyatta, championed the ethic of active civic engagement in education in the form of the self-help ethos *Harambee* ('Let us pull together'), now the country's official motto. Today Kenya is an overperformer in education in terms of most measurable outcomes, notwithstanding the fact that it has a lower per capita income than most African countries and its public spending on education is only one-fifth of that of, for example, South Africa. This is because Kenyans believe that fixing education is not someone else's task or someone else's failure, but involves active citizenship and proactive engagement at every level of society. Their policy is not 'education for all' but 'all for education' (Levy 2018).

When the late Black Consciousness leader Steve Biko published his seminal book *I write what I like* in 1978, it was not about individual self-expression or self-indulgence but a political statement with its origins in the work of the Brazilian literacy activist Paulo Freire. Freire had identified the profound connection between reading, understanding the world and being able to change it. Sadly, in one of Africa's best developed countries, South Africa, the 2016 Progress in International Reading Literacy Study revealed that 78 per cent of Grade 4 learners could not read for meaning in any language. The reasons given for this situation were the absence of a reading culture among many adult South Africans, the dearth of school libraries, the high cost of books and the low quality of training for teachers of reading.

Many parents do not read to their children because they themselves are illiterate and because there are very few children's books available in African languages. Reading is also not a priority among the middle class or new elite, as it is regarded as uncool and nerdy. This is a tragedy, as recent neuroscience research has shown that reading (becoming literate) alters the brain because learning the visual representations of language develops cognitive abilities (such as verbal and visual memory) and reinforces problem-solving skills. Failing to learn to read is therefore bad for the cognition that is necessary to function effectively in modern society (Aitchison 2018).

In 2017, it was estimated that 32.6 million children of primary-school age and 25.7 million adolescents were not attending school in sub-Saharan Africa (Roby, Erickson & Nagaisha 2016). These frightening statistics, combined with the poor quality of education in many African schools, are however partly offset by the digital revolution that is resulting in the burgeoning use of ICT and mobile technology in education. This means that there is huge potential for reaching those who have been excluded from education as well as for creating opportunities to improve the knowledge and skills of teachers. This trend is supported by sharp decreases in the costs of communication and cellphones, increased international connectivity via undersea cables and satellites, and improved intranational connectivity through fibre-optic cables.

M-learning (learning via a connected mobile device) is therefore forecast to be one of the main drivers of a revolution in learner-centred teaching, teacher training and educational data management in Africa. The introduction of MOOCs (massive open online courses) adapted to African countries' needs and capacities has been a further bonus. Furthermore, the widespread distribution of computer hardware such as by the US-based One Laptop per Child project that was launched in several African countries in 2005, together with open-source software, has resulted in a boom in digital teaching and learning on the continent. The Fourth Industrial Revolution (4IR) is spreading so rapidly that the world as we know it will soon change beyond recognition. The future belongs to those who can imagine it and react quickly enough to take advantage of its opportunities. Against this daunting but exciting backdrop, the ways in which we teach and learn must become more agile, interactive and collaborative to ensure that education futureproofs our youth so that they are equipped to deal with a dynamic new world of life and work. The challenges brought on by the Covid-19 pandemic have exacerbated the situation, with 90 per cent of the world's schools closed on 21 May 2020 and 1.5 billion children experiencing a disruption to their education (according to a CNN news broadcast), which has created challenges as well as opportunities.

Lukasa memory board: Cultures around the world have developed different techniques to record and recover memories. One of the most remarkable memory aids from Africa is the *Lukasa* ('the long hand' or 'claw') memory board of the Luba people in the DRC. The *Lukasa*, as well as other memory devices, allowed *bana balute* (court historians) to recall and recite information encoded in the memory aid. This information ranged from mythical stories and societal organisation to sacred texts. Within these small wooden devices (which were adorned with shells, beads, carvings and other material), the *bana balute* could encode and record virtually anything. African traditional knowledge is of course remembered in many other ways as well, including through song, dance and storytelling (Jones 2020).

Ancient and modern libraries: Anthropologists argue that the magnificent collections of petroglyphs (rock carvings), cave paintings and monolithic statues found throughout Africa represent significant archives and libraries of information about the past and should be preserved as such. Just as New Mexico is justly proud of its magnificent outdoor library of over 21 000 rock art engravings by the Jornada Mogollon culture in the Tularosa Basin, so many African countries have ancient libraries to savour and preserve – from the Wadi Al-Hitan palaeontological site in Egypt with hundreds of fossils of extinct whales to the clay towers of the Great Mosque of Djenné in Mali and the Principal Mosque in Timbuktu, the Castle of Gondar and the obelisks at Axum in Ethiopia, the Tellem Tomb Cave in Niger, the Tomb at Kunduchi and the Palace of Husuni Kubwa in Tanzania, the ancient cities of Mapungubwe and Great Zimbabwe in South Africa and Zimbabwe, the Pyramid of Khufu and Great Sphynx at Giza in Egypt, and the magnificent petroglyphs and rock paintings in many African countries. Africa also has an abundance of World Heritage Sites designated by Unesco, comprising 145 sites in 35 countries. South Africa has 10 sites; Ethiopia and Morocco nine; Tunisia eight; and Algeria, Egypt, Kenya, Senegal and Tanzania seven.

The new *Bibliotheca Alexandrina* in Alexandria (Egypt), which opened in 2001, is in the historic centre of the city and was conceived as a revival of the ancient library founded there by Alexander the Great 2 300 years ago but lost to civilisation centuries later. The design of the new library is both timeless and bold, with its vast circular form echoing the cyclical nature of knowledge and its glistening roof recalling the ancient Alexandrian lighthouse. The 20 000-m² open reading room for 2 000 readers (the largest in the world) occupies more than half the library's volume and is stepped over seven terraces. The 11-story library can accommodate 4 million books and could be expanded to 8 million books using compact storage. In addition to the library, the *Bibliotheca Alexandrina* includes a planetarium, interactive science centre, several museums, a school for information science and conservation facilities.

Revolutionary libraries: Reading is one of the most important non-formal learning activities in which people indulge, but many Africans do not have ready access to a library. Innovative libraries have been established in other parts of the world, including telephone booth libraries in the UK, vending machine libraries in Taiwan and China, and a floating ship library in Norway; however, two of the most outstanding examples are in Africa.

To improve access to books by nomadic communities in Kenya, the authorities initiated camel-borne libraries in 1985. The hardy 'ships of the desert' are ideal, as they can negotiate harsh desert and savannah environments, and carry books as well as camping gear for the travelling librarians into the remotest areas. In Burundi, the innovative Library of Muyinga was built by the Belgian company BC Architects and Studies together with local community members. The building was made from locally available materials that are traditionally used in Burundi (such as compressed earth bricks and baked clay roof tiles) and the interior fittings include hammocks made from woven fibres; handwoven bamboo lamp shades; and solid wooden shelves, chairs and tables. The natural look of the building has instilled a sense of pride and ownership in the local community, who make frequent use of it (Nkosi 2020a).

Accra's one-woman library: While Sylvia Arthur from Ghana was working in Brussels in 2011, she sent boxes of books back to her mother in Kumasi. In 2017, when she moved back to her home country following the Brexit referendum, she decided to share her book collection with others by opening a private library and archive in Accra – the Library for Africa and the African Diaspora. Arthur is distressed that many of the archives of African writers are held abroad, for example that of Nigerian author Amos Tutola is at the University of Texas and Chinua Achebe's archive is at Harvard University in Massachusetts.

> We are a library that centres and focuses on the works of Africans and Africa-descended writers from across the diaspora. The reason I started the archive is because there's a big misconception that African literature started in 1956 with the publication of Chinua Achebe's *Things fall apart*. And I knew that African literature goes way before that. Trying to access books by Africans on the African continent is a huge problem.
>
> *Sylvia Arthur, founder of the Library for Africa and the African Diaspora in Accra* (Mallinson 2020)

She runs the library as a business on the basis that people should have sufficient respect for literature to be prepared to pay a small sum to access it. Arthur also launched three school and community libraries in Accra, and planned to initiate a residency programme in October 2020 to allow writers from across Africa and the diaspora to peruse the books and archive (Mallinson 2020).

Global storybooks: Globally about 750 million young people and adults do not know how to read and write, and 250 million children fail to acquire basic literacy skills. As one of the UN's Sustainable Development Goals (SDGs) is to achieve quality education globally by 2030, literacy levels need to be addressed urgently. In 2013, the South African Institute for Distance Learning began developing the African Storybook Initiative, which digitises and makes freely available over 1 000 original children's stories in more than 150 African languages as well as in English, French and Portuguese. This pioneering project led to the development of the open-source, multilingual literacy portal Global Storybooks, which is active in 40 countries on five continents. The aim of the global project is to help democratise global flows of information, facilitate language learning and promote literacy (Norton 2020). Once again, Africa has led the way!

Books from Africa for Africa: Bibi Bakare-Yusuf of Nigeria is determined to challenge stereotypes of what African literature 'should be'. Through her new publishing company, Cassava Press, she aims 'to feed and nourish the African imagination' by providing high-quality, affordable African literature to the largest number of readers (including school learners). The kinds of books that she is promoting include Ayesha Harruna Attah's *The hundred wells of Salaga* (a historical novel about the intrigues of precolonial Ghana) and Elnathan John's

Born on a Tuesday (a coming-of-age story set in the sectarian religious violence of northern Nigeria) (Ayene et al. 2020). She has opened an office in London and plans to target the African diaspora as well as readers on the continent.

Educating through song: The educational NGO Edusong in Lusaka (Zambia) has developed a novel way to teach difficult concepts – song. They argue that music is an effective teaching tool because it lingers longer in the mind of the listener than spoken words, and music and song create a stress-free environment in which to learn. Their 'retention through entertainment' approach is particularly effective for learners who do not grasp concepts the first time or have problems remembering the lessons. 'Edusongs' are rich in content and profusely illustrated, and can be accessed through different delivery channels such as CD, DVD, tablet or mobile app. They are intended to supplement, rather than replace, traditional teaching methods and have been developed to cover subjects in the new Zambian Grades 1 to 12 syllabi but can also be used in other countries.

Decolonising mathematics: The field of ethnomathematics, founded in the 1970s by Brazilian educator and philosopher Ubiratàn D'Ambrosio, recognises non-Western contributions to mathematics and seeks to include indigenous knowledge in mathematics education. For example, Xolisa Guzula (a specialist in multilingual education) points to a traditional southern African game played with stones, *upuca*, that can be used to teach concepts such as number theory and estimation. In this way, she argues, mathematics can be connected to a learner's culture. Teaching in indigenous languages alongside English has also been shown to improve mathematical comprehension by teaching the subject from a different perspective (Lewton 2018).

Those who argue that a pure science such as mathematics cannot be decolonised are wrong. Verhoef and Kruger (2020) – and many others – have shown that by rediscovering and using African examples in the teaching of the subject, it is possible to deconstruct an exclusive Western body of knowledge and get rid of teaching and learning methods that reflect a colonial mindset. They identified a range of examples in agriculture and warfare from Africa that could be used to teach how mathematical techniques and models are used to make decisions. For example, in agriculture the selection of villages for hosting regular market days involves a range of criteria such as availability, cost, population size, facilities and the fair rotation of locations. Another agricultural example is the complex system of planting maize and allowing land to lie fallow so that the soil can regenerate. Other examples are the tactical games of *marabaraba* and *mankala*; string patterns created by Sotho girls; geometric designs in the art of the Ndebele; and the layout of huts and other structures at Mapungubwe, Great Zimbabwe and in many West African villages. Including examples like these in the curriculum would benefit students who have been marginalised by the exclusive use of Western examples (Verhoef & Kruger 2020).

Tiri Chinyoka (who holds extracurricular classes for mathematics undergraduates at the University of Cape Town) sports a black leather flat cap and dreadlocks, and is not a stereotypical mathematician. He is a strong proponent of decolonising mathematics, but it is unclear exactly what this means. Does it involve revising curricula to promote non-Western contributions to the field, introducing new teaching methods rooted in indigenous cultures or a greater openness to ideas outside the academic mainstream? Some, like Chinyoka, would like to go further and challenge the philosophical foundations of mathematics itself. He argues that mathematics should be taught with concrete applications in mind rather than purely theoretically – a luxury that can only be afforded in the West. 'There is no flexibility to try to do things alternatively or to try to open ourselves to interrogating new ways of thinking differently', says Chinyoka, adding that allowing researchers to explore new directions in mathematics would not mean throwing out existing methods (Lewton 2018).

However, unlike the arts and humanities, most people regard mathematics as universal and objective and not subject to cultural interventions. But there is an opposing view that mathematics is an evolving work-in-progress whose truths are dependent on culture and invented rather than universal and discovered. There is also concern that African mathematicians would be disadvantaged in the international arena if alternative methodologies were to be adopted. 'We can't cut ourselves off from the mathematical developments in the rest of the world,' argues Loyisa Nongwa, the vice-president of the International Mathematical Union, 'Our intellectual project would be impoverished' (Lewton 2018).

Mathematics board game: Staff at North-West University in South Africa have developed the innovative board game/app Maths Whartels for preprimary and primary learners to help them improve their understanding of mathematics. The game focuses on both cognitive learning (mathematics concepts, metacognition and information processing) and psychological factors (mathematics anxiety, resilience and motivation), and includes principles used in play therapy such as art, bibliotherapy, board games and puppets. The Maths Whartels game is designed to facilitate self-directed learning and is also suitable for use by parents, therapists and teachers.

Roots & Shoots: In 1991, a group of 12 local teenagers met with Dr Jane Goodall (the world-famous primate researcher and conservationist) on her porch in Dar es Salaam (Tanzania), eager to discuss a range of problems that they had witnessed in their communities. Goodall was so impressed by their compassion, energy and desire to develop a solution to problems that she decided to establish the Jane Goodall Foundation's Roots & Shoots project. This youth-led global community action programme, which now flourishes in nearly 100 countries, helps young people to become the informed generation of compassionate citizens the world needs. It partners with schools, educators and youth organisations to inspire and educate young people to make a difference on an individual level. Roots & Shoots has developed into an unparalleled multiplying force in conservation and service-based learning, giving young people the knowledge and confidence they need to act on their beliefs and make a difference by being part of something bigger than themselves.

Feed the Monster app: The growing alarm over the child literacy crisis caused the MTN SA Foundation, partnering with Bellavista School in Johannesburg and the American NPO Curious Minds, to pilot the children's literacy app Feed the Monster that significantly enhances reading fluency and comprehension. Readers aged six to eight years who use the award-winning app can access reading instructions via a specialised curriculum. By matching letters with sounds, the app helps children to learn that sounds combined make words, and words together make sentences that carry meaning.

> We are pleased to be able to bring the transformative power of technology to South Africa's children. Not only does this bridge the digital divide and prepare our children for a future in the information age, but we are also providing the basic building blocks for early childhood education.
>
> *Kusile Mtunzi-Hairwadzi, general manager of the MTN SA Foundation*
> (Mtunzi-Hairwadzi 2019)

Importance of coding: In his February 2020 Sona address, President Cyril Ramaphosa announced that computer coding and robotics would be introduced in Grades R to 3 in 200 schools in South Africa in 2020, with full rollout to all the schools in the country by 2022. Although this is not a new issue (it was mentioned in the 2019 Sona), it has attracted plenty of debate. To implement this ambitious programme, the Department of Basic Education under Minister Angie Motshekga formed – in partnership with Standard Bank and the NGO Africa Teen Geeks – a team of experts on computer applications technology to develop the new curriculum. In addition, thousands of teachers are being trained.

The Cape Town Science Centre (CTSC), headed by Julie Cleverdon, is a strong advocate for introducing coding to children at a young age. As coding is widely regarded as the 'new literacy' and is an essential 21st-century skill, the science centre became a founding member of the important continent-wide initiative Africa Code Week, which is spearheaded by the multinational software corporation SAP. In its first year (2015), the CTSC conceptualised and piloted Train the Trainer workshops in Cape Town for Africa Code Week and, from 2016 to 2018, it served as the global coordinator for this coding initiative. Year after year, the project has exceeded all expectations. In 2018, it engaged 2.3 million learners and 22 999 teachers in 37 African countries. In addition to focusing on coding, the CTSC has taken the lead in the EduConservation initiative in Africa on behalf of Sabine Plattner African Charities, working closely with education departments in various African countries to develop conservation-enriched curriculum material that can be used in schools – once again with a teacher-centric, Africa-centric approach. Through these various engagements, the science centre has developed substantial networks with organisations, governments and thought leaders across Africa who promote the advancement of STEM among their youth (Julie Cleverdon, personal communication, September 2020).

Drone and robotics technology training: The Malawi-based African Drone and Data Academy, a Unicef-sponsored partnership between Virginia Tech in the USA and the Malawi University of Science and Technology, is pioneering drone technology training in Africa and plans to train about 150 young Africans to pilot drones by 2021, with a master's degree in drone technology planned for the future. The academy accepted its first cohort of students in January 2020. In Accra, the Ghana Robotics Academy Foundation was founded in 2011 by Dr Ashitey Trebi-Ollennu as a non-profit volunteer organisation dedicated to motivating and inspiring the next generation of Ghanaians to develop the requisite skills so that they can pursue careers in STEM. The foundation will pioneer science education through hands-on robotics workshops and competitions throughout Ghana, and promote the development of science and robotics clubs.

Learning during the Covid-19 pandemic: Every catastrophe creates opportunities for entrepreneurs to use their creativity and come up with novel solutions. The Covid-19 pandemic in 2020 led to many schools and universities being closed, which threatened to disrupt the education of millions of African youths. Many educators rose to the challenge and strengthened their e-learning and other distance learning capacities to overcome the problem. Associate Prof. Willie Chinyamurindi of the University of Fort Hare in South Africa recommended that lectures should be made available online using YouTube, Sound-Cloud, Twitch or Audiomack, and that virtual web conferences and meetings should be held with students using Skype and WhatsApp video. He emphasised that online access to library resources was very important (Chinyamurindi 2020).

African Leadership University (ALU): The ALU is a network of world-class tertiary education institutions whose mission is to produce 3 million young

African leaders over the next 50 years. The ALU's first campus, the African Leadership College (inaugurated in September 2015 in Mauritius) offers accredited undergraduate degree programmes through its founding academic partner, Glasgow Caledonian University. The second campus opened in September 2017 in Kigali Heights, Rwanda. The unique ALU learning model helps students to learn new skills and master concepts, as well as discover gaps in their knowledge and skillset. Whereas most universities start speaking to their students about life after graduation in their final year, the ALU's students begin their career development journey from Day 1 and use their four years at the university to lay the right foundation for their future career.

Anti-apartheid university: The University of the Western Cape in Cape Town has transformed over the past 60 years from a minor college that was the intellectual home of the left to an ultramodern, research-led institution that is playing a vital role in helping to decode the novel coronavirus genome and contribute to the development of a vaccine. The foundation of the university, whose motto is *Respice, Prospice* ('Looking back, looking forward'), was built on the struggle against apartheid and the institution played a crucial role in drafting the interim constitution of South Africa and developing a functional democracy in the country. Instead of allowing the Covid-19 lockdown to disrupt its 2020 graduation ceremony, the university pioneered a virtual graduation ceremony. In addition, the #NoStudentWillBeLeftBehind campaign was launched to ensure that they catered for the needs of the 30 per cent of their 24 000 students who did not have access to flexi-learning facilities.

> Under the exceptional leadership of Professor Jakes Gerwel, and then Professor Brian O'Connell, it was transformed from a marginalized 'bush college for Blacks' into a deracialised centre of academic excellence, with an acute social conscience. The courageous commitment of the university community to the parallel values of intellectual growth and justice was second to none.
>
> *Archbishop Emeritus Desmond Tutu, former chancellor of the University of the Western Cape* (Anon 2020c)

Top university in Africa: The University of Cape Town, which was founded in 1829, is the leading university in Africa (according to the QS World University Rankings 2021) and a major centre for research and innovation on the continent. The work of many of its researchers and students is referred to frequently in this book. The university's tradition of research excellence has led to many great inventions and discoveries, and its academics and graduates have been innovators in almost every discipline. Some of their most famous innovators are Prof. Christiaan Barnard, who pioneered heart transplants; Allan Cormack, who co-developed the CAT (computerised axial tomography) scanner; and Sir Aaron Klug, who contributed to the development of crystallographic electron microscopy and furthered our knowledge of nucleic acid-protein complexes. Numerous products based on intellectual property developed at the university have been successfully commercialised and, for some, spin-off companies have been created for commercialisation. These spin-offs include Nautilus Technologies, Impulse Biomedical, Abalobi, Nisonic, Cape Catalytix, CURIT Biotech, Cape Bio Pharms, Dream Haven, Attiri Orthopaedics, Lumkani, HyPlat, DroneSAR, AngioDesign, Elemental Numerics, Tuluntulu, Cape Ray Medical and Cape Carotene.

Using their Wits: Staff and students at Wits University in Johannesburg have pulled out all the stops to ensure that the Covid-19 pandemic has had the least effect on teaching and learning (www.wits.ac.za/covid19). In late April 2020, the university's entire academic programme went online after 5 000 laptops had been delivered to students in need and 30 GB of data had been provided to every student.

> We are aware of the anxiety and uncertainty that online teaching and learning presents for both our colleagues and students. The world as we know it is in flux, and it will take our collective courage, dexterity and commitment to fend off the effects of this pandemic and to adopt new ways of teaching and learning.
>
> *Prof. Ruksana Osman, deputy vice-chancellor academic of Wits University*
> (Bright 2020)

Supporting the youth: *Time* magazine voted Greta Thunberg Person of the Year for 2019, as a representative of the many young people who are taking the lead in environmental, political and other campaigns worldwide, but Africa has its own young heroes. In the Sudanese capital Khartoum, 22-year-old activist Alaa Salah led the fight to oust the country's leader for three decades, Omar al-Bashir, and succeeded. When he was two years old, Eddie Ndopo from Namibia was diagnosed with spinal muscular atrophy and given three years to live. Now Eddie, aged 29 years, says that he has outlived himself by 24 years and plans to make the best use of his good fortune – like Stephen Hawking. Since then, this emancipated 'born free' (South African youth born after the fall of apartheid), despite being confined to a wheelchair, has earned a master's degree in public policy from Oxford University, been selected by the UN as one of its 17 global ambassadors for the SDGs, and hopes to travel in space and address the UN

General Assembly while flying weightlessly above it (https://time.com/collection/davos-2020/5764739/eddie-ndopu-space/).

> I want to be able to use zero G as a stage. If I have five minutes to talk to the world, what would I say to capture humanity's attention? I want to become the first disabled person in space. I defied the odds and challenges that faced me at birth, and now it's time to defy gravity.
>
> *Eddie Ndopo* (Kluger 2020)

Traditionally, space travel has been reserved for the sublimely abled. Ndopu's flight would be a paradigm shift, a dramatic democratisation of space travel. If he manages to fly free from the wheelchair that confines him, it will be a lyrical triumph of physics over physique – like the parabolic flight undertaken by Hawking in 2007 which inspired millions. But Ndopu's long-term aim is to champion the rights of disabled people. He has established a global fund that will encourage public–private investment to address the many obstacles that prevent disabled people from participating fully in society, such as access to buildings, the use of computer keyboards and job training. The fund already has the backing of the UN and the WEF, but he wants more. His attitude is: If the sky's the limit, how come there are footprints on the moon?

African Gong: Dr Elizabeth Rasekoala is the president of African Gong, the Pan-African Network for the Popularisation of Science & Technology and Science Communication, which aims to advance public learning and understanding of science as well as scientific literacy in Africa (www.africangong.org). She is also a member of the AU Monitoring and Evaluation Committee for the 10-year Science, Technology and Innovation Strategy for Africa, and has championed transformative development through advancing diversity, sociocultural inclusion and gender equality in STEM. In addition, she is an outspoken critic of the lack of resolve by African governments to support research and development in a meaningful way.

> We just need our African governments to make the concerted funding commitments to R&D that will show that they take STI research and development on our continent very seriously indeed ... How do we get African governments to place it firmly at the heart of development and the transformation of African societies and countries? ... That core belief, that strategic enrolment, is still not there in many of our African governments. If you

don't believe in something, if you don't see the efficacy, you are not wired into it. You are not going to set the policies in place that will make it an imperative.

So, we end up with very poor, patchy and uneven STI policy environments across our continent. If you don't believe in it, if you don't have the policies and the legislation for it, then you are not going to commit in a sustainable way to the strategic funding regimes and budgets that you need to make that scientific endeavour and the research and its impact on societal development happen in your country.

Dr Elizabeth Rasekoala, president of African Gong (Skupien 2019)

Importance of public–private partnerships: The South African Medical and Education Foundation (SAME) and corporate donors (including Scania, Distell and Toyota) officially launched the Masia Maths and Science Academy (MMSA) in Ha-Masia (Limpopo province, South Africa), in collaboration with the Masia Traditional Council, in March 2020. The new academy (developed in a derelict school) has state-of-the-art learning facilities, including fully equipped classrooms, science laboratories and a computer centre. It will serve thousands of learners and greatly improve the quality of STEM education in the Vuwani region, which is characterised by poverty, high illiteracy rates and many child-headed families. According to Lehlohonolo Mohapi of SAME, the MMSA will offer a mathematics and science programme for seniors, the junior Mathematics Pioneers programme and a homework-assistance programme for primary-school learners (samefoundation.org.za). At the opening ceremony, Chief Nthumeni Masia noted the link between high crime levels and poor education in rural South Africa, and expressed the opinion that education was the only effective way to combat the scourge of crime and corruption. SAME's motto is 'Free to live, free to learn'.

One laptop per child in Madagascar: A study of the way in which laptops affect the behaviour of primary-school learners on Nosy Komba Island in Madagascar revealed interesting results. The laptops were made available by the One Laptop per Child initiative launched by the Massachusetts Institute of Technology in 2005 and were loaded with an operating system featuring free educational software called 'Sugar'. In Madagascar, where 75 per cent of the population live below the poverty line, laptops are high-status objects and also have high symbolic value because parents believe that if their children can master their use, they will become more intelligent and develop useful professional skills.

The survey on laptop usage by Sandra Nogry of the *Université de Cergy-Pontoise* in Paris revealed that they were used differently at school than at home. At school, they were used to learn about word processing; play educational games; support creative activities such as drawing and digital storytelling; and access maps, e-books and calculators. At home, they were mainly used for taking photos, making videos, listening to music, playing games, sharing content and doing homework. A marked trend was that laptops in the home tended to be used collectively rather than individually. Children and their families would gather around a laptop for recreational purposes, and the laptops were used to strengthen social relations among siblings, parents and peers. Although laptops have introduced Nosy Komba children to new tools and resources, the survey revealed that few creative projects were undertaken and that the apps that fostered creativity were underused (Nogry & Varly 2018).

Museums in Africa: According to Unesco (2020), there are about 95 000 museums worldwide, of which about 90 per cent were closed during the Covid-19 lockdowns. Alarmingly, Unesco and the International Council of Museums predict that about 13 per cent of museums may never re-open their doors after the crisis. The number of museums in Africa is unknown but probably exceeds 3 000, as there are over 300 registered museums in South Africa alone (Hannan, Reddy & Juan 2016). The Unesco (2020) report further reveals that only 5 per cent of museums in Africa and Small Island Developing States developed online content for their audiences during the pandemic. With its rich cultural and biological heritage, Africa has a great deal of unique material to collect, preserve, archive, study and display. The continent has some of the greatest museums in the world but, sadly, most of them are underfunded. This is a tragedy, as museums play a vital role in taking people down the information value chain: Information → Knowledge → Wisdom → Changing one's mindset → Changing one's behaviour → Influencing others to change their mindset and behaviour. These are vital steps in bringing about social change.

The leading African museums constantly have to re-invent themselves to stay relevant and many have evolved from curiosity cabinets to agents of social change that actively promote public benefit initiatives such as sustainable living. Their focus has shifted from collecting objects to actively engaging with communities and becoming community hubs, centres for knowledge dissemination and spaces for dialogue. They are placing more emphasis on access, redress and national development, and are moving away from focusing entirely on the past to engaging with current issues and talking about the future.

The new Grand Egyptian Museum, which is under construction near the Giza pyramids in Cairo, is a welcome beacon of hope for museums in Africa. It will display over 50 000 artefacts from ancient Egypt, including the entire Tutankhamun collection, and will become the largest archaeological museum dedicated to one civilisation in the world. Artefacts will be relocated from the Egyptian Museum in Cairo as well as from storage and museums in Luxor, Minya, Sohag, Assiut, Beni Suef, Fayoum and Alexandria. The building is shaped like a chamfered triangle in plan, with the north and south walls lining up with the Great Pyramid of Khufu and the Pyramid of Menkaure. One of the main features of the building is a translucent stone wall, made of alabaster, that makes up the front facade. Construction was set to be completed in 2020 but due to the Covid-19 pandemic, the museum is now scheduled to open in 2021.

African art reclaimed: Hamady Bocoum (the director of Senegal's Museum of Black Civilisations) has big ambitions for his new repository of African art, culture and history. For decades, the most prominent homes for African art were museums in Europe, but he hopes to reclaim some of the continent's lost wealth by demanding the restitution of artworks taken during colonial times. The US$34-million project, which is over 50 years in the making, is designed to be a creative laboratory that will help to shape the continent's future sense of identity. Many of its galleries are deliberately empty, ready to accept newly reclaimed artworks. Some European countries, including France, were prompted to lend pieces for the opening (Baker 2019d).

African art galleries: The Zeitz Museum of Contemporary Art Africa on the V&A Waterfront in Cape Town is the world's largest museum dedicated to contemporary art from Africa and the diaspora, and was hailed 'Africa's answer to the Tate Modern' when it opened in September 2017. Its 6 500 m² of gallery space and a rooftop sculpture garden are housed in a stunning building that was, until recently, a complex of abandoned grain silos. Forty-two 30-m tall concrete silos were artfully sliced at different angles to create unique spaces, including shafts for the lifts, transforming it into a cathedral-like arena bathed in light from over 100 faceted glass windows. The collections include a stunning visual arts display by South African artist William Kentridge, cowhide sculptures of the female form by South African artist Nandipha Mntambo and multimedia installations by Zimbabwean artist Kudzanai Chiurai. At the opening ceremony, Archbishop Desmond Tutu pretended to take a phone call from heaven, chatting with the late President Nelson Mandela: 'Yes,' Madiba said, 'This is what we were fighting for!' (Rockwood 2018).

Dirk Durnez has always been brimful of creative ideas. After developing the company Monex Themed Environments (now MTE Studios), which themed the 128 000-m² Canal Walk shopping mall and its associated theme park Ratanga Junction in Cape Town, he developed a variety of other creative initiatives. Then, in 2016, after 31 years' involvement in developing interactive museums and edutainment experiences, he and Henk van Aswegen established the cutting-edge Art@Africa art gallery on the V&A Waterfront as well as Art@Agency and Cape

Sculpture Technologies – all of which promote the work of African artists. Their stable of artists includes Ndabulo Ntuli, Andries Visser, Barney Bernado, Caelyn Robertson, David Griessel, Eben, Kara Schoeman, Kosie Theart, Lauren Redman, Maureen Quinn and Talita Steyn. However, they are constantly on the lookout for more artists from Africa.

Art@Africa seeks to create a significant positive experience and to address the issues modern African society faces (such as gender equality, racial discrimination, homophobia, African identity and biodiversity loss) through its art. They recognise that artists are often the first to oppose binary prejudices like racism and rather use them as an opportunity to portray the kaleidoscopic continuum between extremes. Durnez and his artists also believe that there is an increasingly strong interface between art and science (especially environmental science) and that art can convey messages and articulate emotions that we are not able to express easily in words. Art therefore transcends cultures, languages and disciplines, and provides a means whereby millions of people who choose not to listen to the scientific message can be informed about the environmental crisis. Art helps us to realise that we are irrevocably entwined in natural processes and that our role is to understand and work in harmony with nature – not to conquer it.

The Art@Africa gallery hosts regular themed exhibitions and events, and actively promotes alternative art forms as well as artworks made from recycled materials. The theme of their mid-2020 exhibition (The Blue Dot) reflected humankind's perspectives on planet Earth and referred to the iconic photograph of planet Earth, the 'Pale Blue Dot', taken by Nasa's Voyager 1 space probe on 14 February 1990 from a distance of 6 billion km. The gallery also prides itself on the development of digital experiences, including augmented reality, that allow its clients to discover and interrogate art in another realm (Haynes 2020).

Sonia Lawson is the director of *Palais de Lomé* in Lomé (Togo), a unique art and culture centre in West Africa. The building, formerly the colonial Governor's Palace, has been restored and transformed into a landmark addition to the cultural scene in West Africa. Since opening in 2019, this ambitious renovation, cultural and environmental project that is funded by the Togolese government has been dedicated to showcasing the diversity of Togolese and African cultural productions in visual arts, design, new media, science and technology, and the culinary and performing arts.

Year of Science and Technology: South Africa's minister of arts, culture, science and technology, Lionel Mtshali, proclaimed 1998 the Year of Science and Technology (YEAST), during which a nationwide, government-funded programme of exhibitions, science shows, publics talks and other science engagement and communication activities took place. One of the most popular interactive exhibitions was the Children's Science Village from the *Cité des Sciences et de l'Industrie* (City of Science and Industry) in Paris. The aim of YEAST was to demystify science and make it more accessible to the public, as well as to promote then Vice-President Thabo Mbeki's vision of an African Renaissance and prepare the country for the new millennium. Following the success of YEAST, the South African government implemented the annual National Science Week, which is ongoing, with the objective of 'taking science, engineering and technology to the people' (Joubert & Mkansi 2020).

Science centres and festivals: Interactive science centres are recognised as one of the most cost-effective ways of strengthening a science culture and supplementing school and university education. Their relatively language-free exhibits are well suited to multicultural and multilingual audiences, and they cater for visitors from a wide range of ages and socioeconomic backgrounds (Joubert & Mkansi 2020). Science centres have the potential to play an important role in Africa by demystifying STEM, but their reach is currently extremely limited. Africa has only 40 science centres (5 per cent of the world total) and they are very unevenly spread, with most in South Africa (30 opened since 1977) and North Africa (five in Egypt [2002] and Tunisia [2002]), and one each in Botswana (2005), Ethiopia (2005), Ghana (2009), Kenya (2012) and Zimbabwe (2016).

South Africa has 700 per cent of the world average of science centres per capita, whereas the rest of Africa has only 7 per cent (Trautmann & Monjero 2019). The vast majority of the 1.216 billion people on the continent therefore have no access to interactive science centres. The umbrella body for science centres in southern Africa is the Southern African Association of Science and Technology Centres, and in North Africa and the Middle East it is the North Africa and Middle East Science Centre Network. African countries are strongly urged to develop their own science centres and to benefit from the services provided by these regional networks. Most science centres in South Africa were started by hardworking and determined individuals rather than by government departments or institutions. They include the Sci-Enza Discovery Centre (started by Lötz Strauss in 1977), the UniZulu Science Centre (started by Derek Fish in 1986), the University of the North's Science Centre (started by Shadrack Mahapa in 1996) and the Cape Town Science Centre (started by Mike Bruton, as the MTN ScienCentre, in 2000). Some African museums have adopted interactive methods of teaching and learning.

Many STEM disciplines (such as physics, chemistry, engineering and biology) can be taught very effectively by using non-formal interactive methods, but mathematics is perhaps the one that most needs demystifying and popularising.

A subject that is widely perceived as boring, difficult and scary can be made exciting, fun and inspiring. The joy of exploring the world of numbers and forms is brought to life in many science centres in Africa, but in general they lack the resources (financial and human) to reach their full potential and to benefit from the very dynamic international science centre movement. While most science centres in Africa have some displays and activities on mathematics and logic, none of them can match, for example, the Museum of Mathematics that opened in New York City in December 2012. Mathematics pervades every aspect of the design of this interactive museum and their exhibits address abstract concepts, like number theory and topology, and reveal how they are used in everyday life.

South Africa launched its National Science Festival (SciFest Africa) in 1996 in the Eastern Cape university city of Makhanda and it has grown strongly over the years. SciFest Africa, which is the largest science festival in Africa, has hosted hundreds of international speakers and workshop leaders (including Nasa astronauts and Dr Jane Goodall) over the years and has welcomed more than 1 million participants of all ages. Now under the leadership of Dr Fredy Mshati, it turned a challenge into an opportunity and went digital in 2020 during the Covid-19 pandemic, thereby reaching a wider and more diverse audience through its theme 'Take root ... nurture'. Other African countries that host science festivals include Namibia, Kenya and Ghana.

Edtech advances: Armed with an Ivy League education, former Google employee Riaz Moola is making vast strides in the Edtech landscape in South Africa. In 2012, he founded Hyperion Development as a national peer-to-peer student support community for learner computer scientists and to provide tuition to those wanting to enter the technology job market. He initially received little support from local companies and investors but succeeded in securing funding from the University of Edinburgh (his alma mater), the Python Software Foundation, Google and Facebook (Prins 2019).

> Finding a scalable and sustainable business model took years. Unlike an accredited provider, we had to build a track record of real outcomes that required every aspect of our programmes to be of the highest quality.
>
> *Riaz Moola, founder of Hyperion Development* (Prins 2019)

In 2018, Riaz created CoGrammar to help improve software code quality and provide free code training through the CoGrammar Career Programme. One of his goals is to help unemployed young people to develop software coding skills and become code reviewers. He believes that technology can play a vital role in

Shell middens, the first technology hubs in
South Africa.
(*Photo: Mike Bruton*)

Ochre engraved artwork, dated at 87 000
years before present (the oldest abstract
artwork in the world), found in Blombos
Cave in the southern Cape, South Africa.
(*Photo: Mike Henshilwood*)

Golden Rhino of Mapungubwe, dated at
about 1200 CE.
(*Photo: University of Pretoria*)

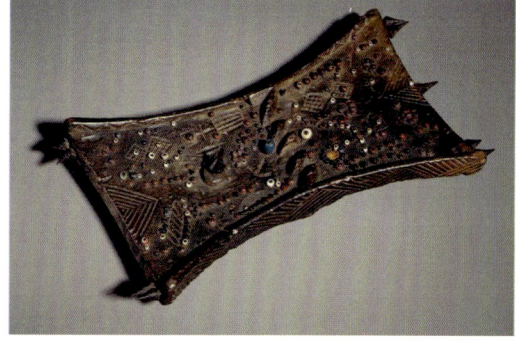

Lukasa memory board of the Luba people
from the Democratic Republic of the Congo.
(*Photo: Brooklyn Museum*)
*(https://commons.wikimedia.org/wiki/File:Brooklyn_
Museum_76.20.4_Lukasa_Memory_Board.jpg)*

Woman fishing with an *isiFonyo* thrust
basket in the Phongola River floodplain,
Maputaland, South Africa.
(*Photo: Bronwyn Jones*)

Traditional fisherman in a *galawa* (wooden
dugout canoe with double outriggers) off
Grande Comore in the Comoros.
(*Photo: Mike Bruton*)

Handmade Madeira fish trap in Zanzibar,
Tanzania.
(*Photo: Mike Bruton*)

Dugout canoes and a wooden plank boat at
a *zimbowera* (floating houses) fishing village
on Lake Chilwa in Malawi.
(*Photo: John Wilson*)

Davis Ndungu from Cape Town,
South Africa, with a coelacanth modelled
from flip-flop scraps.
(*Photo: Mike Bruton*)

Coelacanth modelled from scrap metal by
Chenjerai Mutasa of Zimbabwe at Hout Bay,
Cape Town, South Africa.
(*Photo: Mike Bruton*)

Kente, the indigenous textile from Accra,
Ghana.
(*Photo: Mike Bruton*)

Decorative lampshades based on the
design of Traditional fish dish by Thabisa
Mjo, displayed in the V&A Waterfront,
Cape Town, in December 2019.
(*Photo: Mike Bruton*)

Solar-powered hemp house built in Morocco
by architect Monika Brümmer and students
at the National School of Architecture
in Rabat.
(*Photo: Monika Brümmer*)

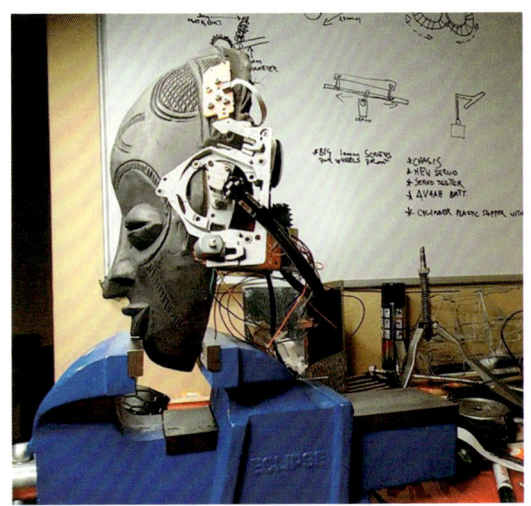

Robotised mask made by David Phume, the
founder of Blackchain, in Johannesburg,
South Africa.
(*Photo: David Phume*)

Sandile Ngcobo with the digital laser
he co-invented at the CSIR in Pretoria,
South Africa.
(*Photo: CSIR*)

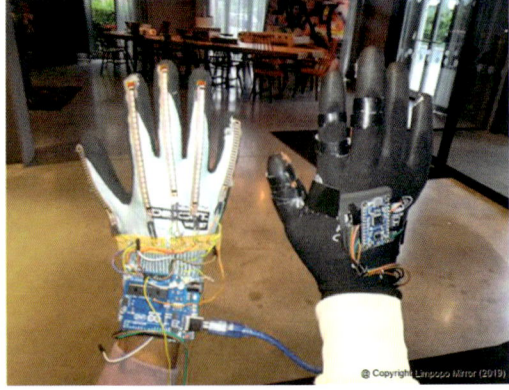

Lucky Netshidzati's Talking Glove that
converts sign language into voice and text.
(*Photo: Lucky Netshidzati/Limpopo Mirror*)

Scrapyard sculptor "Scrapture" artwork
by Vincent Chenjerai of Hout Bay
in Cape Town, made from found
metal material.
(*Photo: Vincent Chenjerai*)

Portrait of Nelson Mandela, made of wood
chips and recycled plastic bottle tops by
Ndabuko Ntuli of Cape Town, South Africa.
(*Photo: Mike Bruton*)

The Dung Beetle project facilitating musical
and environmental education during the
2019 Afrikaburn Art Outreach event at
Pofadder secondary school, in Pofadder
Northern Cape, South Africa.
(*Photo: Jeffrey Barbee*)

Street art symbolising the transition to the
digital age by Stefan Smit in Woodstock,
Cape Town, South Africa, 2020.
(*Photo: Mike Bruton*)

Louis Liebenberg explaining his cybertracker to San
hunters in the Kalahari Desert, Botswana.
(*Photo: Louis Liebenberg*)

Cattle in the Okavango Delta, Botswana,
with eyes painted on their rumps to reduce
the risk of predation by lions in a project
developed by Tshepo Ditlhabang and his
colleagues.
(*Photo: Tshepo Ditlhabang*)

An image of the African beadwork pots.
(*Photo: Mike Bruton*)

Protea, the first production car developed in Africa, in October 1956 by John Myers and his colleagues in Johannesburg, South Africa.
(*Photo: John Myers*)

Wallyscar Iris, one of Tunisia's entries into the lucrative off-road vehicle market.
(*Photo: Wallyscar*)

Tebello Nyokong in her laboratory at Rhodes University in Grahamstown, Eastern Cape Province, South Africa.
(*Photo: Rhodes University, Clint Bradfield/Foto First*)

PGuard robotic policeman and controller
made by Enova Robots in Tunis, Tunisia.
(*Photo: Enova Robots*)

Henry Foretia of Cameroon with his
SaveTheChicken app.
(*Photo: Henry Foretia*)

Muthoni Masinde, Kenyan scientist and
inventor of the ITIKI drought predictor,
which combines indigenous knowledge and
modern science data.
(*Photo: Muthoni Masinde*)

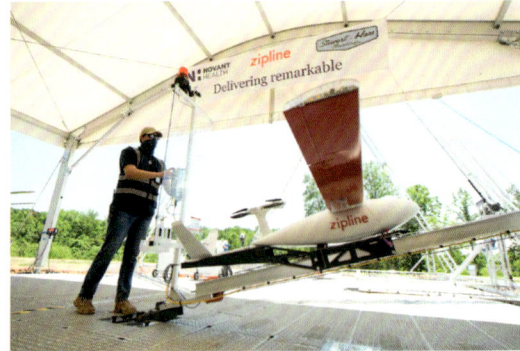

Daniel Marfo, CEO of Zipline Ghana, with
one of his delivery drones.
(*Photo: Zipline Ghana*)

Artist's impression of a MeerKAT satellite
dish at Carnarvon in South Africa.
(*Photo: Ska*)

Skarab, the highly scalable, energy-efficient,
field-programmable gate array developed
in Cape Town, South Africa, for the Square
Kilometre Array (Ska) project.
(*Photo: Mike Bruton*)

Abandoned Vodafone earth satellite station at Kuntunse in Ghana converted
into a radio telescope antenna for the Africa-wide Ska project.
(*Photo: Mike Bruton*)

Egyptian feminist and leading writer
Nawal al-Saadawi prior to a meeting
called by the Union of Egyptian Writers
17 June 2001 in Cairo.
(*Photo: Marwan Naamani/AFP,
Getty Images*)

Nobel Prize laureate Wangari Maathai
addresses the ceremony to mark the
adoption of the Kyoto Protocol at the
Kyoto International Conference Centre
in Kyoto, Japan, 16 February 2005.
(*Photo: The Asahi Shimbun,
Getty Images*)

Rwanda Minister of Agriculture and
Animal Resources Dr Agnes Kalibata,
gestures as she speaks at the African
Green Revolution Forum at the Accra
international Conference Center in Accra,
04 September 2010.
(*Photo: Issouf Sanogo/AFP, Getty Images*)

Aliko Dangote, founder and chairman
of the Dangote group, one of the largest
private-sector employers in Nigeria, 2012.
(*Photo: Photo by Michael Prince/
Forbes Collection/Corbis, Getty Images*)

Dr Matshidiso Moeti, from Botswana, is the
World Health Organisation (WHO) Regional
Director for Africa.
(*Photo: WHO*)

Ethiopia's President Sahle-Work Zewde
speaks during the World Economic
Forum (WEF) Africa meeting at the
Cape Town International Convention
Centre, in Cape Town – South Africa,
04 September 2019.
(*Photo: Rodger Bosch/AFP, Getty Images*)

Moustapha Cisse, head of Google Artificial
Intelligence (AI) Centre Ghana, speaks
during the presentation of the first AI centre
in Africa at the Marriott Hotel in Accra,
10 April 2019.
(*Photo: Cristina Aldehuela/AFP,
Getty Images*)

Professor Mulalo Doyoyo, polymathic
African inventor from South Africa.
(*Photo: Mulalo Doyoyo*)

Rachel Sibande, Malawian technology expert who founded mHub, Malawi's first innovation hub.
(*Photo: Rachel Sibande*)

Mitchell Elegbe is the Founder, Group Managing Director and CEO at Interswitch, a digital integrated payment company in Nigeria.
(*Photo: Interswitch Group*)

Iyinoluwa Samuel Aboyeji, Nigerian entrepreneur who co-founded Andela and was the former managing director of Flutterwave.
(*Photo: World Economic Forum/ Greg Beadle*)

Tony O. Elumelu, leading Nigerian industrialist, philanthropist and developer of the inclusive economic philosophy of Africapitalism, 16 November 2018.
(*Photo: Thomas Imo/Photothek, Getty Images*)

improving education in Africa by using unconventional ideas to access, refine and deploy talent in order to solve global problems (Prins 2019).

Emerging innovators: A computer scientist by profession, Rapelang Rabana of South Africa developed Rekindle Learning in 2013 after realising that there was a learning opportunity in people's compulsion to constantly check their phones. Starting out at age 22 years as the co-founder of Yeigo Communications, Rabana cemented her reputation in the technology industry by pioneering innovations in mobile voice over internet protocol (VoIP) and internet protocol communications. She was named 'Entrepreneur for the World' by the World Entrepreneurship Forum, selected as a Global Shaper by the WEF and was invited to join the annual WEF meeting in Davos – all by the age of 30 years. Paleso Sibeko is the co-founder of Girls Invent Tomorrow, an NGO that empowers girls to explore careers in STEM. Sibeko and two partners developed the NGO in 2013 in response to a global drive to encourage more women to follow science and technology careers, and managed to persuade the global microchip-maker Intel to sponsor some of her projects.

Discussion: Classrooms of the future will be a 'phygital' experience, a combination of the interactive physical world and the power of digitisation that is available through mobile technology and the internet. It is anticipated that there will be a continuous symbiosis between the physical and digital dimensions to ensure that they add value to each another. But, as Clinton Walker (the director of e-learning for the Western Cape Education Department in Cape Town) says, 'The technology tail should never wag the educational dog' (Tobin 2019).

Sahle-Work Zewde (the president of Ethiopia) and Audrey Azaoulay (the chairperson of the International Commission on the Future of Education) have called for a drastic re-think on education, not only what we teach but also how we teach it. They argue that adapting existing educational practices will not fix the severe educational imbalances in the world (especially in Africa) and that a totally new approach is needed. This new approach should ensure that all school leavers have a foundation of basic competencies that equip them for further education, training, employment and civic participation. Their basic education should also equip them to grapple with the major challenges that humankind faces this century (including the need to embrace human diversity, preserve biodiversity, fight climate change and adapt to technological disruptions) and take full advantage of the changed ways in which we live, work, play, communicate, process knowledge and learn in the digital age (Zewde & Azaouly 2020).

Felleng Yende, the award-winning CEO of the Fibre Processing and Manufacturing (FP&M) sector education and training authority (Seta) in South Africa, has risen to these challenges. She believes that much more needs to be done to promote mathematics and science education at school level, especially among girl learners, as these subjects are key to their development as future leaders in the 4IR. She recommends that school curricula should be reviewed and that subjects should be promoted that will make it possible for learners to become big data analysts, scientists, AI and machine-learning specialists, digital transformation specialists and robotics engineers.

> We are also producing too many learners that become dependent on formal employment. There are huge opportunities for learners that have an entrepreneurial flair. Entrepreneurship and business skills must be prioritized at school level.
>
> *Felleng Yende, CEO of the FP&M Seta* (Yende 2019)

Educators are bracing themselves for the onslaught of Generation Alpha (children born between 2010 and 2015), who will soon replace Generation Z as the new kids on the block. Generation Alpha is the first batch of children born to millennials and they are likely to be even more tech-savvy, game-changing and challenging than their parents, having been brought up with screens as their babysitters and toys. 2010 (the year of the generational divide) was also the year when the first iPad appeared, Instagram was invented and 'app' was Word of the Year. Watch out, Generation Alpha's influence in Africa and globally is going to be massive!

13

Sports, games and recreation

'If you want to go fast, go alone. If you
want to go far, go together.'

African proverb

Introduction: Long before African 'screenagers' became addicted to online games on 2D screens, games and sports were popular in Africa. Early Egyptians were keen gamers and their *senet* ('passing') is believed to be the oldest board game because a set was found in the tomb of Queen Nefertan, the wife of Rameses II (1304–1214 BCE), and images of *senet* boards have been found on tomb walls. A game like checkers (called *quirkat*, from the Arabic *qirq* or *qirqa*, meaning 'a pebble' or 'a stone piece') was also played by ancient Egyptians and was important enough to be buried with the dead, presumably for playing in the afterlife.

In South Africa and elsewhere in Africa, considerable effort has been made to preserve indigenous African games that have been played for centuries. The inaugural Indigenous Games Festival, held at the Basotho Cultural Village in the Free State (South Africa) in 2001, included *morabaraba* (*shax* in Somalia and *achi* in Ghana), which is recognised by the International War Games Federation. *Morabaraba* is a two-player strategy board game in which pieces are moved from intersection to intersection along marked lines. Another popular indigenous game is *kgati* (also called *ttimo* or *ugqaphu*), a skipping game in which two players hold a rope on each end and the third skips in a variety of ways while chanting and singing. In 2012, South Africa's *kgati* team won the gold medal at the 5th World Sport for All Games in Lithuania. In the coordination game *diketo* (or *ukugenda*) that is played mostly by young girls, 10 stones are set in a circle and the players try to grab the most stones, toss them into the air, catch them in one hand and quickly replace them in the circle. Another internationally recognised indigenous game is *jukskei*, developed by ox-wagon transport riders during the Great Trek in South Africa from 1835 to 1840, which involves players trying to knock over an upright peg using tossed wooden pins (*skeis*) (Lawson 2017b).

Legendary skateboarder: Capetonian Kent Lingeveldt has been skateboarding since 1994 and was the first black African to compete on the international circuit when he participated in the 1999 Red Bull Downhill Extreme, Africa's first exposure to gravity longboarding. He grew up in the ramshackle townships of the Cape Flats but, 20 years on, he is now a respected board shaper whose longboards are sought-after in Africa, Europe, the Americas and Australasia. Furthermore, he has made the transition from a ruffian street skater to the respected 'old man' of skateboarding without losing his bohemian persona (Bain 2017).

> I spent a lot of time bombing hills in Woodstock. Cops would stop us and lock us up. I spent a few nights in jail – just for skateboarding … I didn't have internet, so I was kind of winging it, reinventing the wheel a bit. I chatted with people in various industries. To learn how to bend wood, I spoke to boat builders in Kalk Bay. And for the fibre-glassing, I spoke to friends who make surfboards.
>
> *Kent Lingeveldt, skateboarding legend* (Bain 2017)

When he started shaping boards in 2000, it was out of necessity because the short streetboards that were fine for tricks like ollies (a skateboarding trick in which the rider and board leap into the air without the use of the rider's hands) and kickflips (an ollie in which the skateboard completes a full rotation along its axis) were not competitive for downhill racing. As he could not afford to buy a longboard, he set about designing and making them, and soon his customised decks were in demand locally and abroad. Since then, longboards have taken off worldwide and Lingeveldt's company, Alpha Longboards, is booming. He has trained a team of young board riders and shapers, called the '52 Crew' after the company's address: 52 Wright Road, Woodstock (Bain 2017).

First surfboards: Surfing is a popular sport in South Africa and West Africa, and is gaining popularity in East Africa. The first surfboard designed and made in South Africa (and probably Africa) was the Crocker Ski made by Fred Crocker in Durban in 1938 using a meranti wood frame covered in canvas, with two drain holes at the back. The 4-m-long skis, which borrowed ideas from early aircraft design in that resin was painted onto the canvas to make it strong and tight, were widely used by early surfers and surf lifeguards in coastal cities in South Africa. In the early 1950s, Les Hillier and Graham Winch adapted the Crocker design using newly available fibreglass, which started a revolution in surfboard design. A few decades later, fishermen in KwaZulu-Natal modified the Crocker Skis into one-person fishing vessels using marine plywood and then fibreglass, with added hatches, rod holders, skegs and transoms, but retaining the original flat-bottomed, banana shape (Hollands 2012).

Les Hillier was a skilled carpenter and a prolific builder of innovative surf craft. He was commissioned by the East London municipality to build paddle boards for surf lifesavers using Sagex foam, a product originally developed for insulation, coated with epoxy resin, with wooden dowels and a central 'stringer' (wooden reinforcement) holding the structure together. These boards were much lighter, faster and more manoeuvrable than the cumbersome Crocker Skis and were far better suited to recreational surfing. Later, the surfing industry relied almost entirely on Clark foam, a polyurethane product that was easier to work (Hollands 2012).

Waves for Change (W4C): W4C is a mental health foundation that uses the power of surfing and close association with the ocean to address trauma in at-risk young people living in unstable communities. Surfing combines aerobic exercise and strength training with a stress-relieving mental health workout as the surfer progresses and develops new skills. W4C (which so far has launched at five localities in South Africa and in Harper, Liberia) provides fun and child-friendly mental health services through 12-month, weekly surf therapy lessons in communities where mental health services are under-resourced and stigmatised. W4C focuses on the needs of young people who are exposed to repeated trauma and adversity, lack supportive adults, feel negatively about themselves and adopt confrontational behaviour patterns. In South Africa, where up to 20 per cent of youth suffer from post-traumatic stress disorder, the project uses the proven therapeutic effects of surfing and ocean experiences to break the cycles of violence that prevent young people from developing normally (Whitby 2019).

> Surfing changed everything about my life. It gave me focus, and purpose. I learnt to have confidence, and how to communicate with people and hold myself. And it gave me the opportunity to become a surf coach, which I absolutely love! … I love having something to offer to people, you know? I can improve their lives, their feelings, and really build relationships.
>
> *Alfonso Peters, ex-gangster rehabilitated through surfing, now a W4C ambassador in Muizenberg (Whitby 2019)*

In Liberia, the aftermath of the civil war (which ended in 2003) and the devastating impact of the EVD epidemic (2014–5) left the country with the enormous task of rebuilding, especially rehabilitating young people who had suffered unimaginable levels of trauma. In 2016, after successfully developing its model in South Africa for five years, W4C partnered with students from Tubman University (a public university located in Harper) to launch the surf therapy programme. Although Harper has no history of surfing, it proved to be an ideal location, as the university was looking for new ways to engage with local youth who suffer from poor education, abusive home lives, poverty and unemployment. Tubman University provided trained counsellors, the NGO Partners in Health gave strategic guidance and logistical support, and two Liberian surfing champions (Peter Swen and Melvin Kabahole) offered expert surf coaching. In collaboration with these partners, a small team of coaches led by Emmed Ross developed the surf therapy programme, which now hosts 60 to 80 children aged 9 to 20 years each day.

The programme not only teaches surfing skills but also life skills (such as making friends, sharing, trust, respect, mindfulness, building confidence and self-esteem, overcoming fear and anxiety, and meditation) and transferring newly acquired skills to the home and classroom 'one wave at a time'. As with all W4C projects, pre-tests are conducted at the start of each programme to provide a baseline for evaluation. These tests revealed that over one-third of the participants had poor coping mechanisms. Away from the beach, the Harper team conducted home visits and spoke to caregivers about the respective participant's progress. The results, after just six months, were overwhelmingly positive: 75 per cent of the parents reported improvements in general behaviour and school performance, and 89 per cent said that their relationship with their child had improved.

> Being in the water, reflecting on the rolling sound of the waves, the kids express the relaxing feelings running through their mind. Their thoughts shift from the stress of daily assignments to the calming movement of the wave – the up and down – riding helping to soothe their anger, upset and stress.
>
> *Emmed Ross, leader of W4C in Liberia* (Whitby 2019)

SharkSafe Barrier: Anti-shark barriers have taken various forms over the years, from gillnets and exclusion nets (which just as often keep the sharks in as out) to drum lines. Technologists at Innovus, the university-industry interaction platform at Stellenbosch University (South Africa), have developed a new technology called the SharkSafe Barrier that works by overstimulating the shark's electroreceptors and repelling them. The barrier comprises tubes containing strong magnets and mimics kelp forests, which sharks naturally avoid. It is an effective method because it does not affect other fish and marine life – or humans – and allows a demarcated area to be kept shark free with a device that is not as inhumane or ecologically damaging as, and is also more cost effective than, previous methods. According to team leader Conrad Mathee, the SharkSafe Barrier was found to be 100 per cent effective (even when food was placed inside the barrier) when tested at Gansbaai on the South African south coast. It was also installed at Réunion Island in the western Indian Ocean.

Kayaking for Africa: Johan Loots of Knysna in South Africa has revolutionised kayaking in Africa with his development of a unique range of sit-in and sit-on paddleyaks. Following an accident that led to the drowning of a paddler, he developed kayaks that are safe, stable, easy to dismount and yet competitive. His kayaks combine the best features of sit-in and sit-on canoes, including a closed hull and cockpit (sit-on elements) and a protective spray cover (sit-in element). His Paddleyak Fusion Classic (2000) was famously used by adventurer Riaan Manser to paddle 5 500 km around Madagascar in 2009, and his Paddleyak Swift Hybrid Adventure Kayak (2002) is widely used by recreational and competitive paddlers in South Africa and Mozambique as well as on Lake Malawi. One of Loots' most innovative designs is the Paddleyak Travelyak, which can be split into three parts for easy transport and is ideal for cross-country trekking (Bruton 2017a).

Telemetric and training cricket bats: Sports scientists at Rhodes University in Makhanda (South Africa) have developed a lightweight cricket bat sleeve that can be placed over the face of any cricket bat. Sensors placed inside the front panel measure the accuracy of ball placement on the bat and the force generated to hit the ball, as well as the angle of the bat swing. From this information, and based on the flight path of the ball, a performance outcome is measured. This information can be displayed on a software app together with digital reconstructions, and used for cricket coaching.

Sports scientist Dr Habib Noorbhai (Mr South Africa 2017) founded the NPO The Humanitarians in 2013 that strives to create a sustainable and innovative society through promoting a healthy lifestyle, education and skills acquisition, sport development, and enhancing sustainable and innovative living. In 2015, Noorbhai was nominated one of the *Mail & Guardian*'s Top 200 young South Africans and, in 2017, he was included in Fast Company South Africa's Top 30 Creative People in Business. His PhD research on backlift techniques in cricket (an essential component of sound batting technique) has underpinned the design of a revolutionary training cricket bat with a racquet-shaped toe-end. The training bat, developed in collaboration with leading sport scientist Prof. Tim Noakes and Russell Woolmer (son of the late Bob Woolmer, a former coach of the South African national cricket team Proteas), has been given the thumbs up by coaches around the world.

Caster Semenya: Caster Semenya was born in a small village in Limpopo province (South Africa) and knew that she was different from childhood. She was born a woman, raised as a woman and is legally a woman but she has unusually high natural testosterone levels and XY chromosome pairings, like a man. Her phenomenal performances as a middle-distance runner (she is a twice Olympic gold medallist and thrice world champion in the women's 800 m) have resulted in her gender identity coming under scrutiny. As a result, in 2018, the International Association of Athletic Federations ruled that she is 'biologically male' and banned her from competing in middle distance events unless she lowered her testosterone levels through medical intervention, which she refused to do. In September 2020, she experienced another legal setback when her appeal against the ban was overruled. She and others argue that male and female are not always binary. Before the Covid-19 lockdowns, Semenya was in her 12th top-flight season and had no plans to retire. Her idol (800-m runner Maria Mutola of Mozambique) ran at top level for 21 years and competed in six Olympic Games, and she wants to emulate that feat. In 2018, Semenya ran the 800 m within a second of the world record time of 1:53.28 (set in 1983) and plans to beat it if she is allowed to run (Gregory 2019).

> There are a lot of athletes who will have the same problem as me, but they cannot fight. If I don't fight for them, no-one will fight for them … I'm not stopping, you understand? I'm here to stay. You'd better get used to it. Or walk away.
>
> *Caster Semenya, champion athlete* (Gregory 2019)

Parkrun: The world's largest provider of free physical activity, Parkrun, was initiated by someone born in Africa. Paul Sinton-Hewitt, who was born in Zimbabwe and educated at Potchefstroom High School for Boys in South Africa, launched the first event in the UK in 2004 – and it has taken the world by storm. Pre-Covid-19, the 5-km Saturday morning run/walk attracted over 350 000 participants at more than 2 000 locations in 22 countries across five continents every week. Furthermore, the Parkrun organisation showed its resilience during the lockdown by organising '(not) parkruns' in which participants recorded 5-km runs on routes of their choice and submitted the time to their parkrun profiles. In recognition of his achievement in launching this highly successful event, Sinton-Hewitt was appointed a CBE (Commander of the Most Excellent Order of the British Empire) 'for services to grassroots sport participation' in 2014 and received the Albert Medal from the Royal Society of Arts in 2019 'for building a global participation movement'.

Detecting the risk of sports injuries: Nneile Nkholise, who is well-known for her medical innovations, established the sport technology company 3DIMO in 2018 that aims to optimise injury detection and prevention using athlete-specific orthopaedic biomechanics data. Rather than just reacting to a costly and debilitating sports injury, 3DIMO aims to prevent the problem from occurring in the first place by using athlete-specific data to predict, analyse and prevent sport injuries. 3DIMO uses performance-tracking devices and stress-sensing wearables to capture data on players during training. Digital models of the athletes are then created that can be compared to parameters measured during training or games to predict the likely onset of injuries, such as torn ligaments, ruptured spleens, dislocated joints or fractured vertebrae – all of which can lead to long-term or career-threatening scenarios. The models can then be used by team managers to design a player's training and playing regimes, and keep them fit. Nkholise was listed 13th by Forbes Africa on the 30 Under-30 technology list in 2019 and her company was selected as one of 22 start-ups from 10 countries to participate in the Cape Town-based start-up boot camp AfriTech Accelerator.

We have noticed that athletes are so determined to be the best and over-train without even noticing that they are over-training. This results in athletes exerting excessive loading on their joints, which is how injuries occur and their performance drops. Our product will help guide them to know when it's enough and to stop.

Coming from a technology background, my biggest hurdle was transitioning into an entrepreneur. The finance jargon used to torment me, but data is a lot of numbers and in sports there are millions of numbers being generated. When we use mathematical modelling, it allows for filtering of data using AI because data can be anything.

The 3DIMO company was born from the lessons and experiences I had in iMed Tech, where we used to do custom prosthesis and surgical planning models. I learned that many of the injuries that people experienced were soft-tissue and the main causes were things such as undetected excessive stress and strain, which in the long run leads to muscle fatigue, over-training, poor nutritional plan, and fatigue or lack of rest.

Nneile Nkholise, founder of 3DIMO (Rahman 2019)

Team Africa Rising: North African cyclists from Algeria, Tunisia and Morocco competed in the *Tour de France* (the world's premier cycling race) from 1913 onwards, with the first being Tunisian Ali Neffati. Later, African cyclists from Eritrea, Rwanda and Ethiopia also competed, with the first Africa-born cyclist to win a stage in the *Grande Boucle* (Big Loop) being Algerian Marcel Molinès in 1938, although he had become a French citizen by then. South African Robert Hunter was a stage winner in 2007 (Quérnet 2020). The first African rider to wear the yellow jersey was Daryl Impey of South Africa in 2013, unless one counts Chris Froome (who was born in Kenya, but had become a British citizen by the time he won the *Tour de France* in 2013, 2015, 2016 and 2017).

There are renewed efforts to get Africans to compete in the top cycle races. History was made in 2015 with the continental MTN-Qhubeka team comprising five Africans (including Eritrean Daniel Teklehaimanot, who wore the polka-dot jersey until the first Pyrenean stage). In 2020, the renamed NTT ProCycling team featured six South Africans and one Eritrean on their 29-man roster, including South African champion Ryan Gibbons.

Few sporting teams in Africa have epitomised the 'Africa rising' narrative better than Team Rwanda Cycling, which was founded in 2007 in a country torn apart by genocide 13 years earlier. A group of cycling industry legends was astounded by the fantastic raw talent of the cyclists who competed in the first Wooden Bike Classic in September 2006 in Kibuye (Rwanda), where wooden bicycles are common, and subsequently developed it into a competitive cycling nation with a top-ranked international race (Tour of Rwanda) and hundreds of young men and women striving to use cycling as a vehicle to escape the poverty spiral.

In 2014, Team Africa Rising was formed to develop the Africa Rising Cycling Centre in Musanze (Rwanda) that houses and trains the national team. Its mission is to recruit and train cyclists as well as coaches, mechanics and nutritionists to compete at the highest level. Team Africa Rising's top rider, Adrien Niyonshuti (who raced for the Team Dimension Data World Tour team throughout his stellar career), is dedicating his time to coaching the next generation of elite Rwandan cyclists. Other prominent Rwandan cyclists include Samuel Mugisha (who has re-signed with Team Dimension Data for Qhubeka's Continental Team) and Joseph Areruya (who races for Delko Marseilles). Areruya was voted Best African Cyclist for 2018, after becoming the first rider to hold three local tour crowns at the same time. Team Africa Rising is planning to grow cycling throughout the continent by inter alia helping Nigeria, Burkina Faso, Ethiopia, Eritrea and Algeria to develop their national teams. Their aim is nothing less than to have a black African rider win the prestigious *Tour de France*! Team Africa Rising is also working towards launching the first pan-African women's professional cycling team.

Most influential African sports star: There are many contenders for the most influential and innovative sportsperson in Africa, on and off the field, including George Weah (Liberia), Caster Semenya (South Africa), Brigid Kasgei (Kenya), Murielle Ahourée (Côte d'Ivoire), Mohamed Salah (Egypt), Pierre-Emerick Aubameyang (Gabon), Sadio Mané (Senegal) and Siya Kolisi (South Africa, the first black man to captain a team to triumph in the Rugby World Cup Final). Kolisi has used his prominence in remarkable ways to encourage African youth to overcome overwhelming odds and reach for their dreams, as he has done, as well as support the needy. Siya and his wife, Rachel, launched the Kolisi Foundation in early 2020 and have been hard at work helping to feed those most seriously affected by Covid-19 in his hometown of Zwide in the Eastern Cape, in partnership with the Imbumba Foundation and the Nelson Mandela Foundation. Kolisi has joined South Africa's top female explorer Saray Khumalo; long-distance athlete Eliud Kipchoge; soccer legend Luis Figo; and the founder of Nigeria's first bobsled team, Seun Adigun, to promote The Bigger Picture campaign to focus attention on malaria during the Covid-19 crisis.

Pierre-Emerick Aubameyang became the first African soccer player to captain a Football Association Challenge Cup (FA Cup) winning side when he scored both goals for Arsenal in the final against Chelsea in July 2020 and a goal in their FA Community Shield victory against Liverpool in August 2020. He also captains the Gabon national side, although he was born in France. Sadio Mané (Senegal), Mohamed Salah (Egypt) and Naby Keïta (Guinea) all played for Liverpool Football Club during its recordbreaking English Premier League cup-winning season in 2019/20. Other top African players in the English Premier League, past and present, include Didier Drogba and Yaya Touré (Côte d'Ivoire), Michael Essien (Ghana), Riyad Mahrez (Algeria), Rigobert Song and Samuel Eto'o (Cameroon), Noureddine Naybet (Morocco), Lucas Radebe (South Africa) and Nwankwo Kanu (Nigeria).

But for sheer audacity and determination, the accolade must go to Eliud Kipchoge of Kenya who did Africa proud by becoming the first person to run the standard marathon in under two hours. His time of 1:59:40.2 equates to eight consecutive 5-km parkruns in just over 14 minutes each! This is not recognised as a world record because he was assisted by a pace car and runners, but his world marathon record of 2:01:30 still stands. Before the race, Kipchoge said that it was more than a race against time; it was a race against history, and about 'making history in this world, like the first man to go to the moon' (*Scroll.in* 2019). The hashtag that he used for the race (#nohumanislimited) summarises his message to the world, especially to the young people of Africa. His accomplishment has been described as 'a giant leap in human endeavour' comparable with those of Edmund Hillary (and Tensing Norgay) in 1953 (ascent of Everest), Roger Bannister in 1954 (first four-minute mile) and Neil Armstrong in 1969 (first person on the moon). We will probably never see his likes again.

Emerging technologies: The Medical Research Council in Cape Town has developed a prototype Sports Injury GeneScreen to identify athletes at increased risk of tendon and ligament injuries. Genescreen is a non-diagnostic DNA test that consists of eight DNA sequence variants within five clinically useful genes associated with the risk of sustaining common tendon and ligament injuries. The results of the genetic test are evaluated in conjunction with physical activity history and medical examination by a healthcare professional, and risk is determined taking into account clinical indicators and lifestyle factors.

Waterbikes, which are pedalled by the rider, are not new (this author remembers riding them in Florida and Hawaii in the 1990s) but their first introduction to Africa took place off Simon's Town in South Africa on 1 August 2020. Waterbikes are clean and silent, which allows riders to watch marine life close-up without disturbing them, and provide a serene and invigorating experience. The two-hour route from Simon's Town passes the naval base and includes the world-famous Boulders Beach penguin colony. Riders may also see Cape fur seals, sea birds, whales (in season) and dolphins. All tours are enriched by knowledgeable crew.

Emerging innovators: At the age of 21 years, Karabo Mathang-Tshabuse noticed that very few women were soccer agents. In 2007, she decided to start her own sports agency business, P Management. Today she is the first African female FIFA-accredited soccer agent and represents about 40 players. She has played a role in the success of several prominent soccer players, including Orlando Pirates' Mbongeni Gumede and Banyana Banyana's Amanda Dlamini, and is currently studying towards an LLB degree with the goal of growing her company into South Africa's largest sports law agency.

James Fernie, the founding director of *Uthando* ('love') in South Africa has initiated a project to install cement table tennis tables in schools, community centres, public parks and old age homes. The tables are brightly painted with indigenous Ndebele designs by Juma Mkwela. The initiative is part of *Uthando*'s goal to transform communities through sport (www.uthandosa.org).

According to Zambia's Ministry of Heath, 10 per cent of men and 25 per cent of women in the country do not get enough exercise, and there has been an increase in diseases relating to inactive lifestyles and poor diet such as hypertension and diabetes. Fitness entrepreneur Makungo Muyembe has launched a new movement of fitness-conscious people through his company Sweat Factory in Lusaka. When the Covid-19 lockdown forced him to close his gym, he started livestreaming fitness classes from his home gym and they have gone viral. Yoga teacher Raine Dunn established the first Goat Yoga classes in Africa at Fairview Farm near Paarl in South Africa that is famous for its goats and cheese. Her classes, which she claims are both lighthearted and uplifting, are a unique combination of yoga and animal therapy that promote relaxation and relieve stress.

Discussion: African athletes (female and male) are prominent on the world stage due to their athleticism and determination to succeed, notwithstanding the lack of sophisticated training and sport science facilities in many African countries. The Sports Science Institute of South Africa (Ssisa) in Newlands, Cape Town, is the leading institution of its kind in Africa and is setting the standard for other African countries to follow. Ssisa, whose motto is 'Enhancing lives through science', was established in 1994 in collaboration with the University of Cape Town's Division of Exercise Science and Sports Medicine and has become a game-changer in sports performance, sports injury prevention and the promotion of healthy living. Ssisa has transformed lives, changed practice, optimised performance, and provided evidence to inform policy and advocate physical activity and sports participation for all South Africans. Its goal is to positively impact lives and to make its services, interventions and expertise as widely accessible as possible.

The initial concept of Ssisa was to develop sporting icons who would promote national pride and unity. This vision was generated during discussions between ex-Springbok rugby captain Morné du Plessis, Prof. Tim Noakes and philanthropist Johann Rupert. Their goal was to develop a facility that would primarily fund research and apply that research to sport so that athletes in all disciplines could improve their performance. They also aimed to build public interest in the country's top athletes whose success would create shared pride in South Africa.

Ssisa's mandate subsequently expanded to include the optimisation of the health of all South Africans. Today the institute has four strategic pillars (wellness, performance, education and research) through which they strive to bring people closer to realising their full potential. Their products and services include education and training, sports and fitness programmes and workshops, the development of centres for high performance sport and general fitness, and conferencing and specialised health services. Ssisa's Community Health Intervention Programme was established in 1997 to address the growing prevalence of obesity and non-communicable diseases (including hypertension, diabetes and high cholesterol) among South Africans and to promote a culture of health, wellness and an active lifestyle through regular physical activity.

Ssisa also actively pursues partnerships that unlock value for their clients and stakeholders. One of their partners is Cape Sports Medicine (CSM), which treats all athletes (from elite to recreational sports enthusiasts) as well as people who have exercise-related medical conditions. The CSM strives to provide the highest levels of care for the prevention, diagnosis, management and rehabilitation of injury and disease. Other South African universities have also developed advanced sports science research and development facilities. For example, the Interdisciplinary Centre for Sports Science and Development at the University of the Western Cape promotes sport as a powerful tool for development, well-being and social change. There is an urgent need to develop a comprehensive network of collaborating sports science institutes throughout Africa so that the enormous potential of African athletes, and ordinary citizens, can be developed to the full.

14

Clothing, accessories and design

'Innovation is the ability to see change
as an opportunity, not a threat.'

Anonymous

Introduction: Clothing styles differ widely from one corner of Africa to the next, and one cannot assume that clothes made in one country will sell in another. When the successful South African clothing retailer Ackermans opened a store in Kenya, selling the same merchandise that it sells in its home country, the store failed because Kenyans are generally more slender and taller than South Africans and culturally tend to dress more formally, as opposed to the more casual clothing that suits the South African market. In many parts of Africa, traditional clothing continues to be made in time-honoured ways. For example: The *shammas* of Ethiopia are colourful toga-like garments or shawls worn by most men and women. They are made mainly from locally grown cotton by the indigenous Dorze, once a warrior people but now farmers and weavers. Using simple handlooms and moving with blinding speed, they can produce up to 9 m of cloth per day (Hancock et al. 1997).

Interestingly, weddings make a major contribution to the economy of Nigeria because they are important events in the country's culture. Large weddings may be attended by hundreds of friends and relatives, and may have a budget of US$250 000. It is a popular tradition to wear matching clothing in identical fabric made from *aso ebi* or *iso eli* ('family cloth' in Yoruba) to weddings as well as parties and funerals. These outfits are intended to display one's close family ties and to reinforce identity and social bonding, but they also have a downside because relatives who cannot afford to wear *aso ebi* may be discriminated against and even excluded. This creates a problem for the hosts of social events whose status is enhanced if they attract large numbers of similarly attired participants. To overcome this problem, *aso ebi* has undergone a transformation by including not only family members (*ebi*) but also co-workers (*alabasisepo*), friends (*ore*), co-residents (*alajogbe*), neighbours (*aladugbo*) and other well-wishers (Tade 2020). *Pantsula* is the latest clothing fad in fashion-conscious Gaborone in Botswana. Some of the old-school-style trendsetters wear expensive brands (Pringle of Scotland sweaters, Brentwood trousers and Converse Allstar sneakers), whereas others don matching chequered shirts and pants or all-white uniforms with extravagant epaulettes and mortarboard caps.

Kente cloth: *Kente* is an indigenous Ghanaian textile made of interwoven cloth strips of silk and cotton, and is worn by almost every Ghanaian tribe as part of their national cultural identity. *Kente* is derived from the word '*kenten*' ('basket') in the Asante dialect of Akan and varies in complexity from a simple design of stripes to abstract designs and the use of words, numbers and symbols. The colours in *kente* cloth have different symbolic meanings and the designs have a variety of names derived from proverbs, historical events, the names of important chiefs and plants. *Kente* patterns may symbolise concepts such as democratic rule (*Obaakofoo Mmu Man*), creativity and knowledge (*Emaa Da*) or hard work (*Owu Nhye Da*). Legend has it that *kente* was first made by two Akan friends who went hunting in a forest and found a spider making its web. They watched the spider for two days, then returned home and implemented what they had seen. *Kente* academic stoles are often used by African-Americans as a symbol of ethnic pride and students hold special ceremonies called 'Donning of the Kente' when the stoles are presented to graduates (https://en.wikipedia.org/wiki/Kente_cloth).

Shweshwe fabric: *Shweshwe* is a printed dyed cotton fabric widely used for traditional southern African clothing. Due to its popularity, *shweshwe* has been described as the 'denim' of Africa and is now popular as far afield as Botswana, Zambia, Kenya and Ghana. The name is derived from the fabric's association with Lesotho's King Moshoeshoe I, who was gifted it by French missionaries in the 1840s after 19th-century German and Swiss settlers imported *blaudruck* ('blue print') fabric for their clothing. It is known as *sejeremane* in Sesotho and *ujamani* in isiXhosa. In South Africa, *shweshwe* is traditionally used to make dresses, skirts, aprons and wraparound clothing. In Ghana, where it is called *mashwete*, it is used to make clothing and shopping bags.

Shweshwe clothing is traditionally worn by newly married Xhosa women (*makoti*) and married Sotho women. Xhosa women have also incorporated it into their traditional ochre-coloured blanket clothing. Besides traditional wear, *shweshwe* is used in contemporary South African fashion design for women and men from all ethnic groups as well as for making accessories and upholstery. In the USA, it is used as a quilting fabric. *Shweshwe* fabric is manufactured in various colours (including the original indigo, chocolate brown and red) in a large variety of designs, including florals, stripes and diamonds, and circular and square geometric patterns. The intricate designs are made using picotage, a pinning fabric printing technique rarely used by contemporary fabric-makers due to its complexity and expense. The trademarked fabric has been manufactured by Da Gama Textiles in Zwelitsha outside King William's Town in the Eastern Cape province of South Africa since 1982. In 1992, Da Gama Textiles bought the sole rights to Three Cats, the most popular brand of the fabric made by the Spruce Manufacturing Company in Manchester (UK), and the original engraved copper rollers were shipped to South Africa.

Da Gama Textiles makes *shweshwe* from locally grown cotton as well as cotton imported from Zimbabwe; however, the local industry is threatened by competition from cheaper, inferior quality imitations imported from China and Pakistan. The genuine product can be recognised by feel, stiffness (from starch), smell, taste (salty) and the trademark Three Cats logo. The fabric was made famous by Jill Scott, who portrayed the main character (Mma Precious Ramotswe) in outfits made from *shweshwe* in the TV series *The No. 1 Ladies' Detective Agency* (https://stylefabrics.co.za/general-prints/shweshwe/history-of-shweshwe/).

In July 2020, a consignment of 1 200 brightly coloured Covid-19 masks made from *shweshwe* cloth was exported to Belgium from the townships of Diepsloot, Alexandra and Tembisa, injecting over R1 million into local businesses. Making these masks was a life-changing experience for 270 township seamstresses working for the Masks4All initiative of the Youth Employment Service. The colourful masks were marketed though LinkedIn and were an immediate hit in the dreary Belgian weather (Nair 2020).

> The vibrant yellow, blue and red *shweshwe* cut through the staid bureaucracy of EU headquarters and old grey buildings. They carry the waft of big blue sky, a story of triumph, a real person who made this by hand and whose life was changed as a result.
>
> *Tashmia Ismail Saville, CEO of YES* (Nair 2020)

In addition to *kente* and *shweshwe*, Africa boasts a variety of other eye-catching fabric designs, including *berber* (Morocco), *bogolan* (Mauritania), *kabyle* (Algeria), *tuareg* (Libya), *ankara* (Gabon, Congo, Chad and Niger), *alindi* (from

Somalia), *aso oke* (Côte d'Ivoire, Liberia, Sierra Leone, Guinea, Burkino Faso and Senegal), *kuba raffia* (DRC), *lamba* (Madagascar), *samakaka* (northern Namibia and Angola) and *chitengi* (Zimbabwe).

Ndebele motif art: Esther Mahlangu from Mpumalanga in South Africa is a living legend in the design world, yet she is hardly known in Africa. She is the foremost exponent of traditional Ndebele motif art, using bright colours and repetitive symmetrical patterns outlined in black to create works of stunning beauty. Her bold art is more than decorative, as the kaleidoscopic patterns represent a visual narrative within Ndebele communities. Traditionally, women painted their houses – inside and out – in the hypnotic but soothing Ndebele patterns to celebrate important events such as weddings (Borrill 2017).

As an artist, Mahlangu deconstructed Ndebele décor art and repackaged it in the form of paintings, painted vessels and tapestries, thereby altering their meaning by introducing them into new contexts, which catapulted her into the global spotlight. However, amazingly, it was a painted car that made her the darling of the international design scene. In 1991, BMW commissioned Esther to convert a BMW 525i into an Art Car using traditional Ndebele motifs – the first African and female artist to be so honoured, following in the footsteps of Andy Warhol, David Hockney and Frank Stella. The project was a huge success and the car was exhibited at the National Museum of Women in the Arts in Washington DC in 1994 and later at the British Museum in London. Twenty-five years after her initial commission, BMW asked her to design the interior trim for the latest BMW 740Li. This project was also widely acclaimed and was exhibited at the 2017 Frieze Art Fair in London. In April 2018, at the age of 82 years, Mashangu received an honorary doctorate from the University of Johannesburg (Borrill 2017).

> To paint is in my heart and it's in my blood. The way I paint was taught to me by my mother and grandmother. The images and colours have changed and I have painted on many different surfaces and objects, but I still love to paint. The patterns I have used on the BMW marry tradition to the essence of BMW.
>
> *Esther Mahlangu, Ndebele artist* (Borrill 2017)

Kaunjika **used clothing:** According to some estimates, 70 per cent of donated clothing ends up in Africa, where it forms the basis of a multimillion-dollar industry that provides a livelihood for millions of people. East Africa alone imports over US$150-million worth of used clothes and shoes, largely from the USA and Europe (Banik & Gresko 2020). But some have argued that this trade, called *kaunjika* ('clothes sold in a heap') in Malawi, has destroyed the local textile industry. This led Rwanda, Uganda, Tanzania and Burundi to consider major tariff increases on imported used clothing in 2016, but these never came about because the merchandise serves a real need in many African communities. Its affordability, higher quality and association with popular Western culture are attractive to many buyers, and entrepreneurs have found that they can launch their *kaunjika* start-ups with minimal funds. Others have argued that the continued importation of used clothing is undignified and erodes African pride, but a study on the used clothing industry in Malawi by Banik and Gresko (2020) found no evidence of this.

Fashion and film: To 32-year-old Senegalese artist and fashion designer Selly Raby Kane, Dakar is a huge source of inspiration because of its fascinating history. Her first official clothing collection in Dakar in 2012 attracted attention for its juxtaposition of the traditional with the futuristic and incorporated designs from military attire to urban street fashion, with fantasy costume and sci-fi influences. Since Beyoncé wore one of her kimonos in 2016, her popularity has soared and her garments are now on sale in Europe, the USA and Nigeria. As her star has risen, Kane has experimented with other art forms. Her 2017 virtual-reality film *The Other Dakar* was a tribute to Senegalese mythology and the city's hidden secrets. In early 2019, when Ikea commissioned 10 African designers to create a range of homeware products, Kane fashioned a remarkable basket designed to look like braided hair. 'It was a beautiful ritual to explore, and the bond it creates between two people was very inspiring to me' (Haynes 2019).

Kane's latest clothing collection was inspired by a landmark report commissioned by France's President Emmanuel Macron that recommended the repatriation of artefacts from French institutions back to their original sites in sub-Saharan Africa. 'It made me realise that in each artefact and archive is encoded a type of knowledge and a small fragment of history that informs you of a past vision of the world' (Haynes 2019). She is now busy sharing her insights and experiences with a trans-African community of young creatives that stretches from Dakar to Lagos, Nairobi and Kigali.

Presidential shirts: One of the most prestigious clothing styles created in South Africa is the presidential range of men's tops based on the colourful hangout shirts worn by late President Nelson Mandela. The patterns on these shirts are based largely on indigenous designs and all sales benefit the Nelson Mandela Foundation. The shirts were made famous by the iconic jazz band Ladysmith Black Mambazo, who wore them at every show. The Presidential range has now been extended to include women's dresses and robes.

African haute couture: For nearly two decades, Cameroonian designer Imane Ayissi has been turning traditional African fabrics into made-to-order womenswear worn by the likes of Zendaya, Angela Bassett and Aissa Maïga. But it was only in January 2020, when he was invited to present his Spring/Summer 2020 collection to the *Chambre Syndicale de la Haute Couture* in Paris – the first sub-Saharan African to do so – that the international press took notice. For his couture week debut, he presented a collection called *Akuma* ('richness' in Beti) to express the idea that true wealth depends on what you do with what you have, be it a little or a lot. On the runway, red raffia from Madagascar covered a strappy dress, strips of Ghanaian *kente* were assembled on a loose coat, the gowns were woven from strips of organic *Faso Dan Fani* (a cotton cloth from Burkina Faso), and *obom* tree bark from Cameroon was shaped into petals and appliquéd onto floorlength silk evening dresses. 'It's about the relationship you have with material things, and the respect you have for other people. It's the way you craft a dress that will give it life,' Ayissi said (Gaboué 2020).

The son of a boxer and a former Miss Cameroon, Ayissi was a dancer with his home country's national ballet before he moved to France in the early 1990s to work with French ballet star Patrick Dupont. He has had no formal design training, but caught the fashion bug while modelling for Dior, Givenchy and Lanvin (brands with which he now shares the runway). He started his eponymous line in 2001. He admits that his early collections were not always successful, but he remained patient, gradually improving his knowledge of textiles and tailoring. In the tradition of haute couture, Ayissi's custom-fitted clothing is handmade from high-quality, often unusual, fabrics and he is best known for mixing ethically-sourced, organic fabrics from Africa with typical couture materials such as silk and taffeta.

> I had to stay true to myself. It takes time to find one's voice. So, I improved my lines, tried to bring new ideas, went further with fabrics, became more audacious.
>
> *Imane Ayissi, Cameroonian clothing designer* (Gaboué 2020)

While Ayissi is glad to have his work embraced by the European and Asian shoppers who make up most of his client base, he would love to see more Africans wearing his work. While some African shoppers regard his clothes as too expensive, he argues that *kente* is a noble fabric and that African designers deserve to be supported at the right level. His ascent is happening at an important moment for African fashion, as young designers such as Thebe Magugu and Kenneth Ize are also showing in Paris and other creatives such as photographer Kristin-Lee Moolman and stylist Ib Kamara are building their brands internationally (Gaboué 2020).

Rwandan fashion to the world: Self-taught fashion designer Cedric Mizero mixes local artisanal techniques with a wide range of media and disciplines, repurposing everyday objects to showcase the vibrancy of life in Rwanda's rural districts. Mizero grew up in a small village and moved to the capital Kigali in 2012. Since then, his goal has been to tell stories about the richness of village life while demonstrating that fashion is for everyone, not just the rich and famous. In 2014, he launched Fashion for All, which uses fashion in an inclusive way to highlight the everyday experiences of people. The pinnacle of his career was attending the 2020 London Fashion Week, where his work was a highlight of the International Fashion Showcase (Appel 2019).

As a result of his village experiences, Mizero sees beauty in a different way from most people – as a quality of inner strength rather than a matter of aesthetics. As one of 12 children, he was always amazed at how happy his mother was despite her heavy workload; to him, that was real beauty. These ideas sparked his 2017 art installation 'Strong Women', with photographer Chris Schwagga, and 2018 fashion collection 'Beauty in the Heart' – both of which highlighted the inner strength of village women like his mother. His favourite form and narrative are that of the butterfly, as it represents the transformative nature of life for him. He is keen to experiment with different media and methods of presentation, and is developing ideas for combining fashion with the performing arts and film (Appel 2019).

Simon and Mary hats: Head gear is popular in Africa, both to shelter from the sun and rain and as a fashion statement. Research has shown that about one-third of the South African population wears hats, and that their use is increasing. Dean Pozniak, a Johannesburg-based milliner, has created the Simon and Mary brand of hats and aims to turn an old-fashioned accessory into a modern-day must-have.

'We tap into all parts of South African society to come up with our designs. For example, our Fez-style range pays homage to the Ndebele culture. The colours – popping yellows, pinks, greens, blues and black – were all inspired by Ndebele beadwork,' he says. Like many other brands, Simon and Mary benefitted when some A-list celebrities such as Maps Maponyane, The Black Eyed Peas' will.i.am and rapper Riky Rick wore their hats. His factory now employs over 70 people and they have their eyes set on the African, European and American markets (www.simonandmary.co.za).

Solar-powered backpack: Thato Kgatlhanye has been publicly earmarked by Bill Gates as one of the entrepreneurs to watch on the African continent. She tackles issues from plastic pollution to education and poverty, but her latest solution addresses a combination of problems African children face today.

In 2013, her company Rethaka launched an innovative backpack crafted from recycled materials and fitted with a solar panel that charges up during the day so that it can be used as a light for studying at night. The light also ensures the safety of children who must walk to or from school after dark. She was listed as one of Forbes Africa's 30 Under-30 technology list in 2017 and as Innovator of the Year by Forbes Women Africa in 2016, and is a CEO Global Leader of Tomorrow honouree. Kgatlhanye aims to continue to drive social change, present sustainable solutions and inspire others.

Versatile children's buggy: Guillaume du Toit of Shonaquip in South Africa has designed the Madiba2GoBuggy, which is a rugged, off-road, posture-support buggy for children who cannot self-propel and/or are unable to sit independently. The unique seating system offers adjustable full body and head support with a reclining option. The award-winning buggy also has a folding frame so that it is easy to transport and store.

South African flag design: When Fred Brownell (a Free State farmer's son) attended a meeting with Cyril Ramaphosa and Roelf Meyer in February 1994 to approve his design for a new South African flag, he explained to them in Sesotho that its three-armed converging cross met the criteria that it should encompass 'unity, interlinking or convergence'. As a result of strong support from traditional QwaQwa chiefs as well as Ramaphosa, Meyer and then President Nelson Mandela, his design was accepted. It has become one of the most recognisable modern national flags. Brownell, who died on 10 May 2019, designed a flag that found its way into the hearts and minds of all South Africans and became a unifying symbol.

Sunglasses: Dave de Wit of Cape Town has made the unusual transition from a skateboarder to a sunglass-maker. In fact, he combines his two passions by making sunglasses from used skateboards. 'My dad was very handy, from plumbing to electrical and carpentry...so I learned a lot from him. I worked for a while for a textile group...and I qualified as a loom tuner,' he says (Waterworth 2020). He also got involved in building skate parks and ramps, and worked for a while on a cruise ship. When his sunglasses broke, he decided to make a new pair from finely veneered skateboard wood. The experiment worked well and Sk8Shades was born. He uses high-quality, UV-protected, shock-absorbing and scratch-resistant lenses (with polarised mirror lenses an option) and stainless-steel spring hinges in his glasses, which have been a hit with both board riders and foreign tourists (Waterworth 2020).

Triggerfish animations: Where does technology stop and design start? Triggerfish Animation Studios (Africa's most award-winning animation studio) was established to upskill animation designers, foster the development of animation networks and bring African animation to global audiences. Their programmes, which support animators from 30 African countries, build on the success of their fruitful partnership with world-leader Disney. Their projects include setting up a digital learning platform to introduce African youth to animation, conducting drawing workshops in schools and townships, and implementing Africa-wide webinars and talent competitions on animation.

The company, in partnership with the Goethe-Institut and the Federal Ministry of Economic Cooperation and Development in Germany, has also created the Triggerfish Academy (a free digital learning platform for anyone wanting to understand more about career opportunities in animation). Their website features 25 free video tutorials, quizzes and animation exercises that introduce animation as a career and explain the principles of storytelling, storyboarding and animation. The platform was created by Tim Argall, the animation director on Triggerfish's third feature film *Seal Team*; Malcolm Wope, the character designer on Netflix's first animated original film from Africa *Mama K's Team 4*; Daniel Snaddon, the co-director of the multi-award-winning BBC adaptations *Stick Man* and *Zog*; and Mike Buckland, the head of production at Triggerfish.

Emerging innovators: After her first entrepreneurial venture (making school bags from recycled plastic) failed, Reabetswe Ngwane decided to convert old car tyres into handbags, seat protectors and masks for the mining industry. In 2017, her idea piqued the interest of the Recycling and Economic Development Initiative in South Africa, which was so delighted with her products that they now run one of her depots. Ngwane has since opened stores in Cape Town and Pretoria, and also sells her wares online. In 2013, the year she graduated from high school, Mabel Suglo launched her Eco-Shoes project aimed at assisting artisans with disabilities to create Afro-themed shoes, handbags and accessories using discarded tyres and recycled cloth. A year later, demand for her products rose sharply. She now aims to produce up to 1 500 handbags annually in addition to her unique shoes.

With a master's degree in architecture but no capital, Lucy Agwunobi and her husband started their made-to-order furniture business in Nigeria in 2010. Today their company, Arredo by TRT, has offices in Lagos and a factory in Abuja, as well as a reputation for innovation and craftmanship. Lucy aims to do nothing less than change the face of 'Made in Nigeria' and take her products to the world. Malian textile artist Abdoulaye Konaté takes inspiration for his striking, multicoloured creations from the colours of traditional Malian clothing styles and from the country's war-torn history. The bright reds of his tapestry 'Touareg Rouge No. 1' reference ethnic clothing, but also the bloody conflicts in the north of his country. He has been commissioned to create a centre piece for the spectacular atrium of the Zeitz Museum of Contemporary African Art in Cape Town.

Discussion: African designers and clothing and accessory makers have made a major impact on the world stage in recent years with their innovative ideas that combine the use of traditional motifs and unusual materials, especially from wild nature. The unparalleled diversity of African cultures is a rich source of inspiration for their work and provides a constant stream of new concepts and ideas, which is in stark contrast to the rather inbred European scene. However, designs of African origin are sometimes so out of kilter with European ideas and conventions that they fail to garner approval from the individuals or organisations that control the international haute couture scene. As a result, many highly talented African designers are either not recognised at all or are only recognised late in their careers.

It has taken a while but the fashion world's ever-watchful eyes have finally started to turn towards Africa and its emerging crop of fashion superstars. South Africa's fashion poster boy, Laduma MaXhosa, has made a major impact with his hugely popular knitwear brand MaXhosa by Laduma, which draws inspiration from the Xhosa rite of passage. Katungulu Mwendwa grew up in Kenya where she spent her childhood observing her late grandmother sourcing materials from local artisans for her shop. After completing a fashion degree in London, she returned to Kenya where she experiments with innovative fabrics, traditional methods and modern techniques to produce timeless casual and semi-formal designs that are fast gaining an international following.

Notwithstanding these and many other success stories, and the advent of social media, African designers still need more international forums to showcase their exquisite work. Individuals such as Grace Bampile, the founder of Haute Afrika, are having a huge impact. Grace came from the Congo (where African prints are worn every day) but found, after travelling around the continent, that people wanted good quality outfits but did not know how to access quality fabric. She realised that creating an efficient supply chain would solve the problem and that combining business acumen with creative design would take African fashion onto the world stage. Many others are following her example.

15

Transport and transport infrastructure

'The fact that wheeled transportation was not used in
sub-Saharan Africa until the early colonial period is paradoxical
because it is well established that African societies knew
about the wheel from the early modern period onward.
They did not have to reinvent the wheel, only adopt it.'

Chaves, Engerman and Robinson (2014: 322)

Introduction: Many reasons have been put forward to explain why the wheel
was not invented – or even re-invented – in Africa, including the availability
of slave porters; the presence of tsetse flies, which discouraged the use of draft
animals that could pull wheeled vehicles; the absence of horses; difficult terrain;
large rivers to cross; and the lack of centralised state organs to finance major
public works such as roads, railways and bridges. While some of these arguments
have merit, they do not explain why, for example, the wheel was not adopted in
areas free of tsetse flies such as the Sahel or southern Africa or why the extensive
paths constructed through forests by, for example, the Asante and Buganda were
not used for wheeled carts (Chaves et al. 2014). While it is accepted that railways
involve far more technology than just wheels (iron smelting and casting, rails,
bogeys, locomotives and carriages) and needed an industrial revolution to come
to fruition, the development of simple, two-wheeled carts seems to be an obvious
and achievable innovation, especially considering the relatively advanced state of
metallurgy (and carpentry) on the continent (Chirikure 2010). The engineering
of the wheel hub–axle interface was probably the main challenge, but this
would not seem to be beyond the competence of artisans who could already
make sea-going sailboats, anchors, axes and hoes – although none of these tools

had moving parts. Real wheels were also not developed precolonially in Latin America, although Mayan children in Mexico had pull-toys with tiny wheels. It remains a paradox.

First cars in Africa: The first 'horseless carriage' to arrive in Africa was probably a Benz Velo imported into South Africa in 1896 by Port Elizabeth businessman John Percy (Bruton 2017a). The vehicle was publicly displayed in 1897 at the Berea Park sportsground in Pretoria in the presence of Paul Kruger, then President of the Transvaal Republic. The publicity hype claimed that the 'motorcar' was the 'invention of the age' which, like the bicycle, had 'come to stay and [would] be the craze of the century'. The hype was right, but it was not until 1903 that cars were generally available in South Africa when the first Ford Model As became available outside the USA. The successor to the Benz Velo, the Benz Ideal (1 100 cc, single cylinder, single gear and three speeds – the personal car of Karl Benz in Germany for many years) reached South Africa in 1901. This car still exists under the ownership of the University of Cape Town and is on loan to the Crank Handle Club in Cape Town. Today most vehicles in Africa are still imported, but they are not necessarily designed for local usage. High import duties compound the problem, typically doubling the price of a vehicle. Across Africa, degraded roads, widely separated communities, low-income levels and poorly maintained vehicles undermine the continent's transportation system and ultimately constrain economic growth.

Vehicles made in South Africa: There have been several concerted efforts to produce a vehicle by Africans for Africans. The first production cars designed and made in Africa were the Protea (October 1956), Dart (September 1957) and Flamingo (1958) – all high-performance sports cars developed in South Africa (Bruton 2010, 2017a; Schwartz 1957; Stuart-Findlay 2019). The Protea was designed and made in Johannesburg by Coventry mechanic and racing driver John Myers, Yorkshireman Roland Fincher and Scottish manufacturing chemist Alec Roy.

They were initially inspired by the fibreglass-bodied Kreft, powered by a Coventry Climax engine mounted on a simple ladder frame, that was launched at the 1954 *Le Mans* 24-hour Race but also took inspiration from Lotus sports cars. Their aim was to produce a car that was modern enough to be competitive and had good road holding capability for daily use, yet be affordable and easy to maintain. The powertrain comprised a Ford Prefect 100E 1 172 cc side-valve, four-cylinder engine with a three-speed gearbox coupled to spoked wheels, brakes and a rear axle. A company called GRP (Glass Reinforced Plastic) Engineering was established, but production ended after only 26 cars and bodies had been produced due to competition with the Austin Healey Sprite (1958) and onerous customs regulations (Stuart-Findlay 2019). In 2019, Myers was still an active member of the Crank Handle Club in Cape Town, but he died on 1 September 2020 at the age of 97 years.

The Dart and Flamingo were developed by mechanical engineer and racing driver Bob van Niekerk and automotive engineer Verster de Witt, who formed the Glass Sports Motors company in Cape Town ('glass' refers to fibreglass). The Dart was a race-bred production sports car designed to provide maximum comfort and performance, and was raced with great success in South Africa and England. It was fitted with a Ford Cortina 1 200 cc engine and was capable of 160 km/h. The Flamingo was a Gran Turismo Coupé sports car that could drive comfortably and safely at high speeds. The production of these cars ended in 1958 (Bruton 2010).

Leyland has produced very successful 55-seater commuter buses, such as the Leyland Kudu and the Leyland Victory, in South Africa. In the 1970s and 1980s, it secured over 60 per cent of the local big bus market, especially in larger cities such as Johannesburg and Pretoria (Putco), Durban and Cape Town (Brian Hogg, personal communication, 20 January 2020). South Africa has also produced over 80 different kinds of military vehicles since the 1970s (see the 'Military and security' chapter of this book).

Several high-performance sports cars have also been produced in South Africa, including the impressive Shaka Nynya (named after the Zulu king) developed by Advanced Automotive Designs in Pretoria since 1995. Other South African car-makers have opted to produce replicas of famous international marques. For example, Birkin Cars in Durban (founded in 1980) is best known for its quality reproductions of the Lotus 7 Series 3 called the Birkin S3 and British-American Shelby AC Cobras have been made in Port Elizabeth by Hi-Tech Automotive, which has become the third largest specialised vehicle manufacturing plant in the world after Lotus and TVR in the UK. Although most of Hi-Tech's output is exact replicas of vintage sports cars from the 1960s, their under-the-skin technology is 21st century, with all brand-new parts. Furthermore, Superformance chassis are the only Cobra replicas on the market that are Shelby-licenced products. Since 2007, Hi-Tech has also produced the sought-after Perana sports car (now called the AC 378 GT Zagato). A tribute car, the Maserati A6GCS53 Berlinetta Pininfarina MM, was built in Caledon from aluminium, with a chassis modified from that of the original 4275 Shelby Cobra and the drive train from a Jaguar XJ6.

Les Hayden of the Dart Engineering Company in Montague Gardens (Cape Town) has made over 50 replica Hayden Cobras and 30 Hayden Darts, and Jack Levy in Paarden Island (Cape Town) has made replica Darts. Harper Sports Cars in Noordhoek (Cape Town) manufactures racing but road-legal Harper Type 5 sports cars conceptualised and designed by Craig Harper and Jim Page. These high-performance cars, which have a 2-litre Toyota Twin Cam engine and weigh only 720 kg, can reach 100 km/h in 3.5 seconds and achieve a maximum speed of 270 km/h. As the vehicle has no windscreen, the driver must wear a full-face helmet and visor when driving at speed. Prospective owners can order Harper sports cars to their own specifications in terms of drive train, level of comfort and other features.

Bailey Edwards is another top replica sports car manufacturer based in South Africa. Started in 2003 by brothers Peter and Greg Bailey, the company builds and customises classic performance cars such as the Porsche 917 and the Ferrari P4 for clients around the world and has a factory in New York to service the North American market. Its signature replicas include the Aston Martin DB6 (familiar as the classic James Bond car) and the Ford GT40, both much sought-after by collectors worldwide.

Bell Trucks (based in Richards Bay, KwaZulu-Natal, South Africa) was established by Irvine Bell in the 1960s. The company initially made simple three-wheeled sugarcane loaders, but has gained an excellent international reputation for making robust and dependable four-wheel drive articulated trucks such as the 25A Bell Arctic Truck (1984), Bell B40 Articulated Dump Truck (articulated dump truck [ADT], 1989) and Bell B50D (2002, the world's largest ADT) – all with Mercedes Benz engines. Bell also produces haulage tractors, timber loaders, forklifts, crawler bulldozers, refuse compactors, road graders, frontend loaders, tracked hydraulic excavators, road rollers and motorised conveyors. The company has formed partnerships with John Deere, Hitachi, Kato and Liebherr, and is a global player in off-highway bulk cartage, mining, forestry, agriculture and road building. It has sold more than 60 000 vehicles in 50 different models in 80 countries and, since 2003, has also manufactured its trucks in Europe. Bell employs over 3 500 people directly and 35 000 indirectly, and has one of the largest in-house accelerated training programmes in its sector (Bruton 2017a). It is one of Africa's most successful vehicle manufacturing companies.

South Africa has also experienced automotive engineering failures, most notably that of the Joule electric car. In 2008, South African engineers designed and built a prototype of the Joule with a power unit comprising Li-ion batteries that needed to be recharged every seven hours. Fuel and service cost savings of 80 per cent and 50 per cent respectively were anticipated compared to an internal combustion engine car. The Joule was produced by Optimal Energy (established in Cape Town in 2004 by Kobus Meiring) and was designed by South African-born ex-Jaguar designer Keith Helfet. Gerhard Swart (the chief technical officer on the project) emphasised that the Joule was not just an electric version of a normal car but a whole new paradigm, similar to the technology shift from a landline telephone to a smartphone (Bruton 2017a).

Five prototype Joules were made by Jimmy Price's HiTech Automotive Company in Port Elizabeth and they earned high praise from experts. With plans to export about 40 000 units per year at R250 000 each, the Joule had the potential to create jobs, make optimal use of local expertise and generate significant foreign exchange. Unfortunately, due to the highly competitive global market for electric cars, the Joule never went into commercial production. At the time, the BMW i3 and i8 and the Nissan Leaf were already available in South Africa, and Tesla and other electric cars would soon enter the market. Local production costs for the Joule would have been too high, and no foreign manufacturer showed an interest in making them abroad (Bruton 2018). It is ironic that another South African-born innovator, Elon Musk, has now taken the lead in this regard.

Although South Africa has the second largest economy in Africa, a sophisticated technological culture, vibrant car industry and high levels of entrepreneurship – and even hosted Formula 1 Grand Prix races on four different circuits intermittently between 1934 and 1993 – it has never produced a successful homebred car, unless one counts the Chevrolet Nomad or Nissan Sani. If India can produce the Tata, Malaysia the Proton, Australia the Holden, Brazil the TAC Stark, Chile the Puma, Indonesia the Wuling, Ukraine the ZAZ, and some other African countries their own cars and off-road vehicles (although none have been a global commercial success – see below), why can South Africa not do so as well? After all, Daimler-Benz assembles its luxury C-Class saloons in East London and exports them worldwide; Ford assembles its Ranger series of compact trucks in Tshwane and exports them to over 100 countries; and BMW, Chrysler, General Motors, Nissan, Toyota and Volkswagen all assemble vehicles locally on a large scale.

The most likely reason is the high cost of tooling up for the production of a single, unique model. Established car manufacturers can retool their assembly lines relatively inexpensively to produce a new model but a newcomer faces enormous costs, which would be exacerbated by the low production runs of a locally made car. Some cars developed abroad have been modified and made in South Africa, including British Leyland's Austin Apache that was a radical redesign of the popular Austin/Morris 1100 with a larger capacity engine – to 1 275 cc (A. Bruton, personal communication, April 2020). However, it seems that South Africa's automotive entrepreneurs decided that it was pointless to compete against established northern hemisphere brands and decided to rather produce small runs of a few high-performance cars. These vehicles are, however, mainly toys for the 'rich and famous' and do not address Africa's real needs. South Africa has not even produced a taxi that meets local demand. Every day millions of commuters ride in *Inyathi* ('buffalo' in isiZulu) taxis made by the China Auto Manufacturer Company. One opportunity that South African automotive engineers have, however, grasped is to make car parts (a modern car consists of up to 60 000 different parts) and South Africa now ranks as the 22nd largest maker of car parts in the world.

Vehicles made in other African countries: Cars, jeeps and trucks designed and produced by other African countries include the Somaka (1959) and Laraki (1999) in Morocco, the Saroukh el-Jamahiriya in Libya (1999), the remarkable Wallyscar (2006) in Tunisia, the URI off-road vehicles (2008) in Namibia, the Mobius vehicles (2011) in Kenya, the Turtle and Kantanka in Ghana, the Innoson vehicles (2013) in Nigeria and the Kiira (2014) in Uganda. Since 2009, Madagascar has flexed its industrial design and manufacturing muscle by producing the Karenja Mazana, a rugged 4×4 vehicle. In their factory in Fianarantsoa, 60 staff members (including six women) have produced 140 Karenjays ('stroll' in Malagasy), with all the parts (except the Peugeot engine) designed and made locally.

Laraki, a car-maker based in Casablanca (Morocco) is owned by the luxury yacht designer Abdeslam Laraki. The company designs and makes its own range of luxury high-performance cars, including the Borac, the Fulgura (which embodies a Lamborghini in look and spirit) and the formidable Epitome (with a V8, 1 750 horsepower engine) that is the only officially recognised African-made supercar. Larakis are strictly concept cars that are custom-build for each customer and were ranked among the most expensive cars in the world in 2015, priced at over US$2 million each. The legendary 'Libyan rocket' (Saroukh el-Jamahiriya), designed in 1999 for dictator Muammar Gaddafi, was the country's pride and joy. It was fast and luxurious, and was pitted against German luxury car market leaders. Unfortunately, it never went into full-time production even though it was a prime example of inventive North African design, with its designers claiming that it was one of the safest cars ever made. It had many innovative safety features, including the ability to drive long distances on flat tyres and a fully electronic safety system with airbags for all four seats.

The small but powerful Wallyscar, manufactured in La Marsa (Tunisia) since 2006, is a relatively new entry into the lucrative off-road market. The company is building a strong reputation for affordable, reliable and powerful 4×4s despite the relatively small size of its vehicles that are similar to Suzuki and Skoda off-roaders. By 2014, Wallyscar was selling over 600 units per year, predominately in Africa and the Middle East but also in Panama and Europe. The company's plans include making its colourful, sporty vehicles more environmentally friendly.

The Nyayo car project of the Kenyan government to plan and make its own cars never came to fruition. The project was initiated in 1986 by then-President Daniel arap Moi, who asked the University of Nairobi to develop the vehicle. Five prototypes were made, named Pioneer Nyayo Cars, and the Nyayo Motor Corporation was established to mass-produce them. However, due to lack of funds, the car never went into production. The Nyayo Motor Corporation was later renamed the Numerical Machining Complex Ltd and now manufactures metal parts for local industries. Mobius Motors was a more successful venture. It was founded in Kenya in 2011 by Joel Jackson to design and manufacture sports utility vehicles (SUVs) specifically for the African mass market. Mobius II was conceived by re-imagining a vehicle around the realities of the African consumer: rough road terrain, high vehicle loading and low average income levels. The result was a tough SUV sold at the price of a used sedan.

The story of the famous Turtle car in Ghana is quite remarkable. Meile Smets (an artist) and Joost van Onna (a sociologist) from Rotterdam in the Netherlands visited Ghana to learn about the informal economy there and follow the track of car part waste in this West African country. They came across the giant Suame Magazine scrapyard, where old car parts are sorted and sold by over 200 000 technicians in 12 000 informal workshops. They realised there is a big potential. Instead of repairing third-hand vehicles or waste, this dynamic car part recycling industry can build 'new' cars from scrap parts. In March 2012, they decided to find partners who could help them to develop cars made from 100 per cent recycled parts from different brands and, with assistance from the NGOs and innovation funds, they partnered up with the local NGO S.M.I.D.O (Suame Magazine Industrial Development Organisation) and and together they started S.M.A.T.I. (Suame Magazine Automotive Technical Institute). Within three months, they developed the first S.M.A.T.I. Turtle car – named after the reptile due to its robust construction.

Turtle 1 became the first Ghanaian car to be seen in the West when it toured motor shows and was gleefully displayed next to the fully electric Tesla S. The car, which has no electronics and is therefore easy to maintain and repair, is arguably the lowest technology automobile in the world and is somewhat reminiscent of the 'pyschobilly Cadillac' that was built from different parts and described in Johnny Cash's 1976 song 'One piece at a time'. The Turtle is a cross between steam punk and disnovation, and is a classic product of the post-industrial era. As it does not comply with international safety and patent standards, it cannot be sold abroad; however, it has the potential to be useful in Africa. It has even attracted the attention of the king of Ghana, Otumfuo Nana Osei Tutu II! Another Ghana-based automobile company, Kantanka (founded by Apostle Safo Kantanka), assembles passenger vehicles (mainly SUVs and pickup trucks) in Gomoa Mpota in central Ghana, but the company has reportedly pushed back the commercial release of its models pending approval from the Ghana Standards Authority.

Namibia can lay claim to the URI off-road vehicles, which are available in two models: the URI Desert Runner (a simple, reliable jeep with civil, military and police modifications) and the URI Mining Vehicle (with two low-profile cab heights of 1.4 m and 1.8 m). The original URI all-terrain truck was designed and built by Ewert Smithis, an Angora goat farmer based in Windhoek (Namibia), in 1995. He named it 'Uri' after the Khoisan Nama word for 'jump'. The vehicle was extensively tested in the Kalahari Desert for rallying as well as for agricultural, police, military and mining applications. Small-scale production started in Witvlei (Namibia) in 2001. Production was subsequently taken over by Uri International Vehicle & Equipment Marketing (UVM) in Waltloo, Pretoria (South Africa). In 2008, UVM became a subsidiary of the South African defence contractor Ivema and, since 2015, the vehicle has been produced by Uri Purposely Built Vehicle of Rustenburg, whose cofounders Andre and Raymond Squire purchased all the intellectual property in URI in 2015. The Land Cruiser 79 4.5D V8 Namib single- and double-cab off-road vehicles have also been developed and sold in Namibia.

The Nigerian Innoson Vehicle Manufacturing Company was originally commissioned by Goodluck Jonathan (the president of Nigeria from 2010 to 2015) and was founded by Innocent Chukwuma. It is based in Anamabra state and is one of Africa's most successful automobile ventures. The company built on its success as a manufacturer of reliable buses and trucks to branch out into building the ubiquitous Uzo minibus taxis, small trucks (IVM 1021A) and SUVs (IVM 6490A), as well as the ambitious Fox sedan released in 2015.

The Kiira EV SMACK, a sedan hybrid electric vehicle originally developed as a group design project by engineering students at Uganda's Makerere University for a project headed by the Massachusetts Institute of Technology, is the first African-made hybrid electronic vehicle. The five-seater sedan is powered by a rechargeable battery and has an internal combustion engine-based generator that charges the battery. The Kiira EV was launched in 2014 and went on sale in 2018. The Ugandan government has invested US$40 million in the Kiira Motors Corporation to create an affordable hybrid car for the African market and turn Uganda into the hub of the automotive industry in East Africa. With the factory in Kampala developing various sedan, off-road and urban variations, the company hopes to go into full production by 2018, employing 10 000 people and making 300 vehicles a year.

Hydrogen fuel-cell vehicles: A South African success story is the hydrogen fuel-cell forklift developed by Impala Refining Services and HySA Systems, funded by Implats and supported by the University of the Western Cape and the Department of Science and Innovation. The prototype has been in operation since October 2015 and is showing promising results. Implats is planning to power most of its mining equipment in future by using hydrogen fuel-cell technology. Fuel cells use electrochemical processes rather than combustion to produce power and are therefore attractive for underground use because they produce no noxious gases or sulphide emissions, and less heat and noise, in mines. As South Africa has about 80 per cent of the world's platinum resources, the potential to develop further platinum-based, fuel-cell-powered vehicles (including scooters and load-carrying tricycles) is enormous. Furthermore, the mining giant Anglo American has announced that it is developing the world's largest hydrogen-powered truck as part of its fleet.

Scientists at the University of the Western Cape, working with HySA Systems, have developed several hydrogen- and fuel-cell technologies that produce zero emissions and are a clean alternative to fossil fuels. Their pilot projects that are currently undergoing commercialisation include membrane electrode assemblies, fuel-cell stacks, solid-state hydrogen storage and compression, and palladium membranes. The fuel-cell industry has the potential to revolutionise the way that power is delivered to our cars, cellphones, computers, homes and offices – and South Africa is in the forefront of these developments.

Solar-powered cars: Solar-powered cars are not as yet a reality in Africa, but the internationally-recognised biennial Sasol Solar Challenge (SSC) which was launched in 2008 is raising awareness of this exciting future mode of propulsion that is ideally suited to the continent's sunny climate. The SSC – whose motto is 'Where innovation meets endurance' – races over 2 000 km from Pretoria to Stellenbosch via Sasolburg, Bloemfontein, Port Elizabeth, Mossel Bay and Cape Agulhas, and pits the world's leading solar vehicle entrepreneurs against home-grown African talent. The SSC caters for solar, hybrid and electric as well as biofuel-powered vehicles and is recognised by the International Solarcar Federation. Teams comprise engineers, technicians, scientists and logistical experts (many of them scholars and students) who showcase their talents over 80 days in this highly competitive 'brain sport'. The challenge is to clock up the most distance in the allocated time, with some teams recording distances of over 4 500 km along loops that they can repeat along the route. The current record of 4 716 km was set by the Dutch team Nuon in their car Nuna in 2016. The SSC also provides a valuable educational service by focusing public attention on cutting-edge science and technology.

Electric superbike: South African motorcycle designer Pierre Terblanche trained under the most famous motorbike designer in history, Massimo Tamburini, and has designed for Ducati and other major marques. His new, all-electric Hypertek from South Africa's carbon fibre specialists BST is a triumph of unabashed futurism with its slim, skeletal monocoque frame and carbon fibre wheels. It is built around a DHX Hawk water-cooled electric motor and has a range of about 300 km, with a 30-minute DC quick-charge capability. A fan at the back of the battery pack draws heat out of the battery and motor and deposits it onto the rear tyre, where it brings the rubber up to temperature. The addition of a clutch, even though the bike is an electric single speed, allows the rider to rev the motor and do wheelies and burnouts! As it is quiet, the bike is even fitted with a sound generator designed to let pedestrians know that you are coming (Blain 2019).

Safe travel for women: A new ride-hail app, ChaufHer (developed by Danielle Wright), was launched in Cape Town in 2019 and in Johannesburg and Durban in 2020. The app connects women drivers with women passengers, and children younger than 18 years, and allows mothers who are running a household and caring for their children to earn money in a safe environment. Women are grossly under-represented in the ride-hail industry because of safety and security concerns, but ChaufHer empowers them to work in a safe environment. The most important safety measure is the comprehensive vetting of both drivers and passengers to ensure that the service is safe for both (Swart 2019).

> Women and children are the most vulnerable members of society, but they also represent our greatest untapped resource. ChaufHer is a way to provide opportunities and freedom of movement to these groups, to help them lead their best lives.
>
> *Dannielle Wright, developer of ChaufHer* (Swart 2019)

Highest bridge in Africa: The Mtentu Bridge on the Wild Coast of South Africa, at a deck height of 220 m, has taken over from the Bloukrans Bridge (217 m) in the southern Cape as the highest bridge in Africa and the southern hemisphere. It will also be one of the longest cantilever bridges in the world, with a total length of 1.1 km and a main span of 260 m. The 580-m long Msikaba Bridge, which is on the same South African National Roads Agency (Sanral) road development, crosses the spectacular Msikaba River Gorge and is the second longest cable-stayed suspension bridge in Africa after the 680-m-long Maputo-Catembe Bridge in Mozambique (Sanral 2017).

According to Sanral Project Manager Gcobani Socenya, the construction of both bridges and their approach roads had been planned in close consultation with local chiefs, especially regarding the relocation of 140 graves and 40 houses that were displaced. The chiefs advised that the following assistance should be provided to the families of the affected graves: one sacrificial goat to appease the ancestors for the bone relocations, a cow to sacrifice and money to buy food for the reburial ceremony. Sanral agreed to these requests, provided services to exhume and rebury the human remains and travel allowances for those who conducted the reburial ceremonies, and performed the archaeological services to protect the graves (Sanral 2017).

Public transport: Two young South Africans – Tyler Hoffman of Pretoria and Unathi Chinco of Westville – developed the public transport navigation app Transit Wise, which enables commuters to discover all they need to know about options to get from A to B using public transport (including trip duration and costs). They believe that their app will encourage more people to use public transport, and use it efficiently, and will reduce congestion on the roads. The pair met at Pretoria Boys' High School, where their passion for programming and apps resulted in them winning an app development competition. Both now work in Silicon Valley in California (USA), where they hope to make an impact in the highly competitive app design industry (Anon 2016).

Rail transport: The CSIR in South Africa has developed an ultrasonic broken rail detector (UBRD) to detect breaks in train rails and remotely communicate this information to rail engineers. UBRDs operate off solar power and scan lengths of rail up to 1 000-m long every three minutes. They are made from robust, rust-free components and can be installed without interrupting rail traffic. The CSIR's next generation UBRD, now under trial, will be able to detect defects in railway lines before they become cracks or rail-breaks and then direct maintenance crews to high-risk areas before a breakage occurs. This will eliminate the risk of derailment and make rail travel even safer (www.railsonic.co.za).

The South African parastatal Transnet has developed its first home-grown locomotive, the Trans-Africa Locomotive (TAL), at a cost of R350 million, specifically for use in Africa. It is the first locomotive to be designed, engineered and manufactured on the continent. The diesel-powered locomotive is suitable for use on branch lines and for shunting, and is also able to travel on old rail tracks originally designed to carry light axle loads (such as the Cape Gauge system) and therefore offers a cost-effective solution for the majority of Africa's railway lines that are currently unused. At the launch of the TAL at Koedoespoort (east of Pretoria) in 2017, Transnet's CEO Siyabonga Gama stated (Van Wyngaardt 2017):

> We are very happy that this is a proudly African locomotive. Made in Africa for African conditions. The locomotive is evidence of the strides we are making in transforming Transnet Engineering into an original-equipment manufacturer

for locomotives, a move designed to restore our position as a catalyst for African innovation, industrialisation and, critically, intra-African trade.

Transnet plans to sell the TALs to other African countries to diversify its income stream and promote rail traffic on the continent. Transnet Engineering also makes and sells railway bogies, passenger coaches and goods wagons suited to African conditions (Van Wyngaardt 2017).

Malawi's waterway: Malawi is a landlocked country even though it has a huge lake in its interior that is large enough for sea-going ships. Its landlocked status places a huge burden on its economy because the cost of imports and exports is high, largely due to the poor quality of its road and rail links to coastal ports in South Africa (Durban), Mozambique (Beira and Nacala) and Tanzania (Dar es Salaam). The search for a solution to this problem has dominated Malawi's foreign policy since its independence in 1964. An exciting option that is now pursued is a waterway to the Indian Ocean through northern Mozambique. Called the Shire-Zambezi Waterway, the project was first mooted in 1891 and then revived during the presidency of Bingu wa Mutharika (2004–12), who claimed that a route from Nsangje in Malawi to Chinde in Mozambique would drastically reduce Malawi's transport costs and facilitate the export of tobacco, tea and sugar and the import of oil, consumer goods and fertilisers.

However, diplomatic blunders on the part of Malawi and intransigence from Mozambique have resulted in little progress being made on the project (Kayuni, Banik & Chunga 2019), which had the support of Malawi's former President Peter Mutharika. The signing of an agreement that allows the Electricity Supply Corporation of Malawi to purchase 200 MW of electricity from Mozambique starting in 2022 suggests that the relationship between the two countries has warmed, which may bode well for the ambitious waterway project.

Tesla self-driving vehicles: On 9 July 2020, South African born Elon Musk announced that Tesla was 'very close' to achieving Level 5 autonomous driving technology (i.e. essentially complete autonomy). Car-makers and technology companies (including Alphabet Inc., Waymo and Uber Technologies) are investing billions in the autonomous driving industry but acknowledge that it will take time, even after the technology has been perfected, for the public to fully trust self-drive vehicles. In June 2020, Tesla sold over 15 000 China-made Model 3 sedans with an autopilot driver-assistance system and later became the highest valued car-maker in the world after its shares surged to record highs and its market capitalisation overtook that of Toyota (Goh & Sun 2020).

Emerging innovators: Sibongile Sambo's pioneering spirit led to the launch of Africa's first 100 per cent black female-owned aviation services company, SRS

Aviation, in 2014. Although she initially started the company (which is based at Lanseria International Airport near Johannesburg, South Africa) as a provider of private aviation services to local and international clients, the company has since expanded its offering to include maintenance, sales and fleet management services to private jet owners. Sambo plans to increase the number of women in aviation through mentorship.

Discussion: Transporting goods, people and now data has always been one of the major drivers of innovation, and the future will be no different. Electric and solar-powered cars, self-driving vehicles, electric jet helicopter taxis, increasingly sophisticated and secure ride-hailing services, bullet trains, and even hyperloops and trips to Mars are all on the cards. A Chinese car-maker now offers to deliver to your door a complete 'four-wheel drive environmental protection small electric car' for you to unpack at your leisure! The German start-up Lilium is even developing a five-seater electric jet that can take off and land vertically, travel at 300 km/h and provide call-up lifts in even the most densely built-up areas (Zorthian 2017).

In 2017, Dr Dieter Zetsche (the chairperson of the Management Board of Daimler AG in Germany) pointed out that, in future, the main competitors of traditional car-makers will not be other car-makers but technology companies such as Tesla, Google, Apple and Amazon. He also predicted that most traditional car-makers will go bankrupt if they adhered to their tried-and-tested formula of making better cars, as they will be out-competed by technology companies that take a more revolutionary approach and build better computers on wheels. The outcome will be products that view the future of mobility from entirely new perspectives and produce exceptional problem-solving solutions. But before African countries can introduce these high-tech innovations, they will first need to fix their basic transport infrastructure. Many international companies complain that it is cheaper and easier to import goods from Europe into Africa than to import the same goods from one African country to another due to breakdowns in supply chain linkages and inefficient and poorly maintained transport infrastructure (Fatunla 2018).

16

Trade, industry, mining, biotechnology and nanotechnology

'How do we thrive in a fast-changing world in which we are torn between the rigid rules of industrial society and the flexible creativity that the future demands, between the certainty and safety of yesterday and the freedom and anxiety of tomorrow?'

South African futurist John Sanei (Sanei 2019a: blurb)

Introduction: Trade has been carried out in Africa for aeons. Egyptian hieroglyphic records show that there was active trade between Egypt, Persia and Ethiopia over 5 000 years ago, and it probably started millennia before that. The pharaohs obtained myrrh, gold, resins and gums from Ethiopia at least 5 000 years ago, and Ethiopia has been exporting ivory to India since time immemorial (Hancock et al. 1997). Minerals and gems have, to a large extent, shaped the course of civilisation and even history. Bronze and iron were so crucial to the development of ancient societies that we have retrospectively named whole eras after them. During the Bronze Age (3300–1200 BC), the production of harder and more durable bronze through the smelting of copper and addition of tin or arsenic gave civilisations a technological advantage. In the Iron Age (1200–600 BC), iron and steel were introduced, and people were able to make tools and weapons superior to their bronze equivalents, which made farming – and conquering an enemy – more efficient.

Africa is extremely rich in minerals, some common and others rare, and many of them would have been known to and used by indigenous people. The first documented discovery of diamonds in South Africa (near Hopetown) was by Erasmus Jacobs (a 15-year-old farmer's son) in 1866, but it was the finding of an 83.5-carat rough diamond by a Griqua herdsman in 1869 (later named The Star of South Africa) that triggered the first diamond rush in Africa. Credit for the first documented discovery of gold on the Witwatersrand Reef in South Africa goes to Jan Gerrit Bantjes, in 1886. But who discovered the world's rarest gemstone, Tanzanite, a blue and violet variety of the mineral zoisite that is only known to be from a single location near the Mirerani Hills in Tanzania?

Manuel de Souza, an Indian tailor and part-time gold prospector living in Arusha (Tanzania), found transparent crystals of the gem near Mirerani and assumed that the mineral was either olivine or dumortierite. The crystals were shown to John Saul, a Nairobi-based consulting geologist and gemstone wholesaler, who was mining aquamarine around Mount Kenya. Saul (who later discovered the famous ruby deposits in the Tsavo area of Kenya) eliminated dumortierite and cordierite as possibilities and sent samples to his father, Hyman Saul, the vice-president of Saks Fifth Avenue in New York. He took the samples to the Gemmological Institute of America, which correctly identified the new gem as a variety of the mineral zoisite. Subsequent research revealed that a Tanzanian government geologist in Dodoma, Ian McCloud, had previously identified the gem correctly. In 1968, the gemstone was named 'Tanzanite' by Tiffany & Co. whose marketing campaign advertised that it could only be found in two places: in Tanzania and at Tiffany's.

In May 2020, a 52-year-old small-scale miner in Tanzania, Saniniu Laize, became an overnight millionaire after selling two rough Tanzanite stones with a combined weight of 15 kg for US$3.4 million – the biggest ever find. In July 2020, he struck 'gold' again when he sold another Tanzanite gem weighing 6.3 kg for US$2 million. Mr Laizer, who has four wives and has fathered more than 30 children, told the BBC that he planned to invest in his community in the Simanjiro district in northern Manyara. 'I want to build a shopping mall and a school … I am not educated, but I like things run in a professional way. So, I would like my children to run the business professionally' (BBC 2020). He also said that the windfall would not change his lifestyle and that he planned to continue looking after his 2 000 cows.

Tunnel vision: James Greathead (a farmer's son from South Africa) studied at St Andrew's College in Makhanda and then at Diocesan College in Cape Town before emigrating to England in 1859, where he became involved in the development of traction and electric railways. A highly innovative engineer, he invented the Greathead shield for underground tunnelling (1869) and the Greathead grouting machine (1891), which were used to develop the underground railways in London and Liverpool. He was also the resident engineer for the first electric underground and overhead railways in the world. His contributions to railway development and tunnelling were so immense that he is regarded as the Father of the London Underground and is commemorated by a bronze statue, unveiled by the lord mayor in 1994, in the City of London (Bruton 2017a).

South African companies are at the forefront of tunnelling technology. In 2018, Master Drilling launched a new 'disruptive' mine tunnel borer that requires no blasting and continuously excavates, supports and removes waste rock in tunnels with diameters up to 5.5 m wide at rates far exceeding those of conventional tunnel boring methods. The company developed the concept in collaboration with Italy's Seli Technologies, which has over 50 years of tunnelling experience (Creamer 2018).

Africa Free Trade Area: In January 2012, agreement was reached among African nations at an AU meeting to establish what is known as the African Continental Free Trade Area (AfCFTA), a trade agreement aimed at creating a single continental market whereby goods and services could move freely. The deal was forecast to generate a GDP of US$2.5 trillion and would encompass a market of 1.27 billion people (Mathe 2020a). The AfCFTA would become the largest free trade area in the world by number of countries since the formation of the World Trade Organisation in 1995. The free movement of goods will be achieved by promoting intra-Africa trade and lowering import duties. The agreement was scheduled to come into effect in July 2020, but the Covid-19 pandemic made it impossible to allow the free movement of people and goods at that time.

At the signing of the AfCFTA in 2018, the then-chair of the AU, Rwandan President Paul Kagame, said that less than 20 per cent of Africa's trade was internal, whereas the level of internal trade in the world's richest trading blocs was three to four times higher. Intra-Africa exports in 2017 were 16.6 per cent of total exports, compared to 68.1 per cent in Europe, 59.4 per cent in Asia and 55 per cent in the Americas. Only five African countries (Eswatini, The Gambia, Togo, Uganda and Zimbabwe) had more exports to African countries than to the rest of the world. In 2019, 27 per cent of South Africa's global exports were to African countries and 12 per cent of its global imports were from African countries. Other large intra-African exporters were Nigeria (9 per cent of all exports) and Egypt (6 per cent). Dr Sithembile Mbete, an expert on African trade at the University of Pretoria in South Africa, has pointed out that the rest of the world is gearing up to trade with Africa and that African countries need to respond quickly to the establishment of the AfCFTA and take control of the process in order to avoid the threat of 'reverse colonisation' (Mathe 2020a). The

AfCFTA agreement will not solve all Africa's financial problems, but it will be a step in the right direction once implemented.

Female mining mogul: Bridgette Motsepe Radebe is a major disruptor-for-good in the mining industry in South Africa. She is the leading black female mining entrepreneur in the country and the founder of Mmaku Mining. Radebe is against the 'capitalist mining model' whereby most of the natural resources in South Africa passed from the white minority to corporate bodies at the end of apartheid. As the president of the South African Mining Development Association, she supports the nationalisation of all mining operations, a state buy-out of mines with dwindling profitability that only exist in the name of black empowerment, and a cooperation agreement between the public and private sectors over the running of South Africa's mines. She argues that this would allow the mining industry, which employs over 170 000 people, to progress for the benefit of all. In 2019, she was appointed as a member of the BRICS Business Council set up to strengthen business, trade and investment between the five BRICS countries: Brazil, Russia, India, China and South Africa (Ayene et al. 2020).

***Hakuna kucheza* in the mining world**: *Hakuna kucheza* means 'no playing around' in Swahili and is the guiding compass for the South African company GeoPoint Africa, a mining conglomerate established by Topman Ngoyama in 2009. His vision is to develop a large, multicommodity portfolio through strategic partnerships throughout Africa and to position GeoPoint Africa as a world-class mining powerhouse (Ngonyama 2020).

> Mining has been the economic backbone of [the] South African economy and the main magnet of foreign investment on the African continent for the past few decades. With slow economic growth, coupled with increased government policy and legislative uncertainties, we have seen a serious decline in Exploration Capital which resulted in few discoveries and [a] shortage of new projects.
>
> *Topman Ngonyama, CEO of GeoPoint Africa* (Ngonyama 2020)

According to Ngonyama, raising capital for new projects has not been their only problem, as uncertainty about changing government regulations and legislation in other African countries has also posed severe challenges. For example, GeoPoint Africa developed a successful copper/lead/zinc project in Tanzania and started to produce raw copper to test the market – only to be confronted by an unexpected government ban on the export of raw materials in favour of beneficiating the product locally, which caused serious financial strain on the company's balance sheet. In response to this setback, one of the company's long-term goals is to set up beneficiation plants that add value to its mined products. Ngonyama points out that many of the projects of the 4IR will not be possible without the minerals that GeoPoint Africa (and other companies) extract from the ground, including graphite, tin, tantalite, tungsten, copper, various rare earth elements, lead and calcined anthracite coal. The company has already carried out mining projects in South Africa, Zimbabwe, the DRC, Mozambique and the Congo, and has new projects in operation in Tanzania and Uganda (Ngonyama 2020).

Hippo slurry pump: Hazelton Pumps in Gauteng (South Africa) exports about 50 per cent of its Hippo submersible slurry pumps and recently increased its range of pumps to include a flameproof, high-voltage, high-volume submersible pump that is the first of its kind in the world. Their pumps have 98 per cent local content, as all castings and manufactured components are made in-house with only bearings and mechanical seals being imported (www.hazeltonpumps.co.za).

Miner's safety system: Technologists at the Wits Commercial Enterprise at Wits University in Johannesburg have invented a system that improves miners' safety by monitoring their location through transmissions of light from LED-based miners' headlamps. These lamps have been modified to emit a unique and imperceptible signature that is monitored by sensors installed every few metres inside a mine. The system facilitates continuous location tracking of miners, which improves rescue efforts in the event of cave-ins and ensures that areas are clear of personnel before blasting. The information transmission systems currently used in mines are not cost effective, whereas the visible light communication and power line communication systems used in the new invention have unique, inexpensive, scalable and compatible capabilities that improve miners' safety. The system is being extended to transmit and log sensor data (such as atmospheric gas and dust content, air temperature and radiation levels) as well as vital signs of the wearer (such as heart rate). This data can be used to trigger emergency evacuations should the need arise.

Pratley Putty: The world-renowned South African industrial products and adhesive manufacturer Pratley Putty has launched yet another unique product – a corrosion-resistant, flameproof compression cable gland that can be installed without specialised tools. The new enviro compression gland, which is certified by the International Electrotechnical Commission, is designed to be used in the most extreme environments, including mines, underwater to 350 m and at temperatures from –20°C to 95°C. The gland is made from high-tensile brass encapsulated in a tough engineering plastic and Pratley's Taper-Tech internal seal design is used, which ensures safe installation even in corrosive environments. According to Pratley's website, 'The primary function of any cable gland is to anchor the cable securely to an electrical apparatus … Strict international testing standards … specify pernicious performance requirements for the certification of such glands' (http://www.pratleyelectrical.com/news/tag/Taper-Tech/).

African polymathic inventor: Mulalo Doyoyo was born in Limpopo province (South Africa) in 1970, and knew poverty and hardship as a youth. He is now a respected researcher in applied mechanics, ultralight materials, green building and renewable energy, and has worked at the interface between academia and industry in the USA and South Africa. He is passionate about using modern technology to fight poverty and develop underprivileged youth. In 2008, he established Retecza, a renewable resource-driven technology concept centre that promotes cross-disciplinary industrial research. His inventions include Cenocell, a 'cementless' concrete produced from fly ash, an industrial waste product; the *Ahifambeni* ('Let's go') hydrogen-powered motorbike; Amoriguard, a non-volatile organic paint coating made from industrial and mining waste; two environmentally friendly chemical binders, Solunexz and Glunexz, that reduce pollution by coal dust; and the Ecocast brick-making machine for making acid bricks. He has also developed a solar-powered flushing toilet that uses nanofiltration and anaerobic digestion, like a miniature waste-treatment plant, that offers a practical solution in places where flush sanitation and water are scarce.

Ngwenya glass: In 1979. the glassblowing factory Ngwenya Glass was set up in Eswatini as a Swedish Aid Project. The Swedes imported the machinery and equipment, built the original factory, and employed and trained Swazis in the ancient art of glassblowing, with two of the most talented glassblowers furthering their education at the world famous Kosta Boda glassworks in Sweden. In 1983, the factory ceased production but was bought by Chas Prettejohn (a marine engineer) and his parents, Alix and Richard – all keen collectors of Swazi glass in South Africa. They restarted production in June 1987 with four former employees, including master glassblower Sibusiso Mhlanga, who now tutors new apprentices (www.ngwenyaglass.co.sz/history). Ngwenya Glass' products (which include tableware, vases, jugs and ornamental African animals) are all handmade from 100 per cent recycled glass, mostly softdrink bottles collected in Eswatini. In 1989, the company launched a successful wildlife conservation fund by donating a percentage of its profits to Mkhaya Game Reserve, a refuge for endangered species in the Eswatini lowveld. Ngwenya Glass is proof that business success and commitment to protecting the environment can be a winning combination.

Oil absorbents: Chris Cooper of Luhlaza Waste Management in South Africa has developed an innovative petrochemical absorbent, Invader Zorb, that is derived from sustainably harvested plants obtained from bush-clearing and wetland rehabilitation programmes. The raw material is treated with an ascomycete fungus which removes cellulose and leaves behind a fibrous, lignin-based material that is strongly hydrophobic and can absorb substantial amounts of petroleum-based pollutants from solid surfaces or dams and rivers. Invader Zorb's absorbent properties are comparable to those of peat moss (the current global market leader), but its manufacture has the additional benefit of creating employment in rural areas and using up organic waste (Ward 2005).

Luhlaza Waste Management also produces EcoFix, a novel and highly effective bioremediation agent that is also derived from plants harvested during bush-clearing operations. EcoFix can degrade a wide range of organic molecules, including highly toxic polyaromatic hydrocarbons and alkanes, that are present in crude oil and its derivatives. EcoFix also contains a range of white-rot fungi which produce enzymes that break down hydrocarbons and ensure that clean soils are produced in a short time (Rhodes 2014).

Graphene: Mohammed Khenfouch, originally from Morocco, is the founder-chairperson of the Africa Graphene Centre at the University of South Africa and one of the leading authorities on the use of graphene materials for opto-electronic applications. His research focuses on the synthesis of novel materials for applications in energy conversion, efficiency and storage, for example in solar cells, LEDs and gas sensors. This work, and that of his colleagues, is of enormous importance in Africa because graphene is regarded as *the* 'wonder material' of the 4IR that will have far-reaching implications for the future of physics, engineering and bio-engineering. It could launch technologies that will change the course of the 21st century.

Although graphite has been used for over 12 000 years, graphene was only discovered at the University of Manchester in 2004. It comprises a single, very thin layer of graphite in the form of a two-dimensional, atomic-scale hexagonal lattice. Graphene is an allotype of carbon and therefore has the same atoms as carbon but they are arranged in a different way, which gives it different properties. It is extremely lightweight and flexible yet 100 to 300 times stronger than steel. Graphene is also a good conductor of heat and electricity, is magnetic, shows uniform absorption of light across the visible and near-infrared spectrum, and is impermeable to most liquids and gases. This unique combination of properties means that it will play a key role in the development of a unique range of products, processes and industries (Fourie 2019). Graphene may also become a battlefield in international power play in future and Africa needs to stake its claim. Like additive engineering, it will hopefully be a major contributor to first-tier technological development on the continent.

Boron nanotube applications: The electronics industry is extremely dynamic because technologies, materials and business models are continuously being restructured to meet rapidly changing evolving consumer demands. Durability, size and performance are major concerns in the electronics industry and manufacturers are focusing on developing advanced materials to improve the performance and life of electronic devices. Carbon nanotubes (CNTs) and graphene have been identified as the candidates with the most potential to replace traditional materials, and printable carbon nanotube inks have already hit the market. There are, however, many challenges in working with carbon nanomaterials and with the large-scale production of CNT materials for the consumer electronics industry. There is also growing interest in boron nanomaterials, as research has shown that they have unique properties that also make them suitable for use in advanced electronics. Unlike CNTs, boron nanomaterials have a much wider band gap (~5.5 eV), exhibit higher resistance to oxidation and show greater thermal stability. Like carbon, boron can be formed into nanotubes, nanosheets (borophene) and spheres.

Boron nanostructures are expected to serve as the key components for the next generation of nanodevices, such as high-temperature semiconductors, field effect transistors, field emission devices and superconductors. As these materials are not available anywhere on the market due to the difficulty of producing them, researchers at the University of KwaZulu-Natal in South Africa have developed a method for producing boron nanotubes that will soon reach the market.

Sun E-Nanobiosensor Platform Technology: The need for the accurate, rapid and sensitive detection of biomolecules is becoming more important in several industries, including clinical diagnostics and screening to assess microbial contamination in water and plant diseases. There are many diagnostic and screening solutions that are already available, but they are all based on expensive technologies, with varying degrees of specificity and sensitivity. In response to this opportunity, researchers at Stellenbosch University in South Africa have developed a versatile, extremely sensitive and specific platform nanobiosensor technology that can be adapted to a range of rapid diagnostic and screening applications.

The Sun E-Nanobiosensor Platform Technology developed by Innovus at the university can be constructed from relatively inexpensive materials, which makes it ideal for incorporation in single-use or disposable sensing, screening or diagnostic tests and in point-of-care devices that require minimal sample input and are easy to use. This allows users who are not skilled in molecular diagnostics to run tests after receiving basic training. The nanobiosensors can also be connected to and exchange data with the IoT, which allows the end-user (such as a nurse) to use the device in rural areas without the need to transport samples to a central pathology laboratory and then wait for the results. The consulting physician would have immediate remote access to the test results and, in consultation with the nurse, could make a quick decision on a course of treatment. In addition, the IoT connectivity of the nanobiosensors would enhance infectious disease surveillance and mapping capabilities. Researchers at Stellenbosch University are working on several different applications of this platform technology, including diagnostic applications for HIV and TB, screening for specific inflammatory biomolecules in human blood, assessment of human platelet function, detection of bacterial contaminants in water and post-harvest disease diagnostics in pomegranates.

Baobab laboratory information management systems (Lims): Lims are an essential component of modern biobanks and laboratories, but many African countries cannot afford them. Baobab Lims, which has been developed at the University of the Western Cape in South Africa, is an Africa-led innovation developed as part of the B3Africa consortium, an Africa–Europe infrastructure project. It is an affordable sample and laboratory management tool for repositories of biospecimens (biobanking), and was developed using standard operating procedures and workflows from a functional human biobank. The Baobab Lims software is an open-source tool with a code that is available free to users, including biobanks, research laboratories, small industries, and equipment manufacturers and suppliers. The Baobab Lims system is continuously updated with additional features and solutions, and can be customised to individual needs. The use of this biobank facility increases laboratory efficiency and effectiveness, enhances quality control and promotes good laboratory practice.

Emerging innovators: Ishmael Msiza, originally from the village of Kameelrivier in Mpumalanga (South Africa), is one of a new generation of researchers in the Biometrics Research Group of the Modelling and Digital Science Unit of the CSIR in South Africa. He has invented a new fingerprint classification and recognition technique, the Structural Fingerprint Classifier, that can quickly and accurately recognise a fingerprint with only partial information – a world first. The old fingerprint recognition system used templates on a huge database that had to be searched each time a fingerprint was identified, whereas the new technique breaks the fingerprint down into parts that can be identified far more quickly (Msiza, Mistry & Nelwamondo 2011).

Discussion: At the end of the Great Depression in the early 1930s, the Italian philosopher Antonio Gramsci lamented that the Old World was dying but the New World (developing world) was not ready to be born – the same may apply after the Covid-19 crisis. Africa came off second best during the Cold War when its importance was relegated to which side it chose in the titanic battle between the capitalist West and the socialist East. Once again, there is a danger that the continent will be marginalised during and after the Covid-19 pandemic. It is also widely predicted that the complex links that maintained globalisation – the global value chains – will become shorter and simpler post-Covid-19, which may also lead to the exclusion of Africa. The European Commission's President Ursula von der Leyen and France's President Emmanuel Macron have both called for the shortening of global supply chains and more inward-looking, European 'economic sovereignty' approaches, which may also reduce trade with Africa. However, China (the main global trader) has prioritised investment in Africa, even if it is based on their self-interest.

Africa will need to be better integrated as a continent if it is to erase the historical barriers currently blocking the easy transfer of goods, services and people within the continent. At present, African countries trade mostly with non-African economies, partly due to the high cost of intra-African trade. The implementation of the AfCFTA was supposed to inject energy into the continent's pan-African ideals, but it has so far failed to do so.

At the same time, African entrepreneurs need to take advantage of new opportunities. Additive engineering (aka digital or 3D printing) offers enormous opportunities for small and medium enterprises (SMEs) in Africa to be competitive in the lucrative space of making conceptual models, functional prototypes, manufacturing tools and production parts, as well as the reverse engineering of complex designs. Additive engineering can now be carried out using thermoplastics, nylon, resin and many other substrates, and the technology offers the advantages of reduced development time and cost, minimal material waste and improved delivery timeframes. It is the perfect 4IR technology for ambitious African start-ups to embrace.

17

Commerce, e-commerce and fintech

'Innovation is the central issue in economic prosperity.'

Michael Porter, economist at Harvard Business School (Jaruzelski 2015)

Introduction: In the spirit of Absa bank's Africanacity ('the distinctly African ability to always find a way to get things done') project in South Africa, Africa has made enormous strides in the development of fintech but it still falls far behind developed continents in this sphere. Only 33 per cent of sub-Saharan Africa's adult population has a bank account (Norbrook 2020b). Furthermore, some African countries still resort to barter or payments-in-kind to settle their debts. In early 2020, in a win-win situation, Chad settled a US$100-million debt with oil-rich Angola by bartering 75 000 head of cattle over a 10-year period. Chad borrowed the money in 2017 but has struggled to pay back the cash, and Angola wanted to replenish its depleted livestock numbers. By late March 2020, 1 000 cattle had already arrived in Luanda by ship (BBC 2020a).

In 2019, Nigeria – for the first time in half a decade – took over as the top venture-capital investment destination in Africa, topping South Africa and Kenya for the biggest share of US$3.1 billion in funding, according to WeeTracker (2020). In response to this development, Victor Asemota (an investor in the Nigerian technology space) stated, 'The important thing to notice is how many seed investments were made. An existing investor will always assist with follow-on funding for selfish reasons. New investments are the canary in the shaft' (Norbrook 2020b). Asemota, the founder of SwiftaCorp (a pioneering software and technology services group with operations in 14 African countries) is also the co-founder of Mfisa (a mobile financial services accelerator). Like Semota, Sunday Folayan (the co-founder and CEO of General Data Engineering Services) believes that fintech is the focus of investor attention at the moment because this is where Nigeria, and many other African countries, will experience most technological growth thanks to innovation and the huge markets that they offer. These and other factors have attracted successful start-ups such as OPay and PalmPay to Nigeria.

Asian investors (such as the Tolaram Group and the HQ Financial Group) and governments (such as the UK government) see big opportunities in the African fintech space. Tolaram is planning to invest in Nigeria, Egypt, South Africa and Ghana. In February 2020, the HQ Financial Group provided US$10 million in debt financing to Aella, a fintech start-up seeking to serve the underbanked population in West Africa and other emerging markets, according to its CEO Akin Jones (Jackson 2020). In January 2020, Flutterware's CEO Olugbenga Agboola announced that his company had raised US$35 million from Worldpay IS to grow its market share, but the biggest investments were expected to come from the South African telecommunications company MTN, Nigeria's largest operator with more than 60 million customers (Olurounbi 2020).

MTN launched its mobile payment platform, MoMo Agent (similar to Safaricom's M-Pesa), in 2020 and it is taking off. Airtel Nigeria (Nigeria's third largest operator) is expected to follow suit and 9Mobile and Globacom (the country's two other major operators) have also been issued licences to become payment-service banks. According to Lagos-based analyst Chukwuemeka Monyei, the development of fintech in Nigeria and elsewhere will lead to the increased adoption of blockchain, cryptocurrencies and AI-powered fintech products – all of which will boost the energy, healthcare and education sectors (Olurounbi 2020). The Central Bank of Nigeria explained its decision to license mobile operators as rooted in its policy goal of boosting financial inclusion, hoping to replicate the experience of countries like Kenya and India. In Kenya mobile money services have helped to lift about 194 000 households out of extreme poverty, according to a World Bank report (Logan 2017).

Blockchain and Bitcoin: Blockchain is widely regarded as an incorruptible ledger of economic and other transactions that can be programmed to record virtually anything of value. Many businesses see it as the most effective way to deal directly with end-users – from e-tail outlets to government departments – in a secure, efficient and cost-effective way. Because information on it is traceable and uneditable, blockchain provides a secure and transparent way to manage tender and procurement processes. Given how trust in governments in Africa is at an all-time low, blockchain can also be used to replace easily altered paper records and databases (Thornton 2018).

John Lombela (the MD of the African investment and technology company Crytovecs Capital) believes, together with many others, that Blockchain technology holds the key to building an inclusive global digital economy that benefits all its stakeholders, but he has concerns about its introduction into Africa. Doing business in Africa is a two-edged sword: despite steady economic growth, there is still an inability to leverage this growth into sustainable development. However, Lombela argues that Blockchain does provide foundational technology that promotes trust, privacy, transparency and stability – all qualities needed in the African economic system – and that its immutable nature and network integrity are major factors that will make it successful on the continent (Lombela 2020).

Blockchain technology allows everyone to participate as an investor, and will encourage African countries to re-invent their financial services and democratise the way that banking and investment are done. It will also stimulate African economies by facilitating the flow of money into projects that are most urgently in need of investment. Although access to and knowledge of technology may be a stumbling block, many African countries (such as Kenya, Nigeria and South Africa) have shown an ability not only to adopt but also to adapt the latest technology to their own needs.

> So, basically, when it comes to investing in technology, Africa needs to realise that the greatest wealth is created by being an early investor in innovation. Making such investment requires believing in something before most people understand it.
>
> *John Lombela, MD of Cryptovecs Capital* (Lombela 2020)

In all three versions of Blockchain (public, private and hybrid), computers (called 'nodes') which are connected to the network validate information that is transmitted before it is permanently recorded. Once this information has been written into the digital 'ledger', it can never be reversed, which introduces an important element of self-governance that potentially eliminates most forms of corruption. But to participate in Blockchain, people need to be educated, which is why Lombela established the Blockchain Incubation Hub in Johannesburg in April 2020 (Lombela 2020).

Bitcoin was launched in 2008 by its founder under the pseudonym Satoshi Nakamoto and it is now just one of over 1 500 different digital currencies that are available. It is different from normal currencies, as it is decentralised and is not backed – or issued – by any country's central bank. In April 2018, the cryptocurrency exchange platform Golix installed the first Bitcoin and Litecoin ATMs in Harare (Zimbabwe), following growing interest in cryptocurrencies there and in other African countries. At the time, Zimbabwe became the second African state (after Djibouti) to deploy crypto-ATMs and did so to stabilise its currency. In 2009, Zimbabwe's currency was destroyed when the inflation rate reached a world record 79.6 billion per cent and the country had to switch to using US dollars. It is hoped that Bitcoin and Litecoin will help to stabilise the economy by providing a trustful means of currency exchange (*Cape Times* 2018). The jury is still out on the future of Bitcoin, but most economists agree that Blockchain is here to stay.

African Bank: Basani Malukele is the first black African woman to be the CEO of a bank in South Africa (African Bank) and, at the age of 42 years, she is already making waves in the industry. Malukele has changed the bank's policy to ensure that its customers, from all levels of society, obtain the best interest rates. She has also established an NGO, Get Me To Graduation NPC, to fund the subsistence needs of university students and launched the MyWORLD scheme that ensures that customers can carry out all their financial transactions (not only loans) with the bank. In 2019, African Bank recorded its third successive year of increasing profit despite less-than-ideal economic conditions (Ayene et al. 2020).

Access Bank and Nollywood: The banking world in Nigeria was shocked in 2019 when Access Bank bought out Diamond Bank to become the largest bank in Africa's leading economy. When Herbert Wigwe and his business partner, Aigboje Aig-Imoukhuede, acquired Access Bank in 2002, it was ranked 65th out of 89 banks in Nigeria. With the acquisition of Diamond Bank, it is now the largest bank and has about 30 million customers. Wigwe, the CEO, admits that he is 'still learning how to navigate such a massive institution'. He is a keen philanthropist and movie buff and, under the aegis of Access Bank, has created the W Initiative to accelerate the development of female entrepreneurs as well as established the Nollywood Fund to help Nollywood stakeholders with the production and distribution of locally made films (Ayene et al. 2020).

Cellulant – Nigeria's goddess of the harvest: Bolaji Akinboro and Ken Njoroge sketched their first ideas for the fintech start-up Cellulant on a napkin, but the idea soon grew beyond their wildest dreams. Their company has helped to build payment architectures that everyone, from smallholder farmers to government agencies, can use and trust. Cellulant now works in 33 African countries, with 94 per cent of the customer base never having used banking before. Their most important intervention in Nigeria was arguably to disrupt the corrupt networks that interfered with the distribution of agricultural subsidies. Their Growth Enhancement Support Scheme created an e-wallet system that gives the subsidy directly to the farmer via his or her cellphone so that it can be cashed out without any intermediaries. They are now developing an app that will connect farmers directly to banks, which will help to fix the financing gap in African farming. Cellulant has even been labelled the 'Nigerian Demeter', after the mythical Greek goddess of the harvest who presided over rich crop yields. Their vision now extends beyond agriculture, as they are also set on disrupting the energy sector (Ayene et al. 2020).

Launch of the Eco: In December 2019, the imminent end of the CFA franc and its replacement by the Eco in West Africa was announced in Abidjan (Côte d'Ivoire) following intensive lobbying by several West African financial experts, many of whom believe that the CFA franc is incompatible with the economic emergence of Francophone Africa. Economist Kako Nubukpo of Togo is an outspoken proponent of the adoption of the Eco and the removal of French influence in the form of the CFA franc from the currency of the eight member states of the West African Economic and Monetary Union. In his book *Sortir l'Afrique de la Servitude Monetaire* ('Get Africa out of Monetary Bondage') and in his articles in '*Jeune Afrique*' ('Young Africa') and '*Le Monde*' ('The World'), Nubukpo (2020) complains about the high interest rates the European Central Bank charge and calls the CFA franc 'monetary slavery'.

Nubukpo also accused the Central Bank of West African States of refusing to debate the issue. This led to the International Organisation of La Francophonie suspending him from his post as its economic and digital director, but he and others eventually won the battle. At the end of 2019, France announced the removal of the CFA franc and the introduction of a new currency in West Africa. The eco will remain pegged to the euro, but West African countries will no longer have to store 50 per cent of their foreign reserves with France (Ayene et al. 2020). The ultimate objective of the Eco lobby is to combine all monetary zones in the Economic Community of West African States.

Interswitch becomes a unicorn: Mitchell Elegbe is an optimist who thinks big. He took on the formidable challenge of weaning Nigerians off cash when he established Interswitch in Lagos in 2002. His company pioneered the development of infrastructure to digitise the mainly paper-ledger and cash-based economy in Nigeria, and today its technology processes over 500 million transactions per month. Interswitch's Verve payment card is the largest domestic debit card scheme in Africa. Its use has extended beyond Nigeria since August 2019 through subsidiaries such as Paynet Kenya Limited, Vanso International Corporation and Interswitch East Africa Limited. Elegbe always envisaged that his technology would eventually facilitate the electronic circulation of money throughout Africa. Visa agreed with him, so the international company bought a minority share stake in Interswitch that raised its valuation to over US$1 billion, thereby qualifying it as Africa's first technology 'unicorn' (Ayene et al. 2020). Venture capitalist Aileen Lee coined the term 'unicorn' in 2013 to earmark any privately held start-up company whose valuation exceeds US$1 billion. She chose the mythical animal to represent the statistical rarity of such successful ventures.

US companies have been active in Africa for a long time: General Electric for more than a century, Coca-Cola since 1928 and Johnson & Johnson since 1932. Facebook's Mark Zuckerberg, after meetings with local techpreneurs, jogged through the streets of Lagos in 2016 and Twitter's CEO Jack Dorsey is planning to set up office on the continent soon following an eye-opening tour from Accra to Addis Ababa in 2019. Former US President Donald Trump's Prosper Africa initiative (which sought to bring under one umbrella the more than 15 US government agencies working on African trade, finance and policy) made some progress but senior-level engagement on the continent by Trump administration officials had been virtually zero (Norbrook 2020b).

Microlending in Kenya: Hilda Moraa (a Kenyan fintech entrepreneur) is the founder and CEO of Pezesha, a peer-to-peer microlending marketplace for Africa that provides access to affordable financial services and credit scores to low-income borrowers. She previously founded WezaTele, which in 2015 became the first African technology start-up to be acquired by another business. In 2016, she was named one of the 30 Most Promising Young African Entrepreneurs by Forbes (Ball 2018).

> · When I think about all the young Africans who don't have jobs, I'm driven to use the knowledge I have acquired to inspire more young people to be entrepreneurs, create jobs and to do even better than me.
>
> *Hilda Moraa, founder of Pezesha* (Africa.com 2020: para. 4)

Moraa's interest in technology started in school, where she excelled in mathematics and physics and realised that she wanted to follow a career in engineering. She enrolled at Nairobi University for a course in electrical engineering – one of a handful of women in a class of about 200 students. She was later awarded a scholarship to Strathmore University in Kenya, where she studied business and information technology, and it was there that her interest in entrepreneurship developed. She started her first business (providing printing and computer services to fellow students) while still studying. It was an immediate success, to the extent that the university soon offered the same services – which put her out of business. The whole experience made her realise that her passion lay in helping to solve challenges and improving the lives of other people.

> One of the mistakes I made was bringing my friends into the business. They agreed to join me because they were my friends but looking back, I realise it

was not really their passion or where they wanted to go with their careers. This ended up creating tension because they were not able to execute the way I wanted and were not motivated about the business … In the start-up environment, with all the uncertainty and limited resources, you really need people who are passionate to be able to push through.

Hilda Moraa, founder of Pezesha (Africa.com 2020: para. 8)

Importance of social entrepreneurs: Milton Friedman's dictum 'There is one and only one social responsibility of business: to use its resources and to engage in activities designed to increase its profits' (Friedman 1967) sounds very old-fashioned in this age of altruism, philanthropy and collaboration. Commercial entrepreneurs who develop their products and services purely for profit are rapidly being replaced by social entrepreneurs whose aim is to heed the triple bottom line while making a profit. Some argue that capitalism as we know it is under fire and is being replaced by 'stakeholder capitalism', a kind of anti-Friedmanite formula that focuses more on the needs and wants of customers, employees, suppliers, activists and the environment. For example, working conditions and environmental impacts in the DRC, which supplies over half the world's cobalt (at an enormous profit) for electric car batteries, are under scrutiny.

According to Patrick Smith (the editor of the online news organisation The Africa Report), globally investments that are motivated by environmental, social and governance concerns are up to US\$30 trillion, with about a third of the funds under professional management, and pension funds (the biggest clients for asset managers in Africa and elsewhere) are increasingly using environmental, social and governance criteria in their decision-making. Furthermore, analysts have found that they are producing higher equity returns than the old shareholder capitalist model (Smith 2020). In addition, the spiralling climate crisis, together with the uncosted global effects of robotics and AI, is undermining faith in established models of trade and market capitalism. The trading environment is ripe for change.

Africa is replete with social entrepreneurs but not all of them achieve their goals, even though they typically offer a double benefit to the economy by creating jobs and helping to resolve social and environmental problems. Furthermore, many young millennial and Generation Z shoppers tend to support enterprises that are aligned with their own social values, and it is no surprise that they lean towards supporting start-ups that have society's best interests at heart. Social entrepreneurs identify a problem, come up with a solution, and then find the resources and funds to implement the solution. It sounds easy, but it can be a hard road to travel because it takes more than a strong work ethic and a healthy dose of idealism to drive a successful social enterprise (Yunus 2020).

The Nobel Prize-winning Indian economist Muhammed Yunus, who pioneered the microfinancing of small enterprises by founding Grameen Bank in 1976, has sound advice for African social entrepreneurs: Do market research on other players in your area of enterprise; continue to study the principles of social entrepreneurship; continuously improve your skillsets; keep in mind that no-one has an obligation to support you notwithstanding your good intentions; be realistic about the financial, human and space resources that you can afford; learn from others' mistakes; and choose your partners carefully (Yunus 2020).

In South Africa, the Gordon Institute of Business Science offers a social entrepreneurship programme at their Gordon Institute of Business Entrepreneurship Development Academy (www.gibs.co.za/programmes). Also, the University of Cape Town's Graduate School of Business' Bertha Centre for Social Innovation and Entrepreneurship is dedicated to advancing social innovation (www.gsb.uct.ac.za/berthacentre) and the Social Enterprise Academy in Cape Town offers internationally recognised qualifications in social entrepreneurship (www.socialenterprise.academy/za). Yet, despite the continent's vast resources, its 54 countries together comprise the world's poorest inhabited continent with the aggregated GDP totalling barely a third of that of the GDP of the USA.

Convergence of telecommunications and banking: Steward Bank in Zimbabwe is one of a few banks in the world that is wholly owned by a mobile network operator. It was established in 2013 following the acquisition of TN Bank by Econet Wireless Zimbabwe, the country's largest telecommunications company, and occupies the space where telecommunications and banking converge. The bank's mandate is to offer simplified, low-cost, accessible and universal banking services to Zimbabweans previously excluded from participation in the formal financial sector. Steward Bank's close relationship with EcoCash (Econet's mobile money subsidiary) has been a key element of their competitive advantage, as it enables them to simplify and scale their business model to reach as many customers as possible without having to build large physical infrastructure as their peers have done. Following the maxim that the customer of the future is more interested in banking than banks, they are developing simplified products that are accessible from smartphones such as savings accounts (EcoCash$ave), microloans (EcoCash loans) and digital group-savings services (Mukando).

Paga, one of the most successful mobile payment companies in Nigeria, is building an ecosystem that enables banked and unbanked people to digitally send and receive money in a stress-free environment. Furthermore, Paga agents earn a commission for every transaction that they perform for their customers. OPay in Nigeria allows customers to use or withdraw savings at any time and offers a 14 per cent interest on savings. The Kudi app facilitates agent banking and allows technoliterate individuals to offer services such as bank transfers, cash withdrawals, airtime purchases and electricity bill payments to their communities while earning an income.

Getting paid: One of the biggest challenges facing entrepreneurs in Africa is getting paid by their customers. Paystack was founded in 2015 by CEO Shola Akinlade and CTO Ezra Olubi (both from Nigeria). The start-up, which has offices in Lagos and San Francisco, enables Nigerian businesses to accept Mastercard, Visa and Verve cards, and is used by over 60 000 businesses to verify the identity of customers and facilitate easy one-time or recurring payments using whatever route is preferred (including credit card, bank account or mobile money). The service automatically routes payments through the optimal channels and endeavours to ensure the highest transaction success rates in the market. In addition, Paystack's combination of automated and manual antifraud systems protects customers from fraudulent transactions and associated charge-back claims.

Nigeria's financial crisis: Nigeria (now the largest economy in Africa) received a stark warning from the World Bank in 2019: Unless policy-makers revive economic growth and increase employment, the country risks becoming home to a quarter of the world's destitute people within a decade. This sombre scenario was caused by many factors, including the Nigerian government's tendency to continuously resort to external borrowing with the consequence that debt servicing takes up between 25 and 50 per cent of its revenue. There is also little coordination between the public and private sectors on macroeconomic policy, resulting in the private sector being marginalised because it has been unable to raise the necessary credit to expand (Tella 2020).

The solution, according to Sheriffdeen Tella (professor of economics at Olabisi Onabanjo University in Ago-Iwoye), is that the government recognises the problems rather than turn a blind eye to them and put a national financial plan in place enveloped in collaboration with the private sector and academics. The government also has to do more to promote agriculture for food and industrial raw materials, and provide more incentives for industrial production to reduce production costs. According to Tella (2020), the government also has to promote the consumption of locally produced food and goods, and invest massively in education and health. He advises, though, that the fuel subsidy should not be removed because this would create hardship and increase poverty levels, but it should be reduced as domestic production increases.

FNB's eWallet eXtra: Millions of people in Africa do not have bank accounts, mostly because of the high fees but also due to lack of access to services, especially in rural areas. First National Bank (FNB) has set out to change this by creating eWallet eXtra, a mobile bank account that allows unbanked and underbanked people to use a smartphone to open and operate a bank account without using a card. Similar services are offered by rival banks in South Africa, such as Bank Zero, Discovery and TymeDigital (backed by billionaire Patrice Motsepe's African Rainbow Capital); and Absa and Nedbank have launched CashSend and Send-iMali respectively, which enable instant mobile banking (Ruzicka 2018). FNB's eWallet exTra has the edge, as it charges no monthly fees and consumers do not need to submit any paperwork to open an account. They cannot, however, process debit orders and there are spending limits of R3 000 per day and R24 000 per month. Throughout Africa, banks are being pressured by online operators and retailers to improve their services, which is good for the consumer, and many further e-commerce innovations can be expected in future.

Can *Stokvel* go digital? A *stokvel* is a home-grown savings institution first developed in South Africa to primarily serve the needs of black people who do not own significant assets and therefore have trouble obtaining loans from banks. *Stokvel* members agree to contribute a fixed amount towards a certain financial goal on a regular basis. The Nedbank *Stokvel* account offers clients additional benefits, such as a discount of up to 10 per cent on groceries or school supplies and R10 000 burial cover for each member. According to research by Nedbank, an amount of over R44 billion is collectively saved in 820 000 *stokvels* around the country annually, with more than 11 million South Africans members (Mathe 2020b). Now, two start-up fintech companies are digitising traditional *stokvels* to make them more transparent and convenient.

Stokfella, an app launched in 2016, is an online savings account that enables traditional savings clubs to manage their administration, monthly payments and claims by using cellphones or tablets. The platform already boasts more than 14 000 users and does more than 1 million transactions annually. Another smartphone app, Franc, allows members to join its *stokvel*; manage their investments; and gain access to high-performing, low-cost investments with the Allan Gray money market fund (with an average return of 7.87 per cent per year) and the Satrix Top 40 (with an average return of 13.6 per cent per year). According to Sebastian Patel, the COO of the Franc Group, the app helps members to make better investment choices and gives them access to better products (Mathe 2020b), a classic melding of traditional and modern financial services. Since it was launched in mid-2018, Franc has been downloaded about 6 000 times, 1 600 people have registered and there are 155 active users with R1.12 million invested.

Lobola calculator: *Lobola* or *magadi* (the African bride price payable to the family of the bride-to-be) was traditionally paid in cattle (a potent symbol of wealth), but can now also be paid in cash or by credit card. In addition to the payment, it is traditional for the prospective husband to buy presents (*uMembeso*) for the bride's closest relatives. These may include knives (symbols of protection), a trunk (wealth), a mattress (rest) or watches (reliability). In addition, there are possible fines for misdemeanours such as making the bride pregnant before the wedding (*inhlawula*) or arriving late for a wedding meeting (Nkosi 2020b).

The negotiation of appropriate *lobola* is a delicate issue. A generally accepted starting point is six cows worth R10 000 each, but this levy can be negotiated up or down depending on the financial status of the groom's family. Late President Nelson Mandela apparently paid 60 cows for his second wife, Graça Machel (Nkosi 2020b). Inevitably, there is now an app, *Lobola* Calculator, that can help you to make this crucial calculation (www.lobola.co.za). App developer Robert Matseneng created it, somewhat tongue-in-cheek, so that brides-to-be can determine how many cows they are worth! He claims to have had 157 hours of consultations with brides, grooms and their families, and meetings with over 100 elders, during the design of the app, which attempts to bring an old-age African tradition into today's technology-driven world. But it is not without controversy. Ponsho Pilane (2015) has dubbed it, 'The thin line between humour and misogyny'.

Factors that are taken into account by the *Lobola* Calculator include the woman's height, weight and attractiveness; her age and level of education; her ability to cook; whether she is a virgin or has children from an earlier marriage; whether she has a driving licence and car; whether she is employed or owns property; and the cost of the wedding. To some, *lobola* is an outdated practice that objectifies and devalues women; whereas, to others, it solidifies relationships between the two families and increases the honour of the bride (Nkosi 2020b). A Nigerian equivalent of the app, Bride Price (released in 2014), uses a more specific barometer that includes the woman's skin complexion, whether she has tattoos and how religious she is (Pilane 2015).

Creating an IoT ecosystem: The rollout of an ecosystem for IoT innovation in South Africa was initiated in November 2016 by SqwidNet, the local Sigfox network operator, in collaboration with Dark Fibre Africa. Sigfox is a global IoT network operator that is unlocking the true potential of the IoT in 45 countries across the Americas, Europe, Africa, Asia and Australasia. By 2019, SqwidNet had reached 85 per cent of South Africa's population, with plans to expand further. SqwidNet links the digital and physical worlds, and will enable the development of IoT solutions through the deployment of long-range networks and access to low-cost and secure connectivity as well as to low-power devices and modules (www.sqwidnet.com).

Nimble: Refresh Creative Media in Cape Town has created Nimble, a web-based, mobile-friendly, desktop learning management system (LMS) that is an effective tool for delivering, administering, tracking and reporting staff training and development programmes. Nimble is designed for the South African market, where bandwidth and access to data can be limited, and can operate in extremely low bandwidth conditions. A subscription model, which caters for unlimited users, is available to suit different businesses and each application can be custom branded to the client's corporate identity. According to Refresh's Director, Craig Bruton, 'It is easy to manage modules as well as audio and video clips, staff assessments and learning pathways. The LMS can also score and track learner progress through the modules' (www.nimbleapp.co.za).

Africa's online marketplace: The ILO has estimated that the informal commercial sector contributes about 41 per cent to the GDP of sub-Saharan Africa. Sprawling, noisy, colourful open markets are as characteristic of Africa as the vast savannahs and herds of game, with the Makola Market in Accra probably the most famous. But that will soon change with the development of hybrid physical/online marketplaces. Africa has over 420 million unique mobile phone subscribers, with a 43 per cent penetration rate, and online marketplaces are likely to grow in importance as more people are digitally connected (Lubele 2018).

Two factors may delay the development of online marketplaces in Africa: educational exclusion and infrastructure deficits. According to Unesco, Africa has the highest rates of educational exclusion, with over 20 per cent of children aged 6 to 11 years, 33 per cent of youths aged 12 to 14 years and 60 per cent of older youth not attending school. As online markets require people to be technoliterate, the introduction of plugged-in markets may leave large numbers of young people behind. Furthermore, a 2017 World Bank Report showed that only 35 per cent of the population of sub-Saharan Africa has access to electricity, with even lower rates in rural areas. Without this access, online markets will not work. Notwithstanding predictions that Africa will be the hub of the 4IR, there are clearly some important structural improvements that will have to be made before this prophecy can be fulfilled (Lubele 2018).

Threat of online betting: The drive for financial inclusion has resulted in hundreds of millions of Africans accessing mobile phone banking services, but one of the unintended consequences of this development has been the proliferation of online betting. The combined size of the online gambling industry in Kenya, Nigeria and South Africa was estimated to be worth US$37 billion in 2018, and in Kenya alone over 2 million people engage in online sports betting (Owuor 2018). Most online betters are young, and many are unemployed. In Nigeria, 60 million people aged between 18 and 40 years spent US$5 million per day on sports betting in 2018, staking an average of US$8.40 daily even though they were unemployed or underemployed. Many of them borrow money to support their gambling addiction, which potentially propels them into a downward spiral of perpetual debt. Victor Owuor (2018) recommends that African governments should recognise that the increased prevalence of online gambling addictions among young people is a serious problem that needs to be addressed.

Cybercrime: As Africa trends towards the connection of 1 billion people by 2022, the continent is increasingly becoming the target of cybercriminals. Cybercrime is already a serious issue in the developed economies of South Africa, Nigeria and Kenya, with half of the top 10 countries in the world that have reported very high levels of economic crime coming from Africa and South Africa (77 per cent of companies) and Kenya (75 per cent of companies) topping the global list. The most prevalent forms of economic crime are asset misappropriation by staff or officials, followed by fraud committed by the consumer and bribery and corruption – which suggests that the entire supply chain may be fraught with criminality (White 2018). As the threat of economic crime continues to intensify, stakeholders (from senior executives to regulators, employees and consumers) are taking a more active interest in their responsibilities and changing their corporate culture to one that respects transparency and adherence to the law. They are also more receptive to the tip-offs of whistleblowers.

According to Anna Collard, the MD of the integrated security awareness training and simulated phishing platform KnowBe4, many African internet users are coming online for the first time and do not understand the risks. From ransomware to phishing, malware and credential theft, users in Africa are not protecting themselves adequately because they mistakenly think that they are well informed. A survey of eight African countries by KnowBe4 revealed that most internet users are vulnerable because they are not aware of what they do not know, which highlights the urgent need for cybersecurity awareness training on the continent. Adding to the problem is the acute shortage of skilled cybersecurity professionals, a lack of legislation and law enforcement related to cybercrime, and inadequate company budgets allocated to cybersecurity (Odendaal 2020b).

> Many criminals consider Africa a safe haven for their illegal operations, as many African governments need to attend to other pressing issues, such as fighting poverty, unstable politics, violent crime and massive youth unemployment, and thus regard cybersecurity as a luxury [and] not a necessity.
>
> *Anna Collard, MD of KnowBe4* (Odendaal 2020b)

Discussion: At the WEF in Davos in 2019, a panel consisting of the presidents of South Africa, Ethiopia and Rwanda discussed foreign investment, industrial growth and good governance, but no reference was made to climate change and the green economy (Swilling 2019). Furthermore, African leaders repeatedly asserted their faith in industrialisation as the means to create the millions of jobs needed to harness the potential of the burgeoning African youth population; however, they ignored the findings of the WEF that while the 4IR would result in 75 million existing jobs being lost, 133 million new jobs would be created. They also argued that a massive increase in foreign direct investment would be needed to fuel industrialisation but they seemed to ignore the fact that, in open economies, the manufacturing sector would have to be highly automated to be competitive – which would result in job losses (Swilling 2019). In this author's opinion, African leaders should have placed far more emphasis on the job-creating capacity of the highly interconnected, youth-orientated Post-Industrial Revolution currently sweeping through Africa and the world. In the past, 'disruption' was a negative concept but now it is an important agent of change.

South Africa's President Cyril Ramaphosa takes recent global developments seriously and set up the Presidential Commission on the 4IR in 2019. However, as Prof. Mark Swilling of Stellenbosch University has pointed out, the real issues will not be addressed if this commission is dominated by technology giants. It would be like asking the arms industry to plan for disarmament. He noted that not even the star of Davos 2020, Ethiopia's Prime Minister Abiy Ahmed, came up with any novel economic thinking appropriate to the African context (Swilling 2019). What no-one doubts is that the economies of all African countries will be turned on their heads over the next 10 years by changing technologies. Robotics, AI, machine learning, nanotechnology, the IoT, Blockchain and other innovative solutions will exert a strong influence, especially in the commerce and fintech sectors. Furthermore, a growing contingent of confident and increasingly daring start-ups will disrupt the established economy and eat away at the large company's revenues, profits and market shares because they offer customer-focused solutions that provide end-users with an experience and not just a product or service. Fintech companies are already disrupting the status quo by adopting different approaches to risk and lending, thereby forcing conventional banks and other lending institutions to re-invent themselves if they want to stay relevant. This trend will address one of the major problems for entrepreneurs – accessing adequate finances because they do not have enough collateral.

While Africa has a large number of mobile phone subscribers, which is predicted to exceed 1 billion by 2022 (Mpinganjira, Brett & Maiden 2018) – second only to the Asia-Pacific region – the lack of a fully functional, continent-wide digital innovation ecosystem is hampering the development of e-commerce at the level of SMEs. Africa also has the most expensive internet costs in the world. In nature, an ecosystem is a community of living plants and animals that have formed complex interrelationships with one another and with their natural environment over millions of years. Similarly, an innovation ecosystem includes all the players, institutions and environmental factors that facilitate innovation such as knowledge generators, idea generators, innovation hubs, skill developers, trading platform developers, entrepreneurs, public–private partnerships, investors, sellers, buyers and those that provide support services. Innovation ecosystems – like natural ecosystems – also have symbioses (both partners benefit), commensalisms (one partner benefits but the other is not harmed) and parasitism (one partner benefits and the other is harmed).

If the players optimise their interactions with one another and with their environment, it creates an innovation ecosystem from which individual players derive value that is greater than what they could generate alone. At this stage, many African countries lack the infrastructure to make this matrix work optimally and to continuously scale it up, with the result that e-commerce does not deliver the economic and social benefits that it has the potential to offer – but this is changing rapidly. i-hub in Kenya, hive-colab in Uganda and AfriLab continent-wide, and other hubs as well as MTN's Entrepreneurship Challenge, are all helping to strengthen the innovation ecosystem in Africa (Mpinganjira et al. 2018).

18

Entrepreneurship and the future of work

'Innovation is the specific instrument of entrepreneurship.
The act that endows resources with a
new capacity to create wealth.'

Peter F. Drucker, Austrian-American management consultant and educator
(Forbes 2015)

Introduction: According to Stanford Prof. Carol Dweck (2008), people display two kinds of thinking: those who operate with a fixed mindset and are more likely to stick to activities where skills that they have already mastered are used rather than risk embarrassment by failing at something new, and those with open minds who are focused on personal growth and make it their mission to understand new things even if they will not succeed at all of them. Entrepreneurs have the 'growth' mindset. Dweck found that one of the most basic beliefs we have about ourselves has to do with how we view and inhabit what we consider to be our personality. On the one hand, a 'fixed mindset' assumes that our character, intelligence and creative ability are static givens that we cannot change in any meaningful way, and success is the affirmation of that inherent intelligence. Striving for success and avoiding failure at all cost become a way of maintaining a sense of being smart or skilled. On the other hand, a 'growth mindset' thrives on challenge, and sees failure not as evidence of unintelligence but as an encouraging springboard for growth and for stretching our existing abilities. From two mindsets (which we manifest from an early age) springs a great deal of our behaviour, our relationship with success and failure, and ultimately our capacity for happiness.

The growth mindset is based on the belief that your basic qualities are things that you can cultivate through your efforts. Although people may differ in many ways, in their initial talents and aptitudes, interests or temperaments, everyone can change and grow through application and experience. People with this mindset do not believe that they can become Einstein or Beethoven, but they do believe that their true potential is unknown and can only become known after years of passion, training, learning and toil. The growth mindset therefore creates a passion for learning rather than a hunger for approval, and its hallmark is the conviction that human qualities like intelligence and creativity can be cultivated through effort and deliberate practice. People with this mindset are not discouraged by failure; they do not actually see themselves as failing in these situations, but rather as learning.

Although 'toil' is an important component of every innovator's toolkit, it is not the only factor that will determine their success. Malcolm Gladwell popularised the concept that 'mastery' of any field requires 10 000 hours of practice (Gladwell 2008). That will normally take about 10 years to achieve if you dedicate a few hours a day to 'practise'. But mindlessly repeating the same thing will not produce the results you are looking for, which is why 'directed practise' (which involves actively engaging in whatever it is you are trying to master) is now recommended. Directed practise is also about identifying activities that you are not good at, practising those things until you are good at them and constantly engaging with the unfamiliar.

Most of the African entrepreneurs and innovators showcased in this book have a growth mindset and have focused on directed practise, but it nevertheless remains one of the major challenges on the continent to provide the conditions in which young entrepreneurs can develop these mindsets and thereby reach their full potential. However, it could be argued that these mindsets are indigenous to Africa because they are deeply embodied in the Nguni concept of ubuntu, which encourages us to be focused and work together to embrace our shared humanity and aspire to create a society imbued with wisdom, respect and opportunity for all.

Innovation Summit: When the SA Innovation Summit canvassed the opinions of its delegates on World Creativity Day (21 April 2020) regarding their reactions to the Covid-19 crisis through their #IamCreative_Imbizo app, the response was illuminating. Forty-three per cent indicated that they were negotiating a better future, 44 per cent that they were in the process of creating a new future, and 13 per cent that they were in denial or depressed. The Innovation Summit itself has pivoted in response to the crisis. Instead of holding a once-a-year summit, virtual meetings and other events are now held throughout the year to inspire innovation, match start-ups with funders, showcase the best talent and grow an innovation-based economy. Their new initiatives include the Founder-Driven Series 2020, the Skolkovo Softlanding Programme, Match and Invest facilitation meetings, Pitching Dens and Inventors Garages.

They also celebrate World Creativity and Innovation Day, a global UN day held on 21 April each year, to raise awareness of the importance of creativity and innovation in problem solving with respect to advancing the UN's SDGs.

mHub: Rachel Sibande is a Malawian computer scientist, STEM educator, social entrepreneur and innovator who founded Malawi's first innovation hub, mHub. She is also the chairperson of GirlEffectMalawi and a board member of GiveDirectlyMalawi. Since the establishment of mHub in 2013, she has championed the development and deployment of innovative technology solutions in elections monitoring; citizen engagement; and agriculture in Malawi, Zambia, Tanzania, Mozambique and Zimbabwe. Sibande has also made an impact on the lives of over 40 000 young innovators by creating an entrepreneurship curriculum and pioneering initiatives to teach software coding to develop the next generation of technology creators. In 2012, she became an alumna of US President Barack Obama's Young African Leaders Initiative and, in 2013, delivered a TED talk on the use of technology in agricultural development.

In 2016, Sibande became Malawi's ambassador of the Next Einstein Forum Initiative that promotes STEM and, in the same year, she was listed by *Forbes* magazine as one of Africa's most promising entrepreneurs under the age of 30 years. Rachel joined the Digital Impact Alliance at the UN Foundation in September 2017 as programme director: data for development, supporting health and food security projects in African countries. She also founded the company Earth Energy that focuses on generating electricity from maize cobs by using a thermal chemical process, for which she won first prize at the Next Einstein Forum global gathering in Rwanda.

Importance of technology hubs: Africa has more than 640 technology hubs, ranging from technology incubators and university-linked start-up support laboratories to co-working spaces, innovation hubs, maker parks and accelerator sites, many of them hybrids of the aforementioned. However, these hubs have a high rate of attrition, with over 100 going bankrupt in the last few years (Soumarè et al. 2020). From Yaba (Nigeria's Silicon Valley in Lagos) to Workshop 17 on the V&A Waterfront in Cape Town, young entrepreneurs are gathering to plot the way forward in 'Africa's century'. About two-thirds of these ventures have fewer than 10 staff members, but they share the same catalytic goals: to help technology entrepreneurs take their first steps towards launching viable and sustainable companies, and to support a growing innovation ecosystem.

Although they mainly provide spaces and facilities where entrepreneurs can work together, offer consultancy services, exchange equity for logistical support or locate venture capital funding, it is now recognised that their primary benefit is to provide 'a safe haven for dangerous ideas' – a place where young people can freely discuss sometimes wild ideas without fear of repercussion (this definition was previously applied by this author to interactive science centres).

At these technology hubs, the participants typically debate issues of critical social relevance to find novel solutions to the thorniest problems of the day. As representatives of *the* best-connected generation in history, they strive to bring together people from sectors that ordinarily would not interact because this multicultural approach is more likely to produce solutions that do not perpetuate the status quo, as they are aware that the problems of the present cannot be solved with the same mindset as the past. Many award-winning African enterprises have emerged from debates held in technology hubs, including Nigeria's BudgIT (a start-up hatched during a hackathon at the Co-Creation Hub in Lagos) and LifeBank (whose founder, Temie Giwa-Tubosun, won the inaugural Jack Ma Entrepreneur Award in November 2019).

Big technology companies that are always on the lookout for the 'Next Big Thing' (such as Microsoft, Facebook, Google, Amazon, IBM, BNP Paribas and Orange) are, not surprisingly, the major supporters of technology hubs and their stakeholders but other entities are showing increasing interest. Visa pumped US$200 million into Nigeria's payment fintech Interswitch; Chinese investors put a combined US$170 million into the Nigerian fintech OPay and US$40 million was raised for another African fintech, PalmPay, by the Chinese telecommunications company Transsion. African investors are coming to the party too, including Liquid Telecom (the data infrastructure company led by Zimbabwean billionaire Strive Masiyiwa), Standard Bank (Africa's largest bank) and MTN (the continent's largest telecommunications company). Nairobi, Lagos, Johannesburg and Cape Town are the top African technology cities, as they are nested in large markets that make widespread use of the internet and have sophisticated networks supported by telecommunications giants such as MTN, Vodacom and Safaricom (Soumarè et al. 2020).

Oumar Cissè, a 42-year-old entrepreneur in Senegal, has been able to extend the services of his fintech company InTouch into several West African markets (including Côte d'Ivoire) through technology hub alliances and new connections. Abidjan, the economic capital of Côte d'Ivoire, now has three undersea fibre-optic cables linking it to the global Web. Furthermore, the presence there of technological giants MTN and Orange has greatly facilitated the development of technology hubs and promoted small-scale entrepreneurs' access to new markets. Another important Francophone market is Morocco, as global giants such as Atos, IBM, Oracle and Sage have set up in Casablanca and are using it as a launch pad into Africa (Soumarè et al. 2020).

blueMoon: Eleni Gabre-Madhin, who launched Ethiopia's first commodities exchange, also established blueMoon (an agritech incubator hub in Addis Ababa). According to Shem Asefaw (the 'chief rain-maker' at blueMoon), the hub invests in innovative and scalable businesses in the agriculture, livestock, fisheries and forestry industries. It is aimed at bridging the gap between early-stage start-ups and sources of capital by taking the first risk itself with seed funding of US$10 000 to enable the start-ups to build their prototype and proof of concept, and then providing several options so that they can obtain the next round of financing elsewhere. At the end of the incubator programme, the start-ups participate in a competitive Lion's Den event during which they pitch their projects to raise equity financing. In addition, blueMoon facilitates investments for start-ups from a network of angel investors on a case-by-case basis. Addis Garage, blueMoon's workspace in Bole (Addis Ababa), is open to any start-ups as a co-working or event space where smart, passionate people gather together and create new synergies (https://www.bluemoonethiopia.com/). blueMoon is now developing an incubation framework that African universities can adopt to build hubs on their campuses (Awosanya 2019).

Doing the Kasi hustle: When Brian Makwaiba had to drop out of medical school after his father's spaza shop (a small shop in a township) had gone bankrupt, he decided to find out what had caused it to fail. He found that it was the pricing of goods, which were far higher than those of foreign-owned shops, and a lack of knowledge on how to grow the business. He founded I Am Emerge to help spaza shops by buying their stock in bulk; his company now services over 4 900 shops in the Alexandra and Soweto townships in South Africa. In addition, Makwaiba has developed an app that allows shop owners to place orders online and he plans to expand his enterprise into the rest of Africa (Peter 2018).

Plastic bricks: Nelson Boateng of Ghana is a problem solver. He noticed that enormous amounts of plastic waste are generated in Accra and clog up landfill sites. He decided to form a company, Nelplast, that shreds and melts down plastic waste and mixes it with sand and red oxide to create plastic bricks, paving slabs and roof tiles. These products are now used to make roads, pavements and buildings in Accra. Boateng says that he aims to recycle at least 70 per cent of the plastic waste in Ghana 'into useful products that that can be used for a lifetime'. The initiative has attracted funding from the Ghanaian Ministry of the Environment and is set to expand into other West African countries. Boateng also offers a consultancy service on how to launch a recycling company (Ayene et al. 2020).

Bio-bricks from urine: A team of researchers at the University of Cape Town, led by Dyllon Randall, has developed a process for making bricks from urine. Their bio-bricks offer crucial advantages over conventional clay bricks, as they use a resource that would otherwise end up in waste-processing plants while creating useful by-products (nitrogen and phosphorus) that can be used to make fertiliser. Furthermore, the bio-bricks are made at room temperature, whereas the baking process used to make clay bricks releases large quantities of CO_2. The bio-bricks are created through a natural process called microbial carbonate precipitation. Urine is collected using a fertiliser-making urinal and the liquid left over is mixed with sand and a strain of bacteria that produces the enzyme urease. Urease breaks down urea (a key component of urine) while producing a compound that binds the sand particles together to create a hard, grey-coloured brick. One of the biobrick developers, Suzanne Lambert, likens the process to the way in which marine organisms have built coral reefs for millennia (Wodinsky 2018; D. Randall, personal communication, 2018).

Flying over Soweto: Wiseman Ntombela has qualifications in marketing and communications and a master's degree in theology, but regards himself as a 'visionary and born entrepreneur' (Peter 2018). In 2015, he founded Fly SA Wise – an aviation tourism company that offers visitors an aerial perspective of the country's largest township, Soweto. Ntombela's company also offers tours to Nelson Mandela's childhood home in Qunu in the Eastern Cape as well as to Pretoria and the Kruger National Park, and introduces school children to careers in aviation.

> You know, when tourists come here, they just see Mandela's house, which takes a few minutes, and then they walk around Vilakazi Street … they were not getting exposed to the biggest township in South Africa. While on the [aerial] tour, we also record the trip and later send a video link of the entire tour for the tourist to take home. In this way, we get to share a part of South Africa with them.
>
> *Wiseman Ntombela, founder of Fly SA Wise* (Peter 2018)

Def Jam Africa: Sipho Dlamini of Johannesburg is the CEO of Def Jam Africa, a recording studio affiliated to the Universal Music Group in the USA. Since it was founded by Rick Rubin and Russell Simmons in 1984, Def Jam has been instrumental in classic releases by LL Cool J, the Beastie Boys, Public Enemy, Method Man, Ghostface, DMX, Ludacris, the Roots, Young Jeezy, Frank Ocean, Logic and many others. According to Dlamini, Def Jam is dedicated to representing the best hip-hop, Afrobeats, and trap talent in Africa. It will follow the blueprint of the iconic Def Jam Recordings label. The label will initially be based in Johannesburg and Lagos, but will identify and sign artistic talent from the entire continent. It is launching with a flagship roster, including well-established African artists such as Boity, Nadia Nakai, Cassper Nyovest, Nasty C and Tshego from South Africa, and Larry Gaaga and Vector from Nigeria (Aswad 2020).

> Many of us in Africa grew up on music from legendary labels under the Universal Music Group umbrella. From Blue Note for jazz fans, to Mercury Records, which was Hugh Masekela's first US label and Uptown Records, the home of Jodeci and Mary J Blige and many more. For those into hip-hop, no label has such cultural and historic relevance as Def Jam. It is a historic achievement that we're now able to bring this iconic label to Africa, to create an authentic and trusted home for those who aspire to be the best in hip-hop, Afrobeats and trap. Together, we will build a new community of artists that will push the boundaries of hip-hop from Africa, to reach new audiences globally.

> *Sipho Dlamini, MD of Universal Music*
> *Sub-Saharan Africa and South Africa* (Aswad 2020)

Tony Elumelu Foundation: The Tony Elumelu Foundation (TEF) is an African NPO founded by Nigerian businessman Tony O. Elumelu in 2010 and is headquartered in Lagos. The TEF is committed to Africa's economic transformation by enhancing the competitiveness and growth of the private sector through strong support for entrepreneurship. So far, it has empowered 7 520 African entrepreneurs in 54 African countries. In 2015, the TEF consolidated its programmes and made investing in Africa's next generation of entrepreneurs a priority. Its vision is to unlock the obstacles that Africa's entrepreneurs face as they grow their start-ups into SMEs, their SMEs into national companies and their national companies into African multinationals.

The guiding principles of the TEF are derived from Elumelu's inclusive economic philosophy of Africapitalism, which promotes long-term, sustained and vibrant African-led private-sector investment in key sectors of the continent's economy. Through its programmes, the TEF (which is led by Ifeyinwa Ugochukwu, a lawyer and passionate advocate of entrepreneurship) seeks to 'institutionalise luck' and create an environment in which entrepreneurship can flourish. The TEF's Entrepreneurship Programme is a US$100-million initiative funded by the Elumelu family with the intent of creating at least 1 million jobs and contributing over US$10 billion in revenue to the African economy. The Tony Elumelu Entrepreneurship Forum held its fifth annual conference in Abuja in July 2019.

Facebook traders in Sudan: The introduction of Shari'a law and public order laws in Sudan in 1983 determined the status of women in Sudanese society and made it difficult for them to follow careers of their own. However, the rise of social media and the availability of smartphones have enabled Sudanese women to transcend traditional gender norms and enter the commercial sector by trading on Facebook (Steel 2018). The *tajirat al-Facebook* ('Facebook traders') use their smartphones to sell cosmetics, clothing, fashion accessories and perfumes from their homes without jeopardising social and cultural expectations. Online trading has allowed the women to combine work and family commitments as well as escape from the constraints of being a *rabbat bayt* ('housewife') and achieve economic and personal freedom without compromising their religious beliefs. Many of them have created women-only 'closed pages' on which they can both trade and interact socially with women without interference by men or the authorities. Communications technologies have thus made it possible for Sudanese women to become entrepreneurs while complying with the gender separation codes of Islam (Steel 2018).

Disruption vs symbiosis: Uber built its reputation on a willingness to flout norms and was previously a guiding light for small start-ups. Now nobody is pitching their company as the next Uber. Furthermore, legions of entrepreneurs previously subscribed to Mark Zuckerberg's one-time motto, 'Move fast and break things', but few follow that guideline today. There has been a shift in mindset from disrupting the status quo to trying to build something that is good for both consumers and employees. There are also signs that market sentiment has turned against high-flying, loss-making technology companies such as Uber, Lyft and WeWork. Investors are happy to live with big losses if they believe that a company can one day turn big profits and weather storms, but increasing concerns over profitability have led to more pragmatic attitudes. WeWork, which specialises in shared workspaces, has shown phenomenal growth and has a presence in 29 countries, but company executives reported in September 2019 that it would delay its initial public offering and retrench 3 000 to 5 000 workers from its workforce of 12 500 employees as well as sell some of its assets (Mathe 2019). Investors are becoming wary of big talk.

In the case of Facebook, there have been serious concerns about the use and abuse of personal data (now the world's most valuable commodity) and analysts are asking: Will Facebook users become the platform economy's masters or slaves? A new phrase has even been invented to describe the battle for market control among subscription services such as Netflix, Hulu and Disney: streaming wars. Algorithms are being developed in ways that allow companies to profit from our past, present and future behaviour – or what Shoshana Zuboff of

Harvard Business School calls our 'behavioural surplus' (Mazzucato 2019). In many cases, digital platforms already know our preferences better than we do and can nudge us to behave in ways that produce still more value for them. But we need to ask ourselves: Do we want to live in a society where our innermost desires are up for sale?

Capitalism has always excelled at creating new desires and cravings but, with big data and algorithms, technology companies have both accelerated and inverted this process. Rather than just creating new goods and services in anticipation of what people might want, they already know what we will want and are selling to our future selves (Mazzucato 2019). Policy-makers need to try to understand how platform algorithms allocate value among consumers, suppliers and the platform itself. Whereas some allocations may reflect real competition (which benefits everyone), others are driven by value extraction rather than value creation (which only benefits the platforms).

Tax reformist: In his capacity as the head of the business lobby *Groupement Inter-Patronal du Cameroun* (Inter-Employers' Group of Cameroon) since 2017, Célestin Tawamba has long championed the development of Cameroon's private sector, often in courageous ways. In January 2020, he made headlines by calling for the sacking of the country's tax chief – a rare example of a business leader criticising a senior government official there. Tawaba also advocates abandoning taxation based on activity in favour of a tax based on profit (Ayene et al. 2020).

Africa Innovation Policy Manifesto: i4Policy, a pan-African community of over 160 innovation hubs in 45 countries, has launched a consultation process to shape the next version of the Africa Innovation Policy Manifesto. i4Policy emerged from a gathering of innovation hub leaders in Kigali (Rwanda) in October 2016 and continued to be developed in May 2018 when a group of 49 innovation leaders from 25 countries met again in Kigali to co-create a manifesto of innovation policies for the continent. The manifesto has already facilitated policy reform at the national level in 11 countries and has influenced important continental policy documents, such as the EU–AU Digital Economy Task Force report on a new AfricaEurope partnership and the draft AU Digital Transformation Strategy 2020–2030 (Jackson 2020).

Following the rapid growth of the i4Policy community and its increasing influence on national and regional policy processes, the movement recently recruited a task force of 21 members from 20 countries to conduct a revision of the manifesto and develop the Africa Innovation Policy Manifesto 2.0. The Innovation for Policy Foundation, a NPO established to sustain i4Policy's activities, has built an open online consultation platform to gather feedback on the manifesto and has partnered with Facebook to build chatbot tools that are integrated into social media platforms which enable more people to make their inputs. In addition, i4Policy has mobilised hubs in 45 countries to host focus group discussions on the manifesto (Jackson 2020).

Future of work: The age of robots is not coming; it is here. There are already 'robot restaurants' where you order your food on a touch screen and it is cooked and delivered by machines, Amazon allows you to shop without queueing and self-driving vehicles are about to enter the market. These innovations may not be taking place in Africa, but they are coming. The WEF has predicted that 41 per cent of all work in South Africa is susceptible to automation, and the figures for Ethiopia, Nigeria and Kenya are 44, 46 and 52 per cent respectively (Smith 2018). However, the transition will likely be gradual – or will it?

Every industrial revolution has made some jobs redundant but also created opportunities that have taken humankind forward exponentially. For example, between 2012 and 2017, the demand for data analysts grew by 372 per cent and that for data visualisation skills by over 2 000 per cent! There is also a huge demand for user interface experts who can facilitate seamless human–machine interactions. To meet these demands, businesses will not only have to take on new staff but also re-skill their existing staff. With the rapid uptake of digital technologies in Africa, this creates a real opportunity for the continent's home-grown digital creators and designers – people who understand the continent's nuances and expectations and speak its languages – to step forward and take the gap.

To take advantage of these opportunities, African countries will have to revolutionise their educational and lifelong skills training programmes. As Jenny Dearborn (SAP's global head of learning) has pointed out, we will have to abandon the old three-stage life model consisting of 'learn–work–retire' and adopt a multistage life model of 'learn–work–change–learn–work–change'. Furthermore, as pointed out by Cathy Smith (the CEO of SAP Africa), we will have to move from rote learning to learning to be critical thinkers, problem solvers, creatives and innovators. We will also have to acquire the ability to unlearn what we learned before. In addition, there will in future be a greater emphasis on uniquely human skills that AI and machine learning will have difficulty replicating, such as collaboration, empathy and communication. If we can achieve this, then the future will consist of humans and robots working together.

Techpoint Africa: Founded by Adewale Yusuf in Lagos in 2015, Techpoint Africa is a technology company that amplifies the best innovations from Africa through its media, data, events and technology-focused platforms. It has grown to become one of the highest profile brands across the continent's technology, start-up and e-business ecosystems, having reached over 11 million investors, entrepreneurs, developers, professionals and technology enthusiasts in its first five years. The company hosts the largest technology and start-up event in West and East Africa, Techpoint Build, and has numerous platforms that provide information on the African technology space. Yinka Awosanya (a technology enthusiast associated with Techpoint Africa) reports that in the first quarter of 2020, Nigerian start-ups raised US$55.37 million, with fintech start-ups accounting for 82.2 per cent of the total even though less than 1 per cent was raised from local funders (Awosanya 2020).

Recognising technological achievements: Many initiatives in Africa recognise technological and entrepreneurial achievements, including the Expo for Young Scientists and the National Science and Technology Forum Awards in South Africa, as well as the newly established Techpoint Africa awards in Nigeria. The latter are administered by Techpoint Build, West Africa's largest gathering of innovators, start-up founders, thinkers, programmers and investors. Shortlisted finalists for the 2020 awards (covering 2015–20) included Paga, OPay, Kudi and Riby Finance in the Most Outstanding Start-up Driving Financial Inclusion division; Paystack, Flutterwave, OPay and Paga in the Most Outstanding Electronic Payments Start-up division; Paga, Bolt, Uber, MAX Okada and Gokada in the Most Outstanding Mobility Start-up division; MAX Logistics, GIG Logistics, Africa Courier Express and Robo360 in the Most Outstanding Logistics Start-up division; and Farmcrowdy, Thrive Agric, Agropartnerships and Payfarmer in the Most Outstanding Agro-Crowdfunding Start-up division.

Other shortlisted finalists included ForLoop Africa, DigiClan, Startup South, Startup Arewa and Tech Skills Hack in the Most Impactful Independent Tech Community division; 54gene, Chaka, PlentyWaka, uLesson and Eden Life in the New Start-up of the Year division; Ventures Platform, Y Combinator, CRE Venture Capital and Omidyar Network in the Most Active Venture Capital Company division; and Carbon, PiggyVest and Cowrywise in the Best Nigerian App division. In addition to these awards, Techpoint recognises the contributions of individuals by inducting them in the Techpoint Hall of Fame. These award-granting organisations need to be supported, as it is important for Africa to recognise its science and technology pioneers and techpreneurs.

Emerging innovations: Bryan Vlok created a novel prepayment platform, Say *Siyabonga* ('Thank you' in isiZulu and isiNdebele), in South Africa during the Covid-19 hard lockdown that allowed customers to pay in advance for vouchers for goods and services at shops and service providers and then claim their purchases after the lockdown. All the participating shops had to do was create a profile on the Say *Siyabonga* website, submit supporting documents, upload their unique digital vouchers, sell their goods and services at full price or discount, and spread the word! The platform made it possible for anyone to provide cash flow for small businesses that were under stress and help them to survive the lockdown (www.facebook.com/pg/SaySiyabonga). Like many Covid-19 innovations, Say *Siyabonga* is likely to continue to offer its services (suitably modified) when the crisis has passed.

Discussion: It is estimated that over 80 per cent of business in Africa is conducted in the informal sector, which means that we must have a formidable entrepreneurial cadre. However, the Covid-19 pandemic has left the economies of many African countries in ruins. South Africa has lost over 1.5 million jobs and, with the oil price at about US$20 per barrel, Nigeria will probably have to halve its budget. How will we rescue African economies from intensive care? Entrepreneurs have a vital role to play in this regard.

Chaos in the entrepreneurial space is the norm – the trick is to learn how to tame it and use it to your advantage. One of the key skills required in this process is the ability to unlearn and re-learn. Companies or investors are no longer interested in what employees already know; they want to know whether they are able to unlearn the outdated things that they know and then re-learn systems and processes that are relevant in today's fast-changing world. In this age of personal empowerment, it is the people themselves and their mindset that are the commodities of value – not what they know or do not know. Their potential and willingness to develop themselves are now the key success factors. Internal qualities such as attitude, passion, ability to communicate and emotional intelligence have become the true values.

Today, success is defined by your self-motivation, your ability to multitask and the value you bring to the team, rather than just your own personal success. Other entrepreneurial traits that employers seek include unique cross-disciplinary perspectives, a tendency to 'live your passion' 24/7, an awareness of new business opportunities, being a problem funder as well as a problem solver, having a learner mindset rather than a 'been-there-done-that' expert mindset, a childlike curiosity, an ability to learn from mistakes, being a good listener, having a keen understanding of customer needs, and of course honesty and integrity. Also, entrepreneurial leaders need to realise that if they are the brightest person in the room, they probably have not hired well.

Many African countries have found that there is a non-alignment between the skills that university graduates are equipped with when they enter the workforce and those that are required in the workplace. For graduates to become successful entrepreneurs, they need to have critical and strategic thinking skills, as well as skills in collaboration and negotiation and a healthy dose of emotional intelligence, cognitive flexibility and resilience. It is widely believed that universities need to place a stronger emphasis on developing the employability of their graduates rather than just their knowledge base. Entrepreneurship plays a vital role in growing the economy of Africa and combating unemployment. The development of an entrepreneurial mindset in the various business schools of African universities is critical in this regard.

19

Essay: African women in science and technology

'For me, strength, resilience and power are all learned and accumulated over time. The more you know about yourself, excel at the work you do and understand the world you live in, the more you realise how nothing worthwhile in life comes without a fight and a lot of patience.'

Sahle-Work Zewde, president of Ethiopia (Iversen 2021)

Introduction: Other than Samia Suluhu Hassan who took over as the head of state in Tanzania in March 2021 after the sudden death of John Magufuli, Ethiopia's President Sahle-Work Zewde is currently (April 2021) the only woman among the 54 presidents in Africa. Is this good for the continent? It cannot be. This conclusion is supported by recent studies that have shown that countries with female leaders at the helm (such as Denmark, Finland, Germany, Iceland, New Zealand and Norway) have been far more effective in handling, for example, the Covid-19 crisis than countries led by men (such as the USA, Brazil and Tanzania), who tend to be more authoritarian and top-down in their management styles. These studies reveal that the characteristics shared by women leaders (such as resilience, pragmatism, benevolence, trust in collective common sense, mutual aid and humility) have all contributed to their success. The researchers also argue that countries led by women tend to be societies where there is a greater presence of women in positions of power in all sectors. They argue that the greater involvement of women results in a broader perspective on the crisis and paves the way for the deployment of richer and more complete solutions than if they had been imagined by a homogeneous group of men (Champoux-Paillé & Croteau 2020).

Prof. Ekanem Braide (the vice-chancellor of the Federal University of Technology in Lafia, Nigeria) became the president of the Nigerian Academy of Science in January 2021 – the first woman to occupy this post in the academy's 43-year existence. Her appointment shows that it is possible to overcome the bias, opinion and attitudes that prevent women from being appointed in leadership positions in science, a domain traditionally reserved for men in many African societies. Braide studied zoology at the University of Ife (now Obafemi Awolowo University in Ile-Ife) and obtained her master's degree and doctorate in parasitology at Cornell University in the USA. She has a distinguished track record of scholarship and leadership in parasitology and epidemiology, and is credited with making major contributions to the eradication of guinea worm in Nigeria. She has also contributed significantly to improving global public health, especially in the area of infectious diseases. Braide was previously the vice-chancellor of Cross River University of Technology in Calabar, Nigeria (Adeyemo 2020).

Eunice Baguma Ball is a Ugandan social entrepreneur whose passion for technology stemmed from her experience working on financial inclusion projects in Uganda, where she saw the impact that technology-enabled access to financial services had on the lives of underprivileged people. She is a strong women-in-technology advocate and the founder of the Africa Technology Business Network, and is now based in London where she facilitates collaboration between British and African start-up communities. Ball is particularly passionate about empowering female entrepreneurs and has launched #HerFutureAfrica, a pan-African, female-focused technology entrepreneurship accelerator. She recently published the book *Founding women. African women who are defying the odds to build businesses in tech* (Ball 2018), which showcases leading African female technology role models.

Importance of women entrepreneurs: A marked trend in entrepreneurship worldwide, and in Africa, is the rise of the 'femmepreneur'. Trend forecaster Dion Chang (the founder of Flux Trends) introduced the term 'the rise of Eve' to describe the way in which women are forging ahead in many previously male-dominated sectors, with a concomitant rise in their spending power (MacAllister 2019). Also, with the emergence of female-centricity, investors are backing female-owned businesses – not only to support gender equality but also because this is where the growth is taking place. Research has further revealed that increasing the diversity of leadership teams leads to more and better innovation and improved financial performance.

> Today's woman is a force to be reckoned with. She is smart, savvy and knows exactly what she wants. Brands need to focus less on the general idea of a female consumer, and more on the specific needs of a massive target market that just happens to be female.
>
> *Dion Chang, founder of Flux Trends* (MacAllister 2019)

According to the World Bank's Director of Strategy and Operations for the Africa Region, Diariétou Gaye, Africa is the only region in the world where more women than men choose to become entrepreneurs. Expanding the opportunities for female entrepreneurs through policies that foster gender equality should therefore have a very positive impact on the growth of Africa's economy. She has found that simple and inexpensive solutions to the barriers that women techpreneurs face have proven effective and should be adopted on a wider scale. While both male and female entrepreneurs face constraints such as a lack of capital, women are specifically impacted by several obstacles such as discrimination and a lack of collateral. As a result, monthly profits of female-owned enterprises are on average 38 per cent lower than those of males. Three factors account in part for their underperformance: lack of capital, choice of business sector and commercial practices.

Information collected in 10 African countries indicates that, on average, male-owned enterprises have six times more capital than female-owned enterprises. The fact that women have less access to assets affects their ability to obtain medium-sized loans and, in turn, impacts the growth of their enterprises. This problem can be counteracted in two ways: by giving women more control over assets through, for example, the granting of joint property rights (as is the case in Rwanda) or by eliminating the need for collateral. In Ethiopia, psychometric testing that measures honesty and willingness to repay loans offers a promising solution (Gaye 2018).

In Africa, female entrepreneurs tend to confine themselves to the traditionally female sectors, largely because of a lack of information about other opportunities. In Guinea and the Republic of Congo, programmes that encourage women to transition toward traditionally male-dominated sectors by providing them with appropriate information and mentoring programmes, among others, are being tested. In Togo, training aimed at fostering proactive behaviours among entrepreneurs, rather than only teaching them basic commercial skills, has had a significant impact. Small-scale entrepreneurs are taught to show initiative, be proactive and demonstrate perseverance. This training has yielded impressive results, as the female trainees have seen a 40 per cent increase in their profits. An example is a female entrepreneur in Togo who rented out wedding dresses. After the training, she expanded her client base by selling dresses as well as veils and gloves, and now owns boutiques in three African countries (Gaye 2018).

Women and the 4IR: Part of the promise of the 4IR is that it will drive economic growth and development, and bring about positive social change. However, there is concern that women, and especially rural women in Africa, will be left behind. Adams (2019) identified several possible reasons for this trend. First, the 4IR looks set to be ruled by science and technology – fields long dominated by men. Second, many of the job opportunities offered by the 4IR are internet-based – another field dominated by men. Third, with advances in robotics and automation, many of the routine-intensive occupations such as clerical, call-centre and care-giving work will in future be carried out by machines. Fourth, research has revealed that women in Africa in general have lower digital literacy than men as well as less access to internet-based technologies, which may exclude them from digital work opportunities. Fifth, the burden of childcare and domestic duties restricts women's ability to further their education and boost their digital skills and employability.

Many African countries have introduced programmes that promote the role and opportunities of women in science and technology. The L'Oréal–Unesco For Women in Science initiative was founded in 1996 and has supported over 3 100 talented young women scientists in 117 countries. Their sub-Saharan Africa regional programme has grown significantly since it was launched in 2010, with the first edition of their regional programme in South Africa in 2017. In Ghana, the initiative STEMbees promotes STEM training for girls and women, and addresses social issues such as digital safety and security. Since 2012, the DSI has held annual Women in Science Awards to recognise and reward excellence by women scientists and researchers, and profile them as role models for younger women. All African countries need to consider how new technologies can be used to empower women rather than exclude them. More research should be carried out on the barriers to female empowerment that the First Digital Revolution might create and on how to overcome them (Adams 2019).

Measuring happiness: The adage that women enjoy the journey while men enjoy the destination is a stereotype that is not always true. Furthermore, while men like to use quantitative measures of well-being such as GDP, women increasingly lean towards more qualitative measures that take non-financial issues into account such as happiness. The term 'gross domestic product' was coined during the Great Depression, when consumption was regarded as a proxy for prosperity: the more we consume, the better off we are and the stronger the economy. But this 'value chain' has many false premises. Even Simon Kuznets, the Belarusian economist who invented it, had doubts about its usefulness and regarded it as a faulty measure of a country's overall well-being.

While GDP (a child of the manufacturing age) is good for measuring physical production, it is useless for measuring the value of service (especially virtual services in the cloud) that dominate modern economies. Wikipedia, which brings human knowledge to millions of people, does not contribute a cent to GDP. GDP is also ineffectual at measuring the many ways in which green technologies have softened our impact on the planet and its life-support systems, which highlights the fact that it is difficult to measure the positive impact of innovation. The good news is that because GDP is poor at capturing innovation, we are probably underestimating the improved quality of our lives. However, we cannot afford to continue this mismeasurement of our well-being any longer.

In June 2019, New Zealand's Prime Minister Jacinda Ardern announced that her country would shift its focus from traditional metrics of national development to a 'well-being budget' prioritising the happiness of citizens over capitalist gain. Although the state-driven pursuit of happiness might appear to be a novel idea, it began in the 1970s with Bhutan's King Wangchuck proclaiming that gross national happiness (GNH) was more important than GDP. Bhutan has used GNH as a development philosophy since 2008. In 2011, the UN General Assembly adopted a resolution entitled 'Happiness: Towards a holistic approach to development' and urged member states to follow Bhutan's example and regard happiness and well-being as a 'fundamental human goal'. Since then, countries in northern and western Europe and North America have regularly occupied the top GNH spots. Africa may not become the richest continent, but it can aspire to be the happiest.

Women in science: Unesco (2017) has reported that only 35 per cent of girls worldwide study STEM subjects and only 3 per cent of female students in higher education choose to study information and communication technologies. This gender disparity is alarming, as many STEM careers are the jobs of the future and the engine of innovation, social well-being, inclusive growth and sustainable development. Africa's ability to harness scientific and technological knowledge is crucial for its socioeconomic development and competitiveness, but this potential cannot be realised without the inclusion and advancement of women in science. Globally, only 30 per cent of women in higher education move into STEM-related fields and only 28 per cent of researchers in all scientific fields are female. According to research conducted by Barbara Tiedeu (2020) at the University of Yaoundé in Cameroon, in sub-Saharan Africa about 30 per cent of scientific researchers are women, although an increase in this percentage was shown between 2011 and 2013 in South Africa, Egypt, Morocco, Senegal, Nigeria, Rwanda, Cameroon and Ethiopia. However, many women in science work at a fairly junior level, with little responsibility and power to make decisions. This means that scientific work is largely missing women's perspectives and contributions, which weakens the science agenda.

Fortunately, there are many initiatives that advance the role of women in science in Africa. The Unesco STEM and Gender Advancement project keeps track of gender data and supports the design and implementation of STEM policy instruments that promote gender equality. The African Development Bank's gender equality index captures data on the appointment of women to posts of responsibility, and Gender Summit Africa maintains a database of talented women leaders and executives and hosted Gender Summits in South Africa (2005) and Rwanda (2014). The AU's Kwame Nkrumah Awards for Scientific Excellence programme honours two outstanding women scientists from each of Africa's five geographical regions each year. The AU declared 2015 the Year of Women's Empowerment and Development towards Africa's Agenda 2063 and adopted the Science, Technology and Innovation Strategy for Africa 2024, which inter alia examines the role that women can play in accelerating the continent's transition to an innovation-led, knowledge-based economy meeting society's needs. Mentorship programmes, such as Coach-Cameroon, have trained hundreds of women in career-building scientific skills (Tiedeu 2020).

20

Communications and telecommunications

ኅበዘ

'I want to put a ding in the universe.'

Steve Jobs, American technology visionary and co-founder of Apple
(Anderson 2011)

Introduction: Language is the main tool that humans use to communicate with one another, but the proliferation of languages and dialects can also be one of the major obstacles to overcome in developing effective communications. People's inability to communicate with one another if they speak different languages is an important real-life challenge. There are more than 6 000 languages in the world, of which about one-third are spoken in Africa. Due to this linguistic diversity, many Africans are multilingual. As a result, it is difficult to determine the most spoken language on the continent. Africa has five language 'families', each of which has a common origin: Afro-Asiatic languages include Arabic, Somali, Berber, Hausa, Amharic and Oromo, which are mainly spoken in North Africa, the Horn of Africa and parts of the Sahel; Austronesian languages include Malagasy; Indo-European languages include Afrikaans, English, French, German, Portuguese, Italian and Spanish; Niger-Congo languages include Swahili, Yoruba, Igbo, Fula and Shona, probably the largest language family in the world with over 1 500 distinct languages; and Nilo-Saharan languages include Luo, Songhay, Nubian and Maasai, mainly spoken from Tanzania to Sudan, Chad and Mali.

Arabic is spoken by over 150 million people in Africa, 62 per cent of the Arabic speakers worldwide. English has 6.5 million first language speakers, with

130 million speakers in Africa. West African Pidgin English (a creole language without a written form) has about 75 million speakers. About 11.5 million people in Africa speak French, whereas about 15 million speak Kiswahili as a first language, with about 100 million speakers in total, making it the most spoken African language in the world (Kaya 2020).

Importance of multilingualism: Being bilingual is not enough in a country such as South Africa that has 11 official languages. As a result, the average South African uses 2.84 languages. IsiZulu is the first language of most South Africans (11.6 million) as well as the most common second language (15.7 million). The least commonly spoken of the national languages are Tshivenda and isiNdebele (1.2 million and 1.7 million respectively). In 2020, President Cyril Ramaphosa added a 12th official language: Sign Language, which is used by 0.5 per cent of South Africans (https://bit.ly/30RZCun; https://bit.ly/3eh6yp6). Swahili has been taught in South African schools since 2020.

Berber and Arabic are the official languages of Morocco (a Francophone country), but the importance of English is emphasised at the London Academy in Casablanca. In Cameroon, where the official languages are French and English, the Anglophone sections of the country feel marginalised by the predominantly Francophone government. Many government notices are only distributed in French and when they are translated into English, the translation is so poor as to be misleading despite the existence in the country of the highly respected Advanced School of Translation and Interpreters. In 2016, these tensions spilled over into protests and conflict, which developed into a civil war. A government initiative to promote bilingualism and multiculturalism did little to quell the violence (Sondo 2020a). Regrettably, language can be divisive.

Unique Ethiopian writing: Ethiopia is the only country in Africa to have its own unique writing. The Amharic script consists of 33 consonants, each with seven different vowel combinations – a total of 231. In addition, 20 symbols are used to express diphthongs and there are seven modifications of the letter 'b' to signify vowel variations of the sound 'v'. Numerals are represented by special characters partially derived from Greek letters (Hancock et al. 1997).

Talking glove: Lucky Netshidzati, 27 years old, of Limpopo province in South Africa could not communicate with his parents because both were born deaf. As he is a problem solver, Netshidzati set out to create a hand glove to translate sign language into voice, text and voice-to-sign language animation so that he could communicate with them. When the Talking Glove sensor is connected to an app on a cellphone, it can also be used to make phone calls. Lucky's innovation will not only help his parents, but will also empower deaf and hard-of-hearing people throughout Africa and beyond (*The Citizen* 2019; Ridovhona 2019).

> Imagine not being able to speak to your own parents because they can't hear you or respond to what you are saying because they only use sign language and you have never been taught sign language at school?
>
> *Lucky Netshidzati, inventor of the Talking Glove* (*The Citizen* 2019)

Mobile app for the deaf: Almost 1 million deaf people in South Africa use South African Sign Language (SASL) as their primary means of communication because they cannot speak, read or read lips in any of the country's 11 official languages. These people are largely excluded from communicating with the hearing majority and their inability to communicate also means that government services, including healthcare, are often not accessible to them. Technologists at the University of the Western Cape have developed the mobile app SignSupport, which has been adapted to a WhatsApp interface, that unlocks prerecorded SASL videos stored on a cellphone and helps deaf people to bridge the communication gap in their preferred language. SignSupport also facilitates interaction with non-deaf users in limited communication scenarios, including pharmacies and International Computer Driver Licence training. When the prerecorded information or the limited interaction is insufficient, a video call with a remote SASL interpreter can be set up to provide further assistance.

Nigerian English in the Oxford English dictionary (OED): Linguists consider that the reason why English has become the dominant spoken language in the world (it is spoken, as a first or second language, by over 1.75 billion people) is because it has kept its 'borders' open and is receptive to new words, allowing it to grow from a minor West Germanic dialect into a world language. English is the national or official language in seven countries or regions of West Africa (Nigeria, Ghana, Sierra Leone, Liberia, Cameroon's Anglophone provinces, The Gambia and Saint Helena) and their use of English has impacted its vocabulary. The 2020 edition of the august OED features 29 new words or phrases from Nigerian English that have officially entered the English language, including *okada* (pesky motorcycle taxis that weave through Lagos' notorious traffic jams), *to chop* (earn money illicitly) and *rub minds* (confer). Other new words and phrases include *bukateria* (an inexpensive, casual restaurant), *danfo* (a privately-owned minibus taxi), *ember months* (the last four months of the year), *put to bed* (give birth to a child) and *tokunbo* (an imported second-hand product).

In a release note, the OED stated: 'By taking ownership of English and using it as their own medium of expression, Nigerians have made, and are continuing to make, a unique and distinctive contribution to English as a global language'. Most Nigerians expressed a great sense of pride in the fact that the unique ways in which they use English were being acknowledged internationally. Kingsley Ugwuanyi, a doctoral student on Nigerian English and consultant working with Prof. Dr Ulrike Gut on the OED project, has keen insight into how Nigerian words were chosen for inclusion in the dictionary. Important considerations included the period over which the word had been used and the frequency and distribution of its use, as the remit of the OED is not to prescribe how languages *should* be used but rather to report how they *are* used. The OED recognises that English – like all languages – is dynamic and ever-changing and that by including new words, it can tell a more complete story. The new words may include borrowings from indigenous languages, new concoctions, blends and compounds. Some words are included, as they have a specific cultural or historical significance. For example, '*Kannywood*' describes the Nigerian Hausa-language film industry based in Kano (Ugwuanyi 2020).

Naija, **Nigeria's pidgin English**: The 'Queen's English' is by no means the universal language of Nigeria. In Lagos, where the country's 500 languages come together in a chaotic medley, the rapid-fire rhythm of pidgin English is the symphony of the streets. Here street hustlers and Harvard-educated politicians alike greet each other with 'How you dey?' or 'How body?' The reply may range from a chirpy 'I dey fine' to a downbeat 'Body dey inside cloth' (literally 'I'm still wearing clothes'). Known officially as *Naija*, Nigerian pidgin is spoken by tens of millions across the country and has become the glue of Africa's most populous nation. Current affairs, English and local languages are brewed together to dish up playful imagery at breakneck speed. Like its variants in Sierra Leone, Ghana and Liberia, pidgin began as a form of broken English that allowed the country's coastal inhabitants to barter slaves and later palm oil with European traders. Later, under colonialism, English became the language of prestige. Not everyone is a fan of pidgin's rise. Some see its continued existence as a glaring symbol of a failed education system and the Nigerian authorities have periodically tried to clamp down on it (Mark 2012).

Naija is used in slogans scrawled on walls and advertising pitches on flashy billboards, and can be heard blaring out on Radio Wazobia (a pidgin-only station played in crammed buses and air-conditioned jeeps and at roadside food stalls). When a parliamentary session descended into chaos, the morning news bulletin included news presenter Steve Onu (aka DJ Yaw) saying politicians 'dey abuse them one by each. Nigeria don become mathematics' (a complicated and difficult problem). His co-anchor agreed, 'We need to reswagger Nigeria'. Rubber farmer Sunday Ayodele, alternating Yoruba (his first language) with pidgin, opined: 'The language of Wazobia is clear. Since Wazobia came, we have been enjoying, it's become like a kind of conversation. Before, it was just blowing grammar,' referring to the traditional English of politicians' speeches (Mark 2012).

> The common man wants to air their views but speaking in English they're scared – they don't want to do *gbagam* (hit something that makes a loud noise). Because we use just pure pidgin, the market women, the bus drivers, the mechanics that are not educated can listen to our news and actually understand what's happening in government.
>
> *Steve Onu, morning news presenter on Radio Wazobia* (Mark 2012)

Cape Town's Gayle: In the past two decades, there has been a greater acceptance of gay men in society and many of them now occupy centre stage on prime-time TV, such as South Africa's Somizi Mhlongo. Some gay men and lesbian women in South Africa, and those who show allegiance with the LGBTQ community, speak a dialect called 'Gayle' that morphed from a campish banter known as '*moffietaal*'. The word '*moffie*', for a gay or effeminate man or transvestite, originates from 'morphy', a slang word of contempt for a delicate, well-groomed youngster used by early 20th-century sailors (McCormick 2019).

'*Moffietaal*' can be traced back to coloured communities in Cape Town that have a long tradition of queer subcultures dating from the 'Kaapse Klopse' (minstrel concerts, first held in 1887) to the 1970s hairdressing salons of working-class suburbs such as District Six. What makes Gayle different from other gay dialects is that it uses a combination of women's names and Afrikaans slang to describe other gay people. For example, a *Polly Papagaai* is a gay man who gossips ('*papagaai*' is Afrikaans for 'parrot') and an *Adele Adder* is a vindictive gay man. Despite the rise of queer subcultures, and the seepage of camp into mainstream cultures, conventionality based on the binaries of man/woman, masculine/feminine and heterosexual/homosexual still rules. Anyone who deviates from this recipe faces the risk of being ostracised or even punished, so Gayle has developed as a means of communication that avoids detection and persecution (McCormick 2020).

Apartheid's court jester: Court jesters (who use humour, irony and exaggeration to highlight the vices, abuses and shortcomings of the political elite) have been popular in Africa since the time of the pharaohs. During the worst years of apartheid in South Africa, satirist Pieter-Dirk Uys and his alter ego Evita Bezuidenhout had no lack of material with which to ridicule the ruling elite. At a time when mentioning the African National Congress (ANC) was a crime, he would carry a bag emblazoned 'CNA' (the name of a stationery company) onto the stage and hold it up to a mirror for the audience to read whenever he 'mentioned' the banned organisation. Evita, whom Uys has portrayed since 1981, has worn many masks during 'her' glittering career, including as a 'member' of the National Party, an ambassador to the fictional homeland of Baphetikosweti, a loving wife and doting mother, and a committed comrade. Uys, who is widely known as the 'doyenne of political incorrectness', has been such a successful satirist that the targets of his sharpest jibes have become his biggest fans. He lives and works in the country village Darling, north of Cape Town, where he converted an old railway station into a cabaret venue called 'Evita se Perron' and created the famous satirical garden 'Boerassic Park', scornfully decorated with apartheid memorabilia and a prominent 'Gravy Train'.

Journalism at its best: At its best, journalism can seek to hold the rich and powerful to account. Through his satirical newspaper *L'Eléphant Déchaîné* ('The Raging Elephant'), modelled on France's *Le Canard Enchaînée* ('The Chained Duck/Paper'), Ivorian publisher Tiémoko Assalée challenges the elite of Coté d'Ivoire. He has reported on a vast network of customs fraud and on the use of tax havens by rich Ivorians but still has time for politics, having been elected mayor of Tiasallé in 2018.

Childhood illiteracy: Although South Africa is one of the best developed countries in Africa by some measures, the levels of childhood literacy are alarmingly low. The 2016 Progress in International Reading Literacy Study revealed that 78 per cent of South African learners in Grade 4 could not read with understanding and 62 per cent did not have school libraries. Moreover, about 60 per cent of South Africans lived in households without a single book (English 2020). Shelley O'Carroll, the director of Wordworks (an NPO specialising in early literacy development) is convinced that childhood literacy is a national crisis that needs to be addressed urgently by all levels of society. O'Carroll argues that investment in childhood literacy should start long before the children reach school. Wordworks develops programmes that focus on literacy development in the home by empowering parents and caregivers, and supports teaching in Grade R and pre-Grade R through various partners.

> The struggle for democracy was won by ordinary people taking a stand against an injustice that could not be tolerated. A generation of young children not able to read is another injustice that we must not accept. We need to encourage and inspire parents and caregivers to see themselves as their children's first teachers and give them ideas for supporting their children's literacy development, in informal settings, anywhere, anytime.
>
> *Shelley O'Carroll, director of Wordworks* (O'Carroll 2021)

Ben Rycroft, the communications manager at *Nal'ibali* (isiXhosa for 'Here's the story'), another childhood literacy NPO, is helping to develop a culture of reading among young children by giving them access to reading content in their home languages. *Nal'ibali* produces a bilingual newspaper supplement in eight languages and broadcasts audio stories in all the official languages on SABC Education radio stations (English 2020).

Zenzeleni Community Wi-Fi Network: Technologists at the University of the Western Cape in Cape Town have developed the Zenzeleni Community Wifi Network Solution, a solar-powered, community-owned telecommunications network that provides affordable communications to remote rural areas. Using wireless mesh networking, users can make free internal VoIP calls, while the cost of breakout mobile calls and data is greatly reduced. The network consists of scattered nodes powered by easy-to-install, modular solar systems from which community members can also charge mobile devices and provide lighting.

Internet shutdowns in Ethiopia: Online connectivity is relatively new in Ethiopia and the country has one of the lowest levels of internet penetration and use. Although Ethiopia experienced steady growth in internet connectivity from a low of 0.02 per cent in 2000 to 22.74 per cent of the population of 115 million people in 2019, these levels are well below those of other African countries such as Kenya (87.2 per cent) and Uganda (40.4 per cent). The internet has nevertheless played a decisive role in transforming the lives of millions of Ethiopians because many private companies rely on it to support technology-driven services, including banking. In the agriculture sector, the Ethiopian Commodity Exchange has launched a gateway for direct online trading of agricultural products among farmers. In the transport sector, Ride (the Ethiopian version of Uber) has simplified the lives of many people by offering affordable and accessible taxi services.

In recent years, internet shutdowns have affected the lives of millions of Ethiopians (Ayalew 2020). Between January and March 2019, residents of Wollega province did not have online access due to a security crackdown in the area. Other reasons given by the government and the national internet provider Ethio-Telecom for recent shutdowns include protecting the country from cyberattacks, quelling civil disobedience and preventing leakage of school examination papers. The shutdown in early 2019 prevented families from communicating with one another and severely affected humanitarian services. Shutting down the internet also had an impact on digital rights, which have been recognised as human rights since 2012 when the UN's Human Rights Council adopted a landmark resolution affirming that 'the same rights that people have offline must also be protected online'. While many conversations on digital rights tend to centre around civil and political rights, internet shutdowns also have an impact on rights to education, housing, health and even social security. Internet shutdowns should not be regarded as the go-to solutions to solve political crises.

WhatsApp in Zimbabwe: Democracy thrives in environments where voices, opinions and views across diverse profiles are respected and given a fair chance of being heard. In Zimbabwe (for many years an authoritarian state), WhatsApp, Facebook and Twitter, combined with radio, have helped to democratise communications by giving everyone a voice. Stanley Tsarwe (a journalism lecturer at the University of Zimbabwe) has found that radio is breaking away from being an analogue communication tool that relays top-down information to one that relies on multiple feedback loops, particularly via WhatsApp, which allows seamless and real-time debate between radio hosts and audiences (Tsarwe 2020). WhatsApp accounted for 44 per cent of all mobile internet usage in 2017 – in a country where 98 per cent of all internet usage is mobile. A downside of this communications bonanza is that not all WhatsApp messages reaching live radio studios are aired, as there is an oversupply of information.

The liberating Web: When Tim Berners-Lee (the inventor of the World Wide Web) visited the Women's Technology Empowerment Centre in Lagos (Nigeria) in March 2019 to celebrate the 30th anniversary of the internet, he encouraged women to use the Web to write about their experiences, hopes and aspirations. He also encouraged them to ensure that their culture, beliefs and local languages were represented on the Web, and to use their coding skills to solve problems specific to Nigeria and West Africa. He announced that he had launched a new contract for the Web as a global plan to ensure that it would serve all humanity. He encouraged companies to think beyond short-term revenue and user growth, and to develop sustainable business practices that people trust (Berners-Lee 2020).

Facebook making the internet safer: In February 2018, Facebook announced plans to create partnerships throughout Africa to create a safer internet. Under the theme 'Create, connect and share respect', Facebook will work with over 20 NGOs and GOs in Africa to raise awareness of emerging online issues and how to deal with them. The initiative includes a specially developed family-friendly animation (available in 50 languages) that directs viewers to the parental portal in the Facebook Safety Centre, where parents can access tips and videos on online safety with teenagers. Partners in the project include the JI Initiative in Ghana, Rudi International in Uganda and Watoto Watch in Kenya (*The Citizen* 2018).

> We recognize the important role we play in creating a better and safer online community for all. And with this year's growing partnerships across the continent, we further demonstrate our ongoing commitment to supporting organizations that raise awareness on these important issues.
>
> *Akua Gyekye, public policy manager at Facebook Africa (The Citizen 2018)*

Airwaves activist: Popular Liberian radio host Henry Costa uses the airwaves to channel public anger over maladministration by Liberia's government into action. The euphoria that followed the election of soccer hero George Weah as president soon dissipated after the quality of life did not improve and allegations of corruption were levelled against his government. Through his radio station, Costa organised a series of public demonstrations, gathering hundreds of Liberians onto the streets of Monrovia, to show Weah the red card. Although Costa has since been arrested for alleged visa violations and has had his radio station shut down, he has persisted with his campaign and organised protests with his Council of Patriots against Liberia's decision to auction off oil blocks in 2020 (Ayene et al. 2020).

Internet in the sky: According to Information Minister Joe Mucheru, in July 2020 Kenya became the first country in the world to have commercial high-speed internet using balloons in the sky. The technology was used before, but not commercially, when US telecommunications operators deployed balloons to connect over 250 000 people in Puerto Rico during the 2017 hurricane (*Cape Argus* 2020). The Kenya project is aimed at providing affordable 4G internet to under-covered or uncovered rural communities in the Rift Valley region and elsewhere. A fleet of 35 internet balloons fitted with solar panels and batteries will float in the stratosphere about 20 km high, well above airplanes and the weather. They will be launched from facilities in California and Puerto Rico, and controlled via computers from a flight station in Silicon Valley, using helium and air pressure for steerage. Each balloon has software equipped with AI to navigate their flight paths. The service is run by Loon (a unit of Google's parent Alphabet) and Telkom Kenya (the nation's third-largest telecommunications operator) – a perfect example of a high-tech solution to a down-to-earth problem (*Cape Argus* 2020).

> We are effectively creating the next layer of the mobile network around the world. We look like a cell tower 20 km in the sky. The floating base stations have a wider coverage, about 100 times the area of a traditional cellphone tower.
>
> *Alastair Westgarth, CEO of Loon (Cape Argus 2020)*

According to Nixon Muganda (2020), the balloons are effectively wireless internet connectivity towers that beam internet signals to Earth-based stations, which then transmit them to users through service providers. A single balloon will provide internet connectivity to an area about 80 km in diameter and serve about 1 000 users on the ground at a signal strength similar to 4G browsing speeds. The Google internet balloons are expected to have a positive impact on e-commerce, e-learning, e-health (including telemedicine), e-agriculture and e-government in remote parts of Kenya, and could result in reduced data rates (Muganda 2020).

Submarine connections: Mohamed Assoweh Bouh of Djibouti Telecom has praised the new 4 854-km Dare1 submarine cable linking Kenya, Somalia, Somaliland and Djibouti because it will be vital for 'easing congestion across existing systems, promoting competition and supplying much-needed capacity in a rapidly expanding region' (Pierce 2017). The project is being built by Subcom and is backed by Telkom Kenya, Djibouti Telecom and Somtel. It was due to be operational by June 2020, with the wider Dare project, including in Tanzania and Yemen. In Nigeria, engineer Funke Opeke's company MainOne plans to invest N25 billion (US$68.5 million) over the next two to three years to add 2 700 km to the existing 7 000-km undersea fibre-optic cable connecting her country to the outside world. Opeke's goal is to drive broadband penetration in Nigeria, and she has already secured N120 billion to achieve her goal (Egbejule 2020).

BRCK and SupaBRCK: With a portable modem that uses SIM cards to beam Wi-Fi to multiple devices, BRCK has revolutionised internet access in Africa. Now the Nairobi company offers SupaBRCK that provides connectivity to up to 100 devices at a time and its Moja public Wi-Fi system uses SupaBRCK hardware to provide free internet access. A platform for sharing content across Kenya and Rwanda with the rest of Africa is also in their sights. Anyone within range of the signal can connect to the internet for free via Moja (dubbed the 'people's internet') and access its stored content to watch shows, listen to music or read books.

Throughout BRCK's history, connectivity has been at the heart of their creativity. Their first product, BRCK v1 (which provided consistent access to the internet even in remote areas), was followed by the educational Kio Kit (which can transform any room into a digital classroom). SupaBRCK built on the capacity of the BRCK v1 for promoting enterprise connectivity, computing and storage, and PicoBRCK was created for rugged IoT needs such as remotely measuring soil moisture content in agriculture. BRCK is the only company in East Africa with staff that include electrical and mechanical engineers, full-stack cloud engineers, and firmware and OS software teams. They also have in-house design, user experience and product management teams who ensure that their products and services meet the needs of their end-users.

> We live in Nairobi, so there's no reason we should use devices designed in London and San Francisco. If products aren't conceptualized and created here, they won't fit our infrastructure needs, where electricity and internet connections can be problematic in both urban and rural areas.
>
> *Erik Hersman, CEO of BRCK* (Shapshak 2017)

Agricultural digitisation: Digitisation is changing the world of agriculture in Africa. In Ethiopia, over 4 million users subscribe to the 80-28 hotline (a free online farmer's advisory service) and similar successes have been achieved in Kenya and elsewhere. In Ghana, online platforms such as Esoko, Farmerline and Tortro Tractor provide farmers with useful advice, not only on how to grow crops or raise stock but also on how to access markets and make optimal use of extension services. In addition, the Centre for Agricultural and Rural Cooperation has launched the Transforming Africa's Agriculture: Eyes in the Sky, Smart Techs on the Ground project that supports the use of drones in agriculture. The number of African farmers who subscribe to digital services has grown by 40 to 45 per cent per year in the last three years (Abdulai, Duncan & Fraser 2019).

In South Africa, Vodacom has partnered with the farming business BPG Langfontein, which employs most of the inhabitants of Wakkerstroom (a town in Mpumalanga). The cellphone company has moved all the farmworkers from 2G to 3G smartphones, making Wakkerstroom the only smartphone-only town in South Africa. The farmworkers are now able to access online e-health and e-commerce services as well as educational and entertainment materials.

> We are helping to remove communication barriers so that citizens in the area can be part of the Digital Revolution and reap the associated benefits. By moving the more than 14 000 farmworkers to 3G devices, this will also free much-needed spectrum and this can be re-farmed to provide faster networks such as 3G and 4G.
>
> *Zakhele Jiyane, Vodacom* (Matumi 2019)

Transparency toolbox: Oluseun Onigbinde co-founded BudgIT to enable Nigerians to fight bad governance and hold the government accountable. An entrepreneur, open-data analyst and fiscal transparency advocate, he was appointed as the technical advisor in the Ministry of Budget and National Planning in September 2019, but immediately came under fire on social media due to his open criticism of the government that had appointed him. He resigned three days later and stated on his LinkedIn profile:

> I believe in a just, transparent and fair society where every citizen within a community has equal access to information about the fiscal position of their society and uses such opportunity to demand accountability as well as efficient service delivery.
>
> *Oluseun Onigbinde, co-founder of BudgIT* (Ayene et al. 2020)

Striving for dominance: In 1978, when Strive Masiyiwa was dismissed from Zimbabwe's guerilla forces and told to help rebuild the country, he took on the challenge with gusto. After a prolonged legal battle to remove the state monopoly on telecommunications, he founded Econet Wireless Zimbabwe, the parent company of the Econet Group, which now has operations in 15 countries worldwide. He decided that he also wanted to get into mobile money, so he bought a bank, rather than wait for changes in legislation. Not all his ventures have been successful, though, as he was forced to place Econet Media under administration in February 2020. Zimbabwe's richest man is also one of Africa's most generous philanthropists. His various foundations educate children, fight disease, invest in rural entrepreneurs and mitigate against climate change (Ayene et al. 2020).

Technology king: When Koos Bekker first joined the media group Naspers in 1985, it was effectively the propaganda arm of the National Party that had constructed apartheid in South Africa. He transformed it into a media giant, founded M-Net (one of the first pay television services outside the USA), was a founding director of MTN (now the largest cellphone operator in Africa), and was instrumental in launching MultiChoice and DStv. Having grown the company's market capitalisation from US$1.2 billion to US$45 billion, he retired as CEO in 2014 and then returned as chairperson in 2015. His successor, Bob van Dijk, has raised the ire of some stakeholders by appointing numerous foreigners to head a company that many South Africans regard as their own (Ayene et al. 2020).

Youngest black woman publisher: Margaret Busby, who was born in Ghana and educated at London University, became the UK's youngest and first black woman publisher when she co-founded Allison & Busby in the late 1960s. She has published books by notable authors (including Buchi Emecheta, Nuruddin Farah, Rosa Guy, CLR James and Jill Murphy) and her radio dramatisations encompass work by Sam Selvon and Wole Soyinka. A long-time campaigner for diversity in publishing, she has won many awards (including an OBE [Officer of the Most Excellent Order of the British Empire] and the Benson Medal from the Royal Society of Literature in 2017).

Famous photojournalist dies of Covid-19: John Bompengo, who covered the Congo's political turmoil as a freelance photographer and video journalist for The Associated Press (AP) for 16 years, died at the age of 52 years from complications due to Covid-19. He had contributed to AP since 2004, including coverage of the EVD outbreak in northern Congo in 2018, and also worked for the UN-backed news service Radio Okapi. Andrew Drake, AP Africa's news director, remembers Bompengo as a stalwart colleague and an impressive storyteller:

> Whether news was breaking in Kinshasa or across the river in Brazzaville, John was always on top of things, fast to arrive on the scene and with a plan to get the best pictures. He was committed to covering the flow of Congo's sometimes violent politics, always to be found at the heart of the action on the streets taking photos and video, but soon after he would be back in his suit covering the president. (Larson 2020)

Jerome Delay, AP's chief Africa photographer, said Bompengo was a valued colleague: 'John was a one-man-band international multiformat news agency — TV, radio, print and photos — he would excel in all fields. We have lost a brother' (Larson 2020).

Emerging innovator: In November 2016, Dawid Jordaan (who works at the Apps Factory of North-West University in South Africa) developed the android app *isiKokhulumayo* ('speaker'), which makes it easier for Sotho and Zulu students to understand academic concepts by quickly providing Sesotho or isiZulu translations of concepts taught in English. The app currently covers the disciplines of accounting, auditing and taxation.

Emerging innovations: With the advent of Covid-19 and the age of social distancing, there has been a surge in digital book clubs, online writing courses and virtual literary festivals. One of the successful African interventions is *Afrolit Sans Frontières* ('Afrolit without Borders'), founded by author and publisher Zukiswa Wanner. Her virtual literary festival is aimed at deepening knowledge of African literature by allowing authors from across the continent and the diaspora to host readings and discussions. In April 2020, Wanner and Maaza Mengiste curated an online session with the theme 'What I wish you would ask me' featuring 16 writers from 14 countries.

Prime Tech in South Africa is developing a device that will allow users to stream and access content for free without the conventional use of internet service provider data. Prime Tech claims that dataless streaming is the future of connectivity and will change the way in which we imagine and define what 'connection' is. They also envisage that this invention will be the foundation for future smart cities in Africa, as it will connect people from the most remote or off-grid regions to the digital world. A platform developed by biNu in Cape Town, #datafree, allows users to implement a reverse billing model so that when they access the website of an advertiser, no data is taken from their airtime or data bundle balance. This is because biNu has arranged with mobile network operators in South Africa, Nigeria, Cameroon, Zambia and Uganda for the advertisers to pay for the mobile data used.

Discussion: The widespread use of English in some African universities is drawing increasing criticism, especially in South Africa. This usage leads to some students whose second or third language is English being alienated and acquiring a feeling of worthlessness that results in them becoming silent observers rather than active participants in their education. It can also result in emotional stress and interfere with the learning process. Neville Alexander (an anti-apartheid activist and active proponent of multilingualism) states that, to an extent, the lack of creativity, spontaneity and initiative that comes with people using a second or third language predisposes the project of developing a distinctly African academy to one characterised by mediocrity or failure (Sigenu 2020). Another proponent of multilingualism, Zimingonaphakade Sigenu at the University of Cape Town, argues that incorporating African languages into higher education teaching curricula will demonstrate that they are meaningful instruments with which to conceptualise, theorise, conduct research and share knowledge, and their use will help to draw on the experiences of African people and on African traditional knowledge, in addition to validating their history and identity (Sigenu 2020).

Prof. Hassan Kaya, the director of the DSI-NRF Centre of Excellence in Indigenous Knowledge Systems in Durban, has emphasised that language is more than a tool of communication because it is also a channel for communicating people's cultural values and life aspirations. Furthermore, African indigenous languages are rich in unique language tools such as proverbs, idioms and witty expressions that make socioeconomic, cultural-political and environmental messages clear to specific cultural and social groups. It is through their indigenous languages that people articulate their worldviews, ways of knowing, value systems and methodologies of knowledge production in order to adapt and respond to societal and environmental challenges. Indigenous languages therefore play a pivotal role in people's lives, not only as a means of communication but also in cognitive development and social cohesion (Kaya 2020).

21

Electronics, computers and smartphones

'Paying for a taxi ride using your mobile phone is easier
in Nairobi than it is in New York, thanks to Kenya's
world-leading mobile-money system, M-Pesa.'

The Economist 2015

Introduction: The *Groupe Speciale Mobile* Association (GSMA) is a global body representing the interests of mobile operators worldwide and uniting more than 750 operators with almost 400 companies in the broader mobile ecosystem, including handset- and device-makers, software companies, equipment providers and internet companies, and organisations in adjacent industry sectors. The GSMA's 2019 The Mobile Economy report states that there were 456 million unique mobile phone subscribers in sub-Saharan Africa by the end of 2018 (representing a subscriber penetration rate of 44 per cent) and predicts that 300 million more Africans will use the internet by 2025. Furthermore, the report states that about 239 million people in Africa (equivalent to 23 per cent of the total population) use the mobile internet on a regular basis. But, despite the continent's accelerated drive towards a digital future, the high cost of mobile data remains an obstacle to millions of Africans because it limits their ability to use their smartphones to access job opportunities via online job portals; source educational material and other information; and connect themselves with customers, other stakeholders, friends and family (Arnesen 2020).

The Competition Commission in South Africa recently found that the two largest mobile operators in the country, Vodacom and MTN, are overcharging their customers by as much as 30 to 50 per cent for data and is in negotiations with them to reduce the costs. The leaders of several African countries, including South Africa's President Cyril Ramaphosa, have made strong appeals for the cost of prepaid data bundles to be reduced so that more low-income households can access these services. A free daily data allocation and free access to educational and other public interest websites have also been widely supported. Opera (the Norwegian Web browsing company founded in 1995) has endeavoured over the past 25 years to help millions of people to access the internet in the most efficient and cost-effective way, and has pioneered innovations such as tabs and speed dial that have become standard on Web browsers. Since 2016, Opera has focused its core business on browsers and AI-driven content discoverability apps. According to Jørgen Arnesen, the head of marketing at Opera, the company has developed an Africa First strategy that will help to bridge the digital divide on the continent and lead to millions more African smartphone users being able to access digital solutions on the internet (Arnesen 2020).

Opera's user base in Africa expanded by 26 per cent in 2019 to nearly 120 million users and they are endeavouring to reduce the cost of data by using data compression technology. This is achieved by preprocessing and reducing the amount of data from a website before it reaches a user's smartphone, which prolongs the lifespan of the user's data. In Nigeria, they have already reduced data costs significantly in collaboration with Airtel and MTN; they are in the process of rolling out this technology in Kenya, Ghana, Zambia, Uganda, Tanzania and South Africa (Arnesen 2020).

Computicket: Percy Tucker (who invented the first computerised ticketing system in the world, Computicket, in 1971) started life as a ticket seller or, as he puts it, 'a purveyor of posterior placements', but went on to become a theatre impresario mingling with stars of the stage and the screen. It was 18 years before another centralised, computerised ticketing system was developed elsewhere. Now 92 years old, Percy has devoted his life to nurturing the live arts and entertainment, particularly promoting multiracial participation in South African theatre. In 1959, he was part of the production team that brought to the stage the first all-black South African musical 'King Kong', directed by Leon Gluckman and co-starring the then legend-to-be Miriam Makeba. It played to multiracial audiences in South Africa before opening in London's West End in 1961 (Tucker 1997).

M-Pesa: Safaricom introduced the mobile money transfer system M-Pesa ('*pesa*' is money in Swahili) in Kenya in 2007, when 85 per cent of Kenyans did not own a bank account. M-Pesa allows registered users to move money safely using their cellphones, irrespective of whether they have a bank account or the collateral to open one. M-Pesa does not require a minimum balance and has no service charges, only pay-as-you-go fees for transactions initiated. Safaricom is no longer the only mobile network operator in Kenya offering M-Pesa (Airtel also offers the service), but it can take the credit for positioning Kenya as a leading contender for top technology innovator in Africa. Furthermore, Safaricom did not just rest on its laurels but addressed another problem in East Africa – the high cost of Wi-Fi. Instead of charging its customers exorbitant rates for the service, it offered Wi-Fi free to passengers on the Safaricom platform who travel on Kenya's *matatus* (taxis) – which proved to be another brilliant technovation in Kenya.

But who invented M-Pesa? Former Safaricom CEO Michael Joseph should probably get the credit for sparking the thought process that led to its invention, as the company at the time focused on building innovative new products. He asked his creative team to come up with ideas that had not been implemented in Kenya and, not surprisingly, the original idea was not M-Pesa but a microfinance loan system issued via cellphones. Soon Safaricom noticed that most people accessed the money only to send it to rural areas, so they shifted their attention to making a loan system that would facilitate the transfer of money from one phone to another, with each Safaricom customer being able to transform liquid cash into electronic money by visiting an M-Pesa agent. The money was stored on their SIM cards until they chose to withdraw it or send it to another Safaricom customer. According to Joseph, M-Pesa's original business plan targeted 350 000 customers in the first year but they soon exceeded 1 million. By December 2016, the M-Pesa service was processing over 6 billion transactions, peaking at 529 per second; by 2017, it was serving over 30 million customers with more than 280 000 workers in 10 countries!

But who came up with the idea that Joseph took to market? One candidate is Nyagaka Anyona Ouko, a student at the Jomo Kenyatta Institute of Agriculture and Technology in Nairobi, who claims to have pitched the concept to Safaricom. But the certificate of copyright that he has presented as proof is dated 2012, by which time M-Pesa was well established. If he was the inventor, why did he not patent his idea or invest in it? We will never know, but there is little doubt that he had no idea how successful it would become. Whoever first invented it, Safaricom now has the right of ownership of M-Pesa and is – legally and technically – the innovation's inventor (https://www.tuko.co.ke/280391-who-invented-mpesa.html).

mConsulting: Researchers at the University of Ibadan in Nigeria have shown that digital and mobile technology can improve access to healthcare during medical emergencies such as the Covid-19 crisis. An important component of this technology is mConsulting (mobile phone consulting with healthcare providers). In Nigeria, 15 functional mConsulting services are operated by groups of private doctors who use Web chats, text messages, specialised apps and video, and audio calls to dispense drug prescriptions, for patient referrals and follow-ups, and to provide information about healthcare facilities. Although mConsulting allows many people to access healthcare advice, this service does have problems, including many rural people's lack of access to online facilities, the danger of misdiagnoses as a result of clients not divulging all the relevant information, and data security and privacy. Funke Fayehun, Akinyinka Omigbodun and Eme Owoaje from the University of Ibadan have urged the Nigerian government to formulate policies to protect users of mConsulting services while making it mandatory for practitioners to receive appropriate training in digital literacy and interaction with patients at remote sites (Fayehun, Omigbodun & Owoaje 2020).

Piano Boost: Music is not immune to the app world. Technologists at the University of the Free State in South Africa have developed Piano Boost, an interactive mobile app that makes learning how to play the piano fun. The notes of the tune to be played are displayed on a cellphone or tablet, and the user must attempt to play the tune on a piano or keyboard. As Piano Boost has instant note recognition software, it can inform the user whether they are playing correctly and keeps a performance score. The app includes a tutorial that provides basic information on piano playing.

Dance analysis and Parkinson's disease: African people love to dance, whether it's Azonto, Rimbaxpaxpax, Alkayida, Skelewu, Bolo, Coupê-Dêcalé, Shoki or Vosho. Babusi Nyoni from Bulawayo in Zimbabwe has taken his passion for dance to another level by creating the smartphone app *Vosho Fo'sho* ('Vosho for sure') that quantifies and analyses the movements of the challenging Vosho dance sequence. Nyoni (a self-taught designer and innovator) launched the app, which was initially created as a gag, in December 2018 because he believes that dance is both a great communicator and a useful rehabilitation tool. The *Vosho Fo'sho* app allows users to document and assess complex movements over time and also to measure the extent to which the dancer is able to complete the complex routines, like a gymnast being judged for her performance of a set floor routine. But Nyoni has found that the app also allows him to assess the difficulty some people may have in carrying out complex or even simple movements, in particular for the diagnosis of Parkinson's disease (a neurological disorder that causes progressive deterioration in the ability of the body to move freely) (Rupiah 2020).

Nyoni solicited the help of the head nurse of the Ekuphumuleni Geriatric Nursing Home in Bulawayo, Sukoluhle Hove, and together they developed a research plan and gathered insights that helped them to develop the app Patana A1 which quantifiably assesses a patient's posture, gait, tremors and movements. Using TensorFlow (an open-source, machine-learning library developed by Google Brain, the deep-learning AI research team at Google), the app has been optimised to work on the most rudimentary smartphones – all the user has to do is point and shoot (Rupiah 2020). In 2019, Nyoni showcased Patana A1 at the TensorFlow World Conference in Silicon Valley and then visited France, Switzerland and Kenya, and attended the Fak'ugesi African Digital Innovation Festival in Johannesburg. He has since quit his day job and immersed himself full-time in Patana A and another app, Sila, which connects Africans to healthcare professionals through popular chat platforms. He brags, 'It is a deliberately African solution to a global problem' (Rupiah 2020).

Nyoni believes that there is an audience for new technologies in Africa, but they must be thoroughly researched and designed with a specific context and user-group in mind. He therefore co-founded the Tripleback Agency 'to make super-interesting tech things that hinge on the African experience' and took a year off inventing to experiment with machine learning and AI so that he could fully understand their potential (Rupiah 2020). He notes that many technologies that reach the continent are hand-me-downs developed in the West, with Africa as an afterthought, giving facial recognition software as an example. Modern computer vision algorithms are better than ever at distinguishing human faces, but only if the skin colour is pale or Caucasian. Race-neutral face recognition software needs to be developed in Africa because this technology is now widely used in security, law enforcement and the retail sector.

Andela and Flutterwave: Twenty-eight-year-old Nigerian Iyinoluwa Aboyeji is already an established innovator. His first successful start-up Andela, which he co-founded in 2014, identifies and develops smart software engineers to help global companies overcome the severe shortage of skilled software developers. He gained global recognition in 2016 when his company received a US$24-million grant from Mark Zuckerberg and now has offices in Nigeria, Kenya, Rwanda, Uganda and the USA. In 2017, Aboyeji co-founded Flutterwave, a platform that makes it easier for banks and businesses to process payments across the continent. In 2019, he launched the Future Africa Initiative, which provides capital, coaching and a collaborative community for start-up founders investing in rebuilding Africa. Through this initiative, he plans to make US$50 000 in capital available to at least 20 start-up founders each year (Ayene et al. 2020).

Emerging innovators: Dr Fundile Nyati (the CEO of Proactive Health Solutions) has taken the lead in broadcasting understandable information on Covid-19 in several of South Africa's 11 official languages, and sign language, on two Facebook pages, a dedicated website and YouTube. All his broadcasts, which are free, are available for social media sharing within five minutes of being aired and his website (www.fundilenyati.com) archives the broadcasts for later access (Harper 2020c).

Born into a family of entrepreneurs, Regina Kgatle won her first award at the regional Eskom Expo for Young Scientists when she was 10 years old, which encouraged her to help her parents expand their games arcade business through the development of computer games. She established Educade in 2013 and has showcased her outstanding work at the Berlin Game Week, Amaze Festival and GDC in San Francisco (considered the Oscars of the gaming industry), where she was named 'International Ambassador' and received an award from Oculus.

Wesley Diphoko, the editor of *The Infonomist*, believes that Africa needs to get onto the billion-dollar video games bandwagon, but in a different way to the rest of the world. Many of the most popular video games feature brutality and killing, and some have been blamed for fueling violence in schools. Diphoko suggests that instead of being a consumer of violent video games from elsewhere, African entrepreneurs should develop games with non-violent storylines such as achieving the UN's SDGs or promoting biodiversity conservation (Diphoko 2019).

Nothando Moleketi noticed that there is a market for reliable, pre-owned, leading-brand smartphones, so she launched ReWare in 2015 to market these highly saleable items. Her goal is to make smart technology available to a wider African audience and thereby make a positive impact on the environment by keeping used smartphones in people's hands and out of toxic landfills. She has entered into a partnership with the South African retailer Edcon, and plans to expand the business through an e-commerce platform and then a flagship storefront, followed by expansion on the continent.

Emerging innovations: During the height of the Covid-19 epidemic in South Africa, Tamir Shklaz and Wisani Shilumani built the Coronapp that provides accurate information on the virus and counters fake news. They noticed that fake news not only created uncertainty but also had the potential to lead to some dangerous practices, such as gargling bleach or washing food in bleach. The twitter section of their app only shows tweets from reliable sources such as the Department of Health, which ensures that there is a reliable, real-time flow of accurate information to users. Shklaz, who has a degree in computer and electrical engineering, also launched the start-up Quillo (which allows students to buy and sell second-hand items from the safety and convenience of their university campus) during his third year of study at the University of Cape Town. Shilumani has a degree in mechanical engineering and has been developing apps and websites since he was a student (Wentzel 2020).

Babusi Nyoni of Zimbabwe, the founder of Sila Health, created the app Sis Joyce that is uniquely suited to the African market. It uses a chatbot powered by AI to provide intermediate access to free basic healthcare information through platforms such as WhatsApp and Facebook Messenger, both of which are light on mobile data. The app, which has been used in Bulawayo to augment Zimbabwe's Covid-19 response, creates automated responses to a user's questions based on accurate medical information.

> Our platform provides online assistance to people who believe they may have coronavirus symptoms, to help keep low-risk patients from clogging up hospitals. We do this by disseminating information on the virus … enabling users to assess their symptoms through our chat-based World Health Organization symptom checker.
>
> *Babusi Nyoni, founder of Sila Health (Mail & Guardian [online] 2020a)*

In South Africa, Praekelt.org has developed HealthAlert, a tool that responds to user's questions about Covid-19 in multiple languages and is able to replicate the triage process (i.e. help doctors to determine the priority of a patient's treatment based on the severity of their condition or their likelihood of recovery with and without treatment). HeatlhAlert also provides real-time data to decision-makers on the prevalence of Covid-19 infections and has been adopted by the WHO (Nash 2020).

Discussion: According to the San Francisco-based podcasting platform Stitcher, the number of podcasts grew by 129 000 per cent from about 350 000 in 2010 to over 7 million in 2019. That is impressive growth for any technology investor! Interestingly, episode length has decreased by about 2.4 minutes, a reflection of our shorter, modern-day attention spans. In Africa, podcasts have given a voice to people who want to tell stories or share opinions, but do not have access to a radio station. They are now more accessible to many people than FM radio stations, as you can use your smartphone to converse with anyone wherever you are.

A plethora of technologies has converged in smartphones, but could they also be used for medical diagnosis? Scientists at the University of Belfast in Northern Ireland think so, as smartphone-based tests have already been developed for detecting HIV, malaria, TB and various food contaminants, and work is under way to use them to detect Covid-19. The tests would be done with specially made labels designed to react to the presence of a particular virus. A throat swab or blood sample would be added to the label and the light, colour or electrical signal that the reaction generates would be detected and interpreted by the smartphone. The test could be used to determine whether someone has had or currently has the virus, and to detect the virus on surfaces such as food packaging in the home and office. With the rapid uptake of smartphone technology in Africa, this test could be a lifesaver for many people.

22

Robotics, machine learning and artificial intelligence

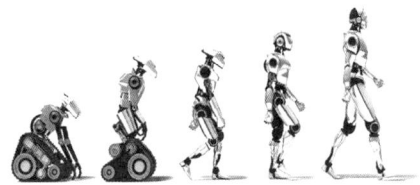

'I would rather be optimistic and wrong than pessimistic
and right when it comes to predicting the future.'

Elon Musk, South African-born innovator (Rogan 2018)

Introduction: The word 'robot' was coined 100 years ago by the Czech writer Karel Čapek in a play about Rossum's Universal Robots taking over the work of people and eventually exterminating them. However, in the past, automated machines did not only exist in science fiction. In the House of Wisdom in Baghdad in the 9th century, the three Banu Musa brothers (mathematicians, engineers and inventors) built 'trick machines' and elaborate fountains that carried out automated functions. They documented 100 of them in their *Book of ingenious devices*. In the 10th century, parks in Baghdad had 'running' camel and lion automatons operated by hydraulic pressure. The Muslim engineer Ismail al-Jazari created elaborate automatic clocks and musical boats in the 12th century that were powered by water and gravity. He documented them in *The book of knowledge of ingenious mechanical devices* (Al-Hassani 2006).

Today robots are an everyday reality, from car assembly lines to remote-operated surgical devices, solar-powered 'wolves' that protect livestock in Japan, sentries that patrol borders, nimble robots that lead exercise classes, robotic buddhas that deliver sermons in Japanese temples, bomb-clearing robots and cleaner robots that kill microbes in hotel beds using ultraviolet-C radiation. Computers are also becoming intelligent much faster than we thought they would. In 2017, a computer beat the best 'Go' player in the world 10 years earlier than expected. Summit, the world's fastest supercomputer, can process 200 000 trillion calculations per second and is about to be knocked off its pedestal by Intel's Aurora. When it is switched on in 2021, Aurora will be able to perform a quintillion calculations per second, five times faster than Summit. When quantum computers arrive, they will be millions of times faster than Aurora (Sanei 2019b).

Robotic policeman in Tunisia: Police robots were deployed in 2020 to patrol the streets of Tunisia's capital Tunis to ensure that people observed the Covid-19 lockdown measures. If the robot spied anyone walking in the deserted streets, it approached them and asked them why they were out. They then had to show their IDs to the robot's camera so that officers controlling it could check them. The four-wheeled robots called PGuards, which were designed and built in Tunisia by Enova Robots, have a thermal-imaging camera and Lidar (light detection and ranging) technology that works like radar but uses light instead of radio waves (Javad 2020). According to Radouhane ben Farhat (the chief sales officer at Enova Robots), the company (which is based in the coastal city of Sousse) also produces a healthcare robot that does preliminary visual diagnoses and uses its sensors to measure certain parameters. One will soon be working at a hospital in Tunis to help diagnose Covid-19, which will minimise physical contact between medics and patients (Javad 2020).

Hospital ward robot: Intensive care frontline workers in Tygerberg Hospital in Cape Town can thank a robot named Quinton for reducing their risk of Covid-19 infection. Quinton does rounds of wards of infected people under the remote control of physicians who are at risk of infection or who are convalescing at home after being infected. According to pulmonologist Prof. Coenie Koegelenberg of Stellenbosch University, 'The odds of at least one of us falling ill are quite high, so we need to realistically plan for what could happen. If any of the specialists gets the virus and is unable to physically go to work, we will be able to function remotely, using the robot from a phone or a laptop'. Charmaine Lambert, the head of the co-working space WorkInProgress (an Absa Innovation Lab), states, 'Robotic process automation can take over manual, repetitive tasks – but instead of making the people in those functions redundant, it frees them up to tackle more important and non-automatable tasks which can improve business operations' (Githahu 2020).

In Netcare hospitals in Johannesburg, Xenex pulsed ultraviolet (UV) robots and Yanex Pulsed-Xenon UV robots have been deployed to use high doses of UV light to decontaminate wards, personal protective equipment and medical supplies. In May 2020, Rwanda deployed four humanoid robots to screen body temperatures and monitor the status of patients in Covid-19 treatment centres to minimise human-to-human contact (*News24* 2020). Students from the Dakar Polytechnic School in Senegal have built a multifunctional robot, Doctor Car, designed to lower the risk of

Covid-19 contamination from patients to caregivers. The device, which is equipped with cameras and is remotely controlled via an app, can move around the rooms of quarantine patients, take their temperatures, and deliver drugs and food.

Blackchain: David Phume, who hails from Bryanston (Johannesburg, South Africa), was born with an entrepreneurial spirit and has never been afraid to dream big. He studied 3D animation in Johannesburg before graduating from a San Francisco-based animation school. In 2005, he founded award-winning Penthouse Motion Pictures (a broadcast design and animation studio) but his dreams were much bigger, so he went on to study robotics and AI online with Udacity. He is determined to become a leader in these fields in Africa and founded the technology company Blackchain that identifies problems and finds solutions through technology. It is founded on four pillars: to inspire, identify talent, nurture skills and innovate. Having spent time in Silicon Valley, it would have been easy for Phume to copy their model but, as an African technologist on a mission, he wanted to start a company that had the right approach to address technology problems in Africa. 'I first needed to shut myself from all the noise and clutter and come to terms with who I am and where I'm from. This helped me find a sense of purpose and mission,' he says. Phume is a strong supporter of start-up accelerators and points out that the accelerator YC combinator in Silicon Valley was single-handedly responsible for a number of unicorns, such as Stripe and Dropbox as well as the Nigerian fintech start-up PayStack.

> I am a big dreamer, so my biggest struggle is my desire to take on problems that are too large at an early stage. Yes, it's a good thing to be ambitious, but without the right steering, it can also be dangerous. Larry Page and Sergey Brin started off by tackling one problem and that was search. They didn't start off wanting to take over the world. Elon Musk didn't start off by building a rocket.
>
> They all started tackling problems their own size then developed a bigger appetite for bigger problems once they conquered the smaller ones … Our goal for the next five years is to have over 5 000 young African technologists enrolled in our Robotics and AI education programme and being a barracks for an army of technologists in Africa.
>
> *David Phume, founder of Blackchain* (Courie 2020)

MIIA robotics kit: Technologists at the National Research Foundation in Pretoria (South Africa) have developed MIIA (a novel build-it-yourself robotics kit and curriculum) to teach some of the core concepts of electronics, programming and robotics to learners across a range of grades. MIIA includes tailored courses that are flexible, scalable and customisable to multiple user groups. The courses are offered as all-inclusive packages that include physical hardware, access to the curriculum and access to enhanced features of a programming app. MIIA has been commercialised by RD9 Solutions and provides packages that are suitable for both in-curricula and extracurricular activities in schools, and can also be used in workshops, hackathons or competitions. Although school learners are the primary target of the kit, it can be tailored to the needs of post-school users. Through its unique mobile app and self-learning driven curriculum, MIIA gradually exposes newcomers to the world of programming, electronics and robotics while simultaneously providing them with the opportunity to obtain hands-on experience with modern technology.

Google Artificial Intelligence Centre and AIMS: Moustafa Cisse from Senegal heads the Google Artificial Intelligence Centre in Accra (Ghana), where his research focuses on the essential requisites of AI: fairness, transparency and reliability. Cisse believes that Africa's technology solutions should be developed within the continent and is developing AI programmes that will help farmers to diagnose blights which might affect their harvests. He plans to translate useful computer software into African languages.

A lack of opportunity on his own continent during Cisse's youth led him to study in Europe but he returned to Africa to establish the African Institute for Mathematical Sciences (AIMS) in 2003 in partnership with the University of Cambridge, University of Cape Town, University of Oxford, *Université Paris-Sud XI*, Stellenbosch University and the University of the Western Cape. AIMS is a pan-African network of centres for postgraduate training, research and public engagement in mathematics and science that will enable the continent's youth to shape the future through STEM education and train Africa's next generation of leaders. It currently has branches in South Africa, Senegal, Ghana, Cameroon, Rwanda and Tanzania. AIMS South Africa, which is based in Muizenberg (Cape Town), provides a one-year structured master's programme in mathematical sciences that will help to develop a critical mass of African academics, researchers and entrepreneurs on the cutting-edge of STEM. The AIMS South Africa Research Centre was launched in 2008 with the goal of conducting and fostering outstanding research and learning in the mathematical sciences, and the AIMS Schools Enrichment Centre is building the capacity of Africa's teachers through training programmes that increase the pipeline of mathematics and science educators.

AI in Madagascar: Madagascar is not an obvious first choice for cutting-edge AI applications, but several companies there are leading the way in finding AI solutions. In particular, Fusion Informatics in Antananarivo is able to assist clients with creating a range of AI solutions by applying natural language handling and machine learning to reduce infrastructure costs by using cloud solutions and mobile apps that enhance business productivity, develop chatbots to facilitate communication with customers, reshape business processes by introducing high-end IoT connections, offer intelligent video surveillance, and create plans that integrate high-tech visual apps and data.

Refresh Chatbot: Refresh Creative Media in Cape Town has created a chatbot that offers quick AI-powered answers to frequent customer questions and requests. The Refresh Chatbot has an easy-to-use menu and can be integrated into a company's online information system. Pick n Pay, South Africa's second largest supermarket chain store, recently installed the Refresh Chatbot to perform tasks such as ordering new SmartShopper cards, obtaining information on store locations and opening times, and even providing answers to Covid-19 FAQs. The chatbot's services can be distributed across WhatsApp, website chats, voice calls or social media channels such as Facebook Messenger (Bruton 2020; www.bit.ly/refreshchatbot).

Novel drones: The US-based company Zipline is using drones to ferry medicines, blood, vaccines and Covid-19 test kits to remote parts of Ghana. According to Daniel Marfo, Zipline's CEO in Ghana, the drones have been exceptionally useful in reaching remote areas quickly and efficiently. 'The government told us that their biggest challenge is that the virus has spread out of the cities, they have suspected cases popping up in the rural areas and the logistics from the rural areas to the cities are very difficult,' he says (*News24* 2020).

Reuben Pillay (a drone enthusiast in Mauritius) spent 18 months developing the project ReubsVision that creates a 3D map of his island home, which is not covered by Google Street View. Using a DJI Phantom 4 Pro camera drone, he shot photos in every direction from points in the sky and then stitched them together into 220 ultra-high resolution 360° images covering the entire coastline of Mauritius. To share his work with the world, Pillay coded up a website that shows overlays of all his photo locations. Clicking on any location marker opens a 360° photo of that site, and viewers can then look in any direction as well as zoom in on the high-resolution image.

> This was actually the first website I ever built, the first domain I ever purchased and the first hosting server I ever 'bought'. I had no prior experience in doing any of that, but it was fun discovering all this, and to be honest, being confined at home during the Covid outbreak really helped in getting me focused on this. All I want for now is that people discover my island and hopefully that'll bring tourists over here once we re-open our airspace and this Covid situation is out of the way.
>
> *Reuben Pillay, creator of ReubsVision* (Zhang 2020)

Nosetsa automatic irrigation system: The Water Research Commission of South Africa has developed a machine-learning-based precision irrigation add-on called Nosetsa that provides autonomous irrigation using real-time sensor data, which allows the user to avoid excessive water losses through evapotranspiration and over-irrigation. Nosetsa can take accurate measurements because of its AI features and records data using record-keeping algorithms while allowing for effective communication through its connection to the IoT. Potential users include commercial and emerging farmers, as well as property owners and golf course and park managers.

Kinematic robotic arm: Scientists at the University of Pretoria in South Africa have invented a hybrid serial-parallel kinematic industrial robotic arm that leverages all the benefits of serial and parallel kinematic machines while minimising their disadvantages. The design is based on the transmission of mechanical power via embedded drivetrains to each joint in the system from actuators that remain stationary in the base. This allows the robotic arm to have a lightweight structure with greater stiffness and stability while still offering high agility and precision. This architecture lends itself to the development of reconfigurable industrial robots, but can also be applied to robotic exoskeletons and prosthetics.

3D printing: The cost of 3D printing has plummeted in recent years, and the printers have become infinitely faster and more accurate. Shoes, building materials, car and airplane parts and spare parts for the International Space Station are already being made using 3D printers, and soon smartphones will have a 3D-scanning capacity that will allow you to scan your feet and print your new shoes at home. It is predicted that by 2027, 10 per cent of everything made will be 3D printed, and this is probably an underestimate. Mehul Shah from Ultra Red Technologies (a 3D printing company in Nairobi, Kenya) has 3D-printed anticoronavirus face masks and is also printing components that will allow ventilators to be used on more than one patient. In Nairobi, electrical engineering student Fidel Mukatia and colleagues from the Chandaria Business Innovation Incubation Centre at Kenyatta University, in collaboration with medical expert Dr Gordon Ogweno, have produced a low-cost ventilator at a 10th of the price of imported machines. In Ghana, the Academic City College in Accra and Kwame Nkrumah University of Science and Technology in Kumasi have produced a ventilator that costs between US$500 and US$1 000 and takes only an hour to assemble.

Emerging innovators: In Somalia, 21-year-old public-health student Mohamed Adawe invented a simple automated resuscitator that pumps oxygen from an air tank into the patient's mask, to replace the conventional Ambu bag (*News24* 2020). In Benin, the digital fabrication start-up BioLab is also producing

3D-printed face masks. FabLab, an innovation hub in Kisumu, has developed an app called *Msafari* ('*safari*' means 'journey' in Swahili) for the contact tracing of passengers on public transport. Passengers entering a *matatu* (minibus taxi) input a simple code on their smartphones with the vehicle registration number. According to Tairus Ooyi (the lead app developer at FabLab), if one of the passengers tests positive, they can trace all the other contacts that travelled in the same vehicle (*News24* 2020).

Discussion: Crises such as wars, extreme climatic events and pandemics have always promoted technological innovation. The Covid-19 crisis is no exception and it is widely predicted that the pandemic will be a major incentive to automate workforces (Semuels 2020). A slew of robots has been developed worldwide to remotely or automatically carry out routine duties such as measuring body temperature, blood pressure and oxygen saturation, reviewing documents, carting supplies around hospital wards, delivering meals and prescriptions to retirement homes or quarantine hotels, or disinfecting indoor and outdoor spaces. In China, quadcopter drones are used to ferry test samples to laboratories and provide light at night while temporary hospitals are built, or use thermal imagery to identify infected people and enforce social-distancing restrictions.

Novel four-winged drones (ornithopters), which flap their wings to generate forward thrust, have been developed at the University of South Australia by reverse engineering the aerodynamics and biomechanics of birds and insects. These drones can take off, fly, hover and land with the agility of swifts, hummingbirds and dragonflies, and outperform existing drones with static wings or propellers (Chahl 2020). They will have wide application, not only in medicine but also in agriculture, pest control and environmental monitoring. Robots are rolling through public spaces broadcasting messages, 'manning' call centres and encouraging social distancing. In Singapore, a robotic dog (Spot) is promoting social distancing; in India, they are using AI-powered myth-busting chatbots; in Japan, students used robots to walk the stage during their virtual university graduation ceremony; a man in Cyprus used a drone operated from his balcony to walk his dog without violating stay-at-home restrictions (Murphy, Adams & Gadudi 2020); in the USA, FluSense (an AI-powered monitor) is used to estimate the size of crowds and the risk of Covid-19 infection.

Machines have made jobs obsolete for centuries. Gutenberg's printing press displaced writers and the spinning jenny replaced weavers, push buttons ousted lift operators, ATMs have replaced many bank tellers and the internet drove many travel agencies to bankruptcy. In many cases, robots (including drones) have not replaced people but have performed tasks that a person cannot safely do or routine tasks at which a robot is more efficient, thereby freeing up workers to cope with their increased workload. Many robots are tele-operated, which enables healthcare workers to apply their expertise and compassion remotely to sick and isolated patients; others are autonomous, like those used to decontaminate hospital wards with UV light (Murphy et al. 2020). While some new robots are being developed, for example, to take blood samples, perform mouth swabs or carry out tests, the speed at which the Covid-19 crisis unfolded has necessitated that existing robots be repurposed so that they can be used immediately. For example, agricultural drones used for spraying insecticides have been adapted to spray decontaminants in public spaces in China and India, and thermal imaging drones have been repurposed from the thermal drones used at bomb disposal sites and nuclear power stations.

The Covid-19 crisis has made us realise that robots are not only useful for repetitive grunt work, but also for sophisticated tasks. Also, prior to the pandemic, robots were mainly seen as devices to improve efficiency, precision and productivity, but we now realise that they will also improve health and safety. Furthermore, technological breakthroughs have allowed engineers to include powerful data-crunching tools in robots, and better digital communications have allowed them to connect simple robots to external brains through the IoT so that they have a 'collective intelligence'. Increasingly, robots are moving away from being human lookalikes to an adaptive radiation of novel 'electronic species' with different abilities.

The future of robots will be determined by factors such as their ability to detect defects and self-repair, and the availability of new, lightweight materials that can conduct electricity. In 2017, when China announced its ambition to become the world leader in AI, world leaders began to talk about a global AI arms race and expressed concern about the growing reach of China's authoritarian surveillance state. However, many of China's AI programmes may benefit humankind, especially in Africa, as their approach is pragmatic and orientated towards finding applications that solve real-world problems. Applications under investigation include 'AI Doctor' chatbots that help to connect remote communities with experts via telemedicine; machine learning to speed up pharmaceutical research; and using deep learning for medical image processing, which could help with the early detection of cancer and other diseases (Elliott 2020).

23

Military and security

'Imagine all the people, living life in peace.'

John Lennon, English singer, songwriter, musician and peace activist
(Anon 2021)

Introduction: Modern warfare in Africa has largely been carried out using imported weaponry, although South Africa developed a formidable armaments industry from the 1960s to the 1980s. Until the importation of guns and cannons from Europe, military battles in Africa were carried out using traditional weapons, especially spears and bows-and-arrows (sometimes wielded by warriors on horseback), as well as knobkerries, battle axes, swords and long-bladed daggers, with shields made from giraffe or buffalo hide. Small swivel-cannons were first introduced on the West African coast late in the 18th century. For example, the city-state of Lagos deployed medium-sized canoes carrying up to 25 men armed with swivel-cannons and muskets that they used to provide covering fire during landings. On the East African coast, sea-going dhows were used by both pirates and military personnel for attacks on shipping and land bases.

On Lake Chad in the early 19th century, the piratical Buduna fielded a fleet of about 1 000 reed canoes with spears and shields for armament; in the east, local kingdoms vied for supremacy using large numbers of war canoes on the African Great Lakes. In Nigeria, large war canoes with up to 20 pairs of swivel-guns mounted on crossbeams were reported in the mid-19th century. In 1841, the ruler of Abo could muster about 300 canoes, many armed with muskets and bow/stern cannons, although some still relied on traditional weapons. In the Niger Delta, the huge *Itsekiri* war canoes (propelled by 40 slave rowers) carried multiple cannons and swivel-guns and over 100 warriors. On several occasions, they fended off British warships by blockading narrow creeks and waterways and, in disputes with the colonial regime or European merchants, they could shut down trade on the Benin River for several months.

Lessons from Shaka: The war strategies of Shaka kaSenzangakhona Zulu, King of the Zulus from 1816 to 1828, had a remarkable similarity to operational strategies used during World War II: the optimal assignment of people to projects was adopted to ensure that the total cost or time to perform a task was minimised. Shaka divided his warriors into *impis* ('regiments') based on their age, skills and physical abilities. The *impis* approached the enemy in a claw formation (*impondo zenkomo*), with the younger warriors at the points of the claw and the veterans in the middle. Shaka also invented a short-shafted, broad-bladed stabbing spear (*iKlwa*) that encouraged his warriors to engage in close combat and a large cowhide shield (*isHlangu*) used to push aside the enemy's shields (Bruton 2017a).

South African military vehicles: South Africa produced over 80 kinds of military vehicles in the past. SAMIL (South African Military) trucks were based on German Magirus Deutz trucks, British Bedford trucks and/or French Panhard vehicles – all of which underwent extensive modifications to suit local conditions. In addition to troop-transporters, a variety of SAMIL vehicles was developed for specific purposes, including the *Ratel* ('honey badger') Light Tank, Cactus Missile Launcher, *Buffel* ('buffalo') Personnel Carrier, Jackals Jeep, *Shongololo* ('millipede') Tank Transporter and the Oliphant Mk 1 Tank. One of the most innovative – and infamous – SAMIL vehicles was the Casspir, a four-wheel drive, armoured personnel carrier designed by the CSIR to provide protection against landmine explosions. The V-shape of the hull, which was unique at the time, could withstand an explosion of 14 kg of TNT under the hull or 21 kg under a wheel. Casspirs were built in many different configurations, including the *Rinkhals* (ringnecked spitting cobra) Field Ambulance, *Gemsbok* (gemsbuck or oryx) Recovery Vehicle, Mechem Mine Recovery Vehicle, Plofadder Mine Clearing System, and a police and internal security Riot Control Vehicle. They even inspired the US Marines' Mine-Resistant Ambush Protected vehicle project. Casspirs were used in combat in Mozambique, Namibia, Angola and Iraq, and on peacekeeping missions in India, Croatia, Afghanistan, Indonesia, Nepal, Peru and Uganda. Although the military stopped making SAMIL vehicles in 1998, modified versions have continued to be sold to farmers, construction and mining companies, and foresters under the names Samag (South African Magirus) and Magirus Deutz (S. Camp, personal communication, November 2018).

Unmanned aerial vehicles: Drones are not 21st-century inventions. The first drone, the Ruston Proctor Aerial Target, was made in 1917. It was a radio-controlled, pilotless airplane based on technology developed by the brilliant Serbian inventor Nikola Tesla that was meant to act as a flying bomb during World War I but was never used in combat. Long before drones came into the public domain, the military used unmanned aerial vehicles (UAVs) for surveillance and other operations, and South Africa was at the forefront of these developments. The first prototype in South Africa was developed by the CSIR in 1977 and delivered to the South African Air Force (SAAF) in 1978. Known as the Champion, it was a technology demonstrator and trainer. Four Champions were built, and some were supplied to Rhodesia (now Zimbabwe) for use in their bush war. In 1986, Denel (the South African state-owned aerospace and military technology conglomerate) delivered the Seeker UAV to the SAAF. It was used operationally in Angola in 1987 for surveillance, reconnaissance and artillery spotting, and to lure Soviet surface-to-air batteries out of hiding so that long-range G5 and G6 cannons could destroy them. The Seeker was also exported to Algeria and the USA (S. Camp, personal communation, 2019).

The Seeker II UAV had a range of 250 km and could reach altitudes of 6 000 m. In 1990, Denel developed the Skua high-speed target drone (which has been exported since 1999) and the Vulture Tactical UAV for target acquisition, surveillance and fire correction for South African-designed G5 and G6 Howitzer artillery cannons. A unique feature of the Vulture is that its launch, flight and landing are all fully automated. It transmits real-time data/video footage to a ground station using a C-Band communication system developed by Tellumat, the original producers of the famous Tellurometer distance measuring device (Bruton 2017a). In 1994, South Africa was the first country to use drones in commercial airspace for civilian purposes when the first open general elections were monitored using UAVs. In 1995, Denel developed the Seraph Stealth UAV with a low radar detection cross-section to reduce its detectability. This remarkable mini-aircraft is powered by a Microturbo turbojet and can achieve a maximum speed of Mach 0.85 (1 041 km/hr) and an altitude of 12 000 m while carrying a payload of 160 kg (S. Camp, personal communation, 2019).

Helmet-mounted display: The first aircraft with a simple pilot's helmet-mounted display to aid in targeting heat-seeking missiles flew experimentally in the mid-1970s. At the time, Denel Optronics and the SAAF developed a novel helmet-mounted sight for Mirage F1AZ pilots that enabled them to make attacks without having to manoeuvre into the optimum firing position. The SAAF was the first air force to fly the helmet-sight operationally and South Africa subsequently emerged as one of the leaders in helmet-mounted sight technology. After the South African system had been proven in combat, the Soviets embarked on a crash programme to counter the technology. In 1985, they produced the

MiG-29 jet with a helmet-mounted display and off-boresight weapon (R-73), which gave them an advantage in close-in manoeuvring engagements.

From 1985 onwards, Denel Optronics developed a high-speed head-tracker system for the Eurofighter-Typhoon (NATO's latest jet fighter). The design comprised three cockpit sensors (essentially tiny video cameras) that detect a series of LEDs embedded in the pilot's helmet. A computerised head-tracker processor captures data from these sensors and rapidly calculates the angle and position of the pilot's head. This information is used to correctly position the display of symbology on the pilot's helmet-mounted display.

The helmet-mounted display projects vital flight, instrumentation, navigation and mission data, together with weapons and countermeasures status, directly onto the pilot's visor. With this information directly in front of his eyes, the pilot never has to take his eyes 'off the road' by glancing down at physical instruments. This is a significant safety enhancement, especially for pilots flying high-speed, low-level missions. The head tracker processor also drives external sensors and missile seekers, keeping them aligned with the pilot's line of sight.

Denel Optronics developed this unique, world-leading optical tracking technology entirely in South Africa. The latest version of the system (for Eurofighter-Typhoon) tracks and processes data three times faster than previous versions, making it ideally suited to the latest generation of fighters. In 2007, the German company Carl Zeiss Optronics GmbH acquired a 70 per cent stake in Denel Optronics as part of Denel's restructuring process.

Eloptro: Eloptro was established in South Africa in 1974 with funding from Armscor (the acquisition agency for the Department of Defence) and technical assistance and the transfer of key personnel from the CSIR's Optical Sciences Division; it is now part of Denel Optronics (www.eloptro.co.za). Its original product range included image intensifiers, riflescopes, observation sights, driverscopes and night vision devices for fire control; its present-day products include submarine periscopes, laser rangefinders and laser target designators. Eloptro also has a submarine periscope upgrade facility that can carry out a complete redesign of the optical system, facilitate integration with a passive rangefinder, add television display capability (day and night) and/or attach a still camera to the periscope, as well as offer continued logistics support. Eloptro has formed an alliance with Zeiss for the design and manufacture of new periscopes for U-109 submarines (McDowell 2006).

Eloptro developed the Nightowl laser designator for the Rooivalk attack helicopter. It has designed and manufactured laser rangefinders for integration into the gunner and commander sights of various armoured vehicles and airborne, naval and land-based optronic systems. The company also produces a range of handheld eye-safe laser rangefinders with a range of 80 to 20 500 m and a resolution of 5 m. One of their most advanced laser devices is the Eagle Eye target-acquisition binocular, which has an integrated eye-safe laser, built-in laser filter, digital compass, GPS, digital camera and voice recorder. Eloptro also has the capability to produce a wide range of spherical lenses, prisms and mirrors for civil society, including lenses for camera obscuras (McDowell 2006).

Parachute fabrics: The South African company Gelvenor Textiles has researched and produced a wide range of aeronautical textiles since 1965. Gelvenor was founded as a joint venture between the Dutch company Gelderman & Zonen and the South African finance house Grosvenor Holdings. Their fabrics for paragliding, skydiving, hot air ballooning and military canopies are world class and have been used internationally for over 25 years. Gelvenor Textiles also developed the first flame-retardant fabric in Africa and, since 1994, has been Africa's largest weaver of aramid (aromatic polyamide) heat-resistant fibre for body and vehicle armour, which is in use in war zones around the world. In 2002, they developed the first microfibre low-bulk parachute fabric, which has become internationally known for its durability, lightness, flexibility and high tear resistance, and is now the industry standard. Most recently, they produced re-usable anticoronavirus face masks with optimal air permeability, pore size, comfort and barrier effectiveness.

DNA test kits for forensics: Crime scene investigators use DNA technology on a regular basis. In rape or sexual abuse cases, male-specific DNA can be targeted for forensic identification or for the exclusion of suspects. However, current DNA test kits were not designed to consider the genetic diversity of African males and perform poorly on the continent. The University of the Western Cape in South Africa has therefore designed a new male-specific DNA identifier, Uniqtyper Forensic DNA Kit, that is tailored for Africa. The technology includes a Y-chromosome short tandem repeat forensic kit that offers higher discriminatory power and is faster and more cost-effective than leading commercial products for processing sexual assault DNA evidence. The prototype has been designed using male-specific short-tandem repeat DNA and is used for the resolution of sexual offences evidence and for kinship analysis, especially paternal lineage. The use of this technology is crucial for improving conviction rates, exonerating innocent men accused of rape and alleviating the backlog of casework in African countries.

Discussion: Mozambican artist Malangatana Ngwenya was active in the nationalistic Frelimo (Liberation Front of Mozambique) guerrilla movement in his youth, but later became famous for his epic paintings and murals imploring people to find a peaceful solution to the Mozambican civil war. When Frelimo came to power, he was instrumental in founding the Mozambican Peace

Movement and later became a Unesco Artist for Peace. Dismas Nkunda, a Uganda-born journalist and activist who has covered the war in South Sudan extensively, became frustrated at not being able to 'name and shame' the politicians who had delayed the peace process in this war-torn African country. In response, he invented the Spoilers of Peace awards, intended to ostracise rather than praise the recipients and expose their corrupt ways of preventing Sudanese people from enjoying peace and prosperity. The first annual awards were in February 2020 and Nkunda plans to expand them to cover the entire continent. Potential repercussions for awardees include banning them from travelling and purchasing property, blocking their financial transactions and preventing them from using their corrupt networks to promote suffering of any kind. The awards will be overseen by the civil society organisation Atrocities Watch Africa, also founded by Nkunda (Allison 2020c).

In December 2016, a South African Casspir armoured personnel carrier covered with colourful beadwork in traditional patterns created by Ndebele women was displayed outside the Iziko South African National Gallery in Cape Town as a symbol of the country's transformation to a peaceful democracy. The decorated vehicle has since travelled extensively in Africa and abroad to spread the message of peace.

24

Astronomy and space science

'One of the greatest challenges of astronomers in Africa
today is being able to make local African world views
more scientific, to link them to other world views,
and to demystify the "mysterious heavenlies".'

Urama and Holbrook (2009)

Introduction: Currently, 19 African states have national space programmes and there is little doubt that the most significant 'blue sky' research and innovation on the continent is in the fields of astronomy and space science. South Africa's contributions to these fields are covered in detail in my book *What a great idea! Awesome South African inventions* (Bruton 2017a); in the many publications on SALT, Ska and Sarao initiatives, such as Addison (2004, 2005a, 2005b), Buckley et al. (2005), Wild (2013) and Raynard (2015); and in SALT, Ska and Sarao annual and special reports. They will, therefore, not be repeated in detail here. Rather, I will try to focus on other African countries' contributions to astronomy and space science.

    Afronaut: On the basis that there is no harm in dreaming, Edward Mukuka Nkoloso (a member of the Zambian resistance movement; high-school teacher; and self-professed founder of the Zambia National Academy of Science, Space Research and Philosophy) initiated a 'space programme' in Zambia in 1960. His goal was to launch a rocket that would send a girl, 17-year-old Matha Mwambwa, and two cats to the moon. To train the afronauts, Nkoloso set up a makeshift facility on an abandoned farm 11 km from Lusaka where the trainees were rolled down a hill in a 200-litre drum and swung around a tree on a tyre to simulate weightlessness. His rocket (D-Kalu 1) was a 3 m × 2 m drum-shaped vessel, to be launched on the day that Zambia gained its independence, 24 October 1964. However, the launch never took place.

A documentary film on Nkoloso's endeavours, *Afronauts* (directed by Frances Bodomo), premiered at the Sundance Film Festival in 2014. However, the Zambian government distanced itself from his efforts and, in 2016, former Zambian President Kenneth Kaunda (after whom the rocket was named) reportedly said of the 'space programme' that it wasn't the real thing and was more for fun than anything else'. Nkoloso Road, at the foot of the long-gone 'weightlessness' slope in the suburb of Matero (Lusaka) serves as a reminder of Nkoloso and his ambitious dream (https://en.wikipedia.org/wiki/Edward_Makuka_Nkoloso). Since then, African participation in space science and astronomy has advanced in leaps and bounds.

African space agencies: South Africa's involvement in space research extends back to the 1950s, 1960s and 1970s when lunar and interplanetary missions conducted by Nasa were supported by a tracking station at Hartebeesthoek, where the first images of Mars were received from the Mariner IV spacecraft during the successful flyby of the planet. South Africa launched its first satellite (Sunsat) from Vandenberg Air Force Base in the USA in 1999 and its second satellite (SumbandilaSat) from the Baikonur Cosmodrome in Kazakhstan in 2009.

The South African National Space Agency (Sansa) was established in December 2010 to promote and develop aeronautics and aerospace research, and advance the peaceful use of outer space. Sansa's main focus is to use data from satellite remote sensing to provide assessments of flooding, fires, resource management and environmental phenomena in South Africa and elsewhere in Africa. Sansa is a key contributor to the South African Earth Observation Strategy and hosts the only Space Weather Regional Warning Centre in Africa. In early 2015, the Kondor-E satellite (which provides all-weather, day-and-night radar imagery for the South African military) was built for South Africa and launched in Russia.

Kenya is an important space science pioneer in Africa. Due to its equatorial location and proximity to the Indian Ocean, the country is ideally located for a spaceport that launches satellites into geostationary and other orbits. The Uhuru satellite was launched from Kenya in 1970 to study celestial X-ray astronomy. Facilities that have been developed since then include the Italian-owned Broglio Space Centre, a tracking station near Malindi, and two sea platforms near Ngomeni and Ungwana Bay. The Kenya Space Agency was launched in 2017 and, in collaboration with Sapienza University in Rome (Italy), staff and students at the University of Nairobi developed the country's first nanosatellite, 1KUNS-PF, which was launched in May 2018 from the International Space Station (ISS).

The Mauritius Radio Telescope (MRT, a T-shaped array commissioned in 1992) is a joint project of the University of Mauritius, the Indian Institute of

Astrophysics and the Raman Research Institute in India. The MRT maps the southern skies and is also used to study pulsars, solar radio emissions and supernova remnants. Mauritius will shortly launch its first nanosatellite, MIR-SAT1, to the ISS in collaboration with the Japan Aerospace and Exploration Agency.

In 2001, Nigeria established its National Space Research and Development Agency, based in Abuja. According to Felix Ale, the communications head of the agency, Nigeria aims to develop a world-class space industry whose focus will be economic development. The country launched its first satellite, NigeriaSat-1 (an Earth observation satellite), in 2003 and another four satellites that have helped to improve agricultural practices, collect climate data and track down hostages kidnapped by Boko Haram. Nigeria cooperates in space technology development with the UK, China, Ukraine and Russia, and launched its satellites from Russia and China.

The Algerian Space Agency was established in 2002 and has flown five satellites, four launched in India (for monitoring natural disasters, agricultural development and other remote sensing applications) and one (the communications satellite Alcomsat-1) launched in China.

Ageos (the Gabonese Studies and Space Observations Agency) was established in Gabon in February 2010, one of the few Francophone countries in Africa to have an active space agency. As over 80 per cent of Gabon is covered by forest, forest exploitation and conservation are vital to the economy of the country and Ageos plays an active role in this regard. One of the innovative methods that the agency has devised to partially fund its satellite programme is to charge a small levy for all activities impacting the integrity of its forests such as mining, oil palm cultivation, logging and agriculture. In exchange, Ageos is responsible for monitoring exploited forestry sites and contributing to the fight against illegal deforestation, mining and other practices by using its satellites. Ageos also monitors forests and their utilisation in 16 Central and West African countries (https://africanews.space/all-about-ageos-gabon-space-program/).

Namibia established its National Space Science Council in 2014 to promote space policy development, provide strategic direction on space science, identify space activities relevant to the country, and promote space research and development. Malagasy Astronomy and Space Science, which was founded in 2016, is the agency that coordinates and promotes astronomy and space science in Madagascar. Angola's space agency, Angolan National Office for Space Affairs, launched its first satellite (AngoSat-1) in 2017 but it suffered a power failure. It plans to launch a second communications satellite, AngoSat-2, in 2022. Both projects were supported by Russia (*SpaceinAfrica online*, 23 April 2019, pages 1–2).

The Ghana Space Science and Technology Centre was opened in May 2012 as the country's first space science and astronomy agency with the aim of promoting private spaceflight and space research commercialisation. The centre will also allow scientists and astronauts to conduct research on natural resource management, weather forecasting, agriculture and national security. The agency's first flagship programme was the Ghana Radio Astronomy Project, which involved converting an abandoned Vodafone Earth satellite station at Kuntunse near Accra into a radio astronomy telescope that will become part of the Ska array. Ghana launched its first satellite in 2018.

In 2015, the multimillion-dollar Entoto Observatory and Research Centre was opened near Addis Ababa; in December 2019, Ethiopia joined the space science community with the launch of the Ethiopian Remote Sensing Satellite-1 (ETRSS-1) on a Long March 4B carrier rocket from the Taiyuan Satellite Launch Centre in China. Twenty-one Ethiopian aerospace engineers (including five women) and 60 doctoral and master's students, working with Chinese aerospace engineers, designed and built the 70-kg satellite in China (Kiruga 2019). Botswana International University of Science and Technology, which plays an important role in promoting space science and astronomy, hosted the Space and Planetary Science in Botswana workshop in 2018. Botswana's primary interest in space science is the use of satellite data to monitor land use, infrastructure development, wildlife distribution, earthquakes, mineral exploration, plant and water resources, and disasters, as well as to revise maps and forecast crop harvests and rainfall.

Egypt launched its first government-owned communications satellite (Tiba-1) in November 2019 to support civil and military communication. In 2019, it was endorsed by the AU as the official host of the headquarters of the AU African Space Agency (ASA) to be established in 2023. The ASA will have four thematic foci: Earth observation, communication, navigation and positioning, and astronomy. Its goals will include reducing the fragmentation of space activities in Africa and representing African member states in the international space arena.

Since 1998, over US$3 billion has been spent on space projects in Africa, driven by Earth observation programmes in Algeria, Egypt, Nigeria, Gabon and South Africa, and investment in satellite telecommunications in Angola and Congo. Earth observation is likely to be the primary focus of the African space programme for the foreseeable future, as this application is viewed as having the most potential to address the socioeconomic challenges of the continent. Further details on Africa's space agencies and programmes can be found in 'Space in Africa', a regularly updated online directory of all space-related institutions in Africa.

South African Astronomical Observatory (SAAO): The SAAO began with the establishment of the Royal Observatory at the Cape of Good Hope in 1820, the first scientific institute in sub-Saharan Africa. The Royal Observatory initially focused on mapping the southern skies and providing accurate time signals for maritime navigation. By the 1950s, as Cape Town grew, light and air pollution made it difficult to perform astronomy there so a new, remote site was established at Sutherland in the Karoo (chosen for its clear, dark skies and predictable weather). Since then, 15 light and infrared telescopes have been installed at Sutherland, in collaboration with various international partners, including the giant SALT (the largest single optical telescope in the southern hemisphere).

SALT, Africa's 'giant eye in the sky', was installed between 2000 and 2005 in a collaborative venture involving six main international partners. With its giant, 11-m wide hexagonal mirror, SALT can record distant stars, galaxies and quasars a billion times too faint to be seen with the naked eye. SALT allows astronomers to explore the scale and age of the universe, the earliest galaxies and quasars, the life and death of stars, planets orbiting other stars and much more. This flagship project has demonstrated that the frontiers of science are not entirely reserved for developed nations. SALT, and its sister Ska, provides a world-class facility for fundamental research in Africa in a field in which South Africa has a long history of excellence. It also provides opportunities for young scientists and engineers from Africa to develop their skills in a stimulating, high-tech environment.

Planetariums in Africa: There are at least 14 permanent planetariums in Africa (seven in Egypt, three in South Africa, two in Algeria, and one each in Ghana and Tunisia) as well as many mobile, inflatable planetariums that travel to schools. One of the most famous African planetariums is at the historic *Bibliotheca Alexandrina* in Alexandria, which includes a science centre and features 3D films on astronomy, space exploration, the natural world and Egyptian history. In Ghana, a small planetarium with a thatched roof and laptop-generated images serves the community of Accra. In May 2017, the Minolta mechanical star projector at the planetarium in the Iziko South African Museum in Cape Town was replaced with a state-of-the-art digital dome offering immersive, multisensory edutainment in the form of 180°, 3D experiences where art, science and entertainment meet. The shows are not only about the night skies but also about the origin of the universe, space travel, Ska, the inner workings of the human body and the intricacies of atomic structure.

First African in space: In April 2002, aged 28 years, Mark Shuttleworth became the first African in space when he blasted off on Soyuz mission TM34 to the ISS in a Soyuz-U rocket from the Baikonor Star City Cosmodrome in Russia. Along with Russian cosmonaut Yuri Gidzenko and Italian astronaut Roberto Vittori, he orbited the Earth and docked at the ISS, where he conducted scientific experiments related to AIDS and genome research for eight days and had a radio conversation with then President Nelson Mandela. On returning to South Africa, Shuttleworth shared his space experience and enthusiasm for science with over 100 000 learners during a country-wide roadshow.

Shuttleworth was born in Welkom, South Africa, and attended school in Cape Town. As a student at the University of Cape Town, he was involved in the installation of the first residential internet connections at the university. In 1995, he founded Thawte Consulting (a company that specialised in digital certificates and internet security) and then sold the company for US$575 million to VeriSign in December 1999. In September 2000, he formed HBD (Here be Dragons) Venture Capital, a business incubator and venture capital provider now managed by Knife Capital. In March 2004, Shuttleworth formed Canonical Ltd for the promotion and commercial support of free software projects, especially the Ubuntu operating system. In 2001, he formed the Shuttleworth Foundation, an NPO dedicated to social innovation that also funds educational, open-source software projects such as the Freedom Toaster. In 2004, he funded the development of Ubuntu, a Linux open-source system based on the Debian operating system that he had helped to develop.

Sadly, another South African, Mandla Maseko, who was poised to be the first black African in space, was killed in a motorcycle accident in 2019 at the age of 30 years. Maseko, an officer in the SAAF who was nicknamed 'Afronaut' by the media, beat 1 million people to become one of 23 candidates who won a seat sponsored by the Axe Apollo Space Academy for an hour-long suborbital space trip. He visited Nasa's Kennedy Space Centre, where he was briefed, trained to skydive and undertook G-force training, but the suborbital flight never took place. As a public speaker and community worker, Maseko worked tirelessly to inspire African children to pursue careers in science.

> He was a larger-than-life figure … He really thought that if he went up to space, he could inspire young African children that they could do anything. He used to always say that the sky was no longer the limit. He put a lot of people first and was an ambitious person with big dreams.
>
> *Sthembile Shabangu, friend of 'Afronaut' Mandla Maseko*
> (Space in Africa 2019)

CubeSats: The advent of shoebox-sized, specialised satellites called CubeSats is set to revolutionise our ability to monitor human impacts on the planet, including mapping the activities of town planners and farmers. CubeSats, which are made from off-the-shelf components and sensors, are relatively cheap to build and launch because they are sent into orbit either as companion payload or on small, re-usable rockets. They orbit at heights of 350 to 500 km and send back multiple high-resolution images at frequent time intervals. They typically relay data back to Earth for a few weeks to a year before ultimately burning out on re-entry into the atmosphere (McCabe et al. 2017).

Africa's most advanced nanosatellite, ZACube-2, was launched on a Soyuz rocket on 27 December 2018 to monitor natural and humanmade disasters such as fires, as well as shipping movements. The miniaturised satellite, which weighs 4 kg, was designed and built at the CPUT in Cape Town in collaboration with Sansa and partners in the French–South African Institute of Technology (F'Sati). The CPUT has also developed a range of compact S-Band and X-Band transmitters for CubeSat missions that require a high data rate downlink. ZACube-2 is South Africa's second nanosatellite, after TshepisoSat, a student-developed 1U CubeSat from the CPUT whose main payload was a high-frequency beacon transmitter that was used to help characterise the Earth's ionosphere and to calibrate Sansa's radar installation at the Sanae-IV base in Antarctica. TshepisoSat, which was developed in collaboration with F'Sati and the Electronic Systems Laboratory at Stellenbosch University, was launched from Russia in November 2013.

Square Kilometre Array (Ska): The Ska radio telescope project that is being developed by Sarao and its partners in Africa is the largest collaborative science and technology project on the continent. It is an internationally collaborative effort to build the world's largest radio telescope with 1 km² (1 million m²) of collecting area. Ska will eventually include over 3 000 radio telescopes, in three unique configurations, that will enable astronomers to survey the entire sky thousands of times faster than any system currently in use. The Ska telescope array will be located in nine countries in Africa (South Africa, Botswana, Ghana, Kenya, Madagascar, Mauritius, Mozambique, Namibia and Zambia) and in Australia. The Ska Organisation has its headquarters at Jodrell Bank Observatory near Manchester (UK) and currently has 11 member countries: Australia, Canada, China, Germany, India (associate member), Italy, New Zealand, South Africa, Sweden, the Netherlands and the UK.

South Africa has already demonstrated its excellent science and engineering skills by designing and building the 64-dish MeerKAT radio telescope array near Carnarvon as a pathfinder to Ska. The first seven dishes (KAT-7) produced superb images and attracted considerable interest internationally, with more than 500 international astronomers and 58 astronomers from Africa submitting proposals to carry out research using the pathfinder telescope. MeerKAT allows scientists to monitor radio waves that have been travelling through space since the beginning of time and will help to answer key questions about the origin of the universe. The first images received by the completed MeerKAT array in July

2018 provided the clearest view ever of the centre of the Milky Way (Philander 2018). The technology developed for MeerKAT is cutting-edge and the project is creating a significant cohort of young scientists and engineers with world-class expertise in ultra-fast computing and data transport, sensor networks, software radios and imaging algorithms, which will be crucial in the next 10 to 20 years.

MeerKAT will be integrated into the first phase of Ska-Mid, which will comprise about 200 dishes spread around the Karoo. The goal is to expand into the African partner countries to carry out observations on gravitational waves and pulsars, and to search for signatures of life in the galaxy. In July 2017, Ghana became the first African country to successfully convert a redundant telecommunications antenna into a functioning very long baseline interferometry (VLBI) radio telescope that will be integrated into the African VLBI Network in preparation for the second phase of the Ska project.

Since 2005, the African Ska Human Capital Development Programme has awarded over 1 000 grants to students in astronomy and engineering from undergraduate to postdoctoral level while also investing in technician training.

Three Ska stations will probably be located in Namibia, which has already gained valuable experience hosting the world's leading gamma ray telescope, the High Energy Stereoscopic System (Hess), near the Gamsberg Nature Reserve south of Windhoek. Botswana will potentially host four Ska stations and has initiated several undergraduate programmes in astronomy and astrophysics at the University of Botswana.

The University of Mauritius, which already has significant expertise in radio astronomy through the construction and operation of the Mauritius Radio Telescope, has been teaching undergraduate courses in astronomy for many years and has extended its existing postgraduate programme with the help of Ska South Africa. Ghana maintains a strong position within Africa's astronomy and astrophysics community, particularly through the presence of the African Physical Society headquartered in Accra. In Madagascar, the University of Antananarivo has launched courses on astronomy and astrophysics, and will make use of the observatory at Ankadiefajoro to develop practical skills. Mozambique has initiated courses in astronomy at Eduardo Mondlane University in Maputo.

Skarab: The Ska project has already spun off some significant technologies, including the Skarab Processing Platform. Skarab is a highly scalable, energy-efficient, rack-mounted field-programmable gate array (FPGA) computing platform. At the heart of the platform is a motherboard that provides unparalleled bandwidth (1.28 terabits per second) to four high-performance sites. An advanced reconfiguration interface allows for extremely fast reconfiguration of the FPGA to take place, which enables computer clusters to rapidly change function with minimal downtime. A rich board support package is also available to enable users to take full advantage of the platform's features and facilitate rapid customisation for specific applications.

Hydrogen Intensity and Real-Time Analysis eXperiment (Hirax): South Africa has become one of the world's most important radio astronomy hubs thanks largely to its co-hosting of Ska. Now a new radio telescope, the Hirax, has been added to the array in Sutherland. Hirax, an internationally collaborative project led by scientists from the University of KwaZulu-Natal under the leadership of Prof. Kavilan Moodley, is an interferometer array consisting of 1 024 m × 6 m dishes that (like Ska) combine signals from many telescopes to provide the resolution of a larger telescope (Moodley 2018).

Hirax has two main goals: to study the evolution of dark energy by tracking neutral hydrogen gas in galaxies and to detect and localise mysterious radio flashes called 'fast radio bursts'. Dark energy is a mysterious force driving the accelerated expansion of the universe. Hirax will map the distribution of neutral hydrogen, a useful tracer of the distribution of matter in the early universe, to unravel the properties of dark matter. The telescope's huge field of view will also allow it to observe large portions of the sky daily and record fast radio bursts at a frequency far higher than ever before – research that will be supported by eight-dish outrigger arrays in other southern African countries. Like Ska, Hirax will also promote research on the design and manufacture of high-precision dishes as well as smart ways to compress, store and analyse vast amounts of data (Moodley 2018).

Meerlicht telescope: Meerlicht is a fully robotic optical telescope that will provide a simultaneous, real-time optical view of the same part of the sky as the MeerKAT radio telescope array, to which it will be permanently linked. It is located at the same site as SALT in Sutherland and was developed through collaboration among institutions in South Africa, the Netherlands and the UK. The telescope will be used to study star explosions, neutron stars and black holes, and is one of several projects aligned to the Multi-Wavelength Astronomy Strategy (whose aim is to forge closer working ties between radio, optical and gamma ray astronomers).

Discussion: While the main goals of astronomy and space science are to explore and understand extraterrestrial space bodies and phenomena, another important objective is to recalibrate our view of Earth and try to achieve the right balance between development and the sustainable use of natural resources. The 'lifeboat concept', whereby humans would have to live on another planet or moon in order to survive in the long term as a result of planet Earth becoming unlivable due to climate change, has been proposed by some futurists. However, it is regarded by others as an unethical mindset that takes the focus away from the need to live sustainably on Earth, a fragile but also powerful and elegant home. We cannot afford to give up on our planet.

Space science has yielded a cascade of benefits, spin-offs from space, that are now embedded in the life of earthlings. They include satellites, GPS, miniaturised digital technologies, memory foam, freeze-dried food, advanced firefighting equipment, cloud computing, airplane winglets, emergency space blankets, infrared thermometers and cameras, DustBusters, cochlear implants, LZR racer swimsuits, advanced lubricants, wireless headsets, scratch-resistant lenses and advanced image sensors. Although space is now congested, competitive and highly commercialised, it is still an exciting scientific and technological frontier – and African countries have a strong role to play in its exploration.

But the benefits of space science and astronomy go far beyond that. Many nations, in both developed and developing countries, realise that data and knowledge have become the key resources that promote economic development and have increasingly embraced knowledge generation as a means of securing their competitive advantage. They have done so by investing, not only in education but also in 'big science' projects with the aim of producing highly skilled graduates who thrive in an increasingly technology-driven world. International examples of these 'big science' projects include Nasa in the USA; Cern (the European Council for Nuclear Research) in Switzerland; the Large Synoptic Survey Telescope in Chile; and, of course, Hess, Sarao, SALT, Hirax and Meerlicht in Africa. Breakthrough technologies that resulted, often serendipitously, from these 'big science' projects include integrated circuits, improved robotics and satellite imagery from Nasa, and the World Wide Web from Cern. It is anticipated that there will be many spin-offs from MeerKAT and Sarao, including improved antenna design and manufacture, data processing, digital signal processing, the development of reconfigurable open architecture computing hardware for other applications and the suppression of electromagnetic 'noise' (Davidson 2012).

Nishina Bhogal (2018) of the University of Cape Town has identified four factors that will ensure that 'big science' projects will lead to strengthened knowledge economies: robust, world-class institutions; collaborative relationships across international boundaries; a strong innovation culture that commercialises some of the technology spin-offs; and the availability and training of highly innovative individuals. She anticipates that Africa's space science projects will achieve these goals.

25

Essay: Future of technology

'For tomorrow belongs to those who prepare for it today.'

African proverb

Luckily for humans, in Africa and elsewhere, the future is not a head-to-head competition with robots and AI – an 'us-or-them' scenario. Just as we have done before, we will learn to live with the new technologies and benefit from them. The first controlled use of fire by prehumans, about 1 million years ago, was an equally transformational event. We used fire to keep warm, cook food, fend off enemies, bake clay and melt metals – and thrived. The first ventures in horticulture and agriculture 12 000 years ago disrupted hunter-gatherer societies, but they moved on to the next level – as humans did after they had harnessed wind- and hydropower to do work. The controlled generation, storage, transport and use of electricity led to an industrial revolution, and the generation of nuclear power and renewable energy has further increased our options. Most recently, computers, the internet, robots, AI and machine learning have propelled us into another dimension. Each leap has enabled us to climb to the top of the technology 'food pyramid' and improve our quality of life, at least theoretically.

The robot revolution is different in one important way: the potential for our machine-like creations to become part of us until eventually we might become cyborgs that are inseparable from our machines. Until now, all our tools (except our hands) have been separate from us, and there has been a clear distinction between where we end and where machines begin. However, as we integrate technology and physiology, this distinction will gradually disappear and there is a fear that we will end up as 'humachines' – like Darth Vader, pseudo-people powered by robotics, immortal until we run out of energy to maintain and upgrade our 'bodies'. However, as John Sanei (2019a, 2019b) points out, even if we make a near-perfect 'humachine', no amount of engineering, coding or AI technology will be able to reproduce the undefinable, unprogrammable qualities that make us human – love, empathy and compassion. After all, the word 'robot' comes from the Czech term for 'slave'.

Beyond silicon-based changes, there could also be massive carbon-based changes to humans. Medical advances may extend our lifespans to 150 years, big data may find a cure for cancer, nanotechnology will transform drug delivery and targeted therapy, and 3D printing will make prosthetics affordable and liberating. Through computer-assisted learning, massive open online courses and internet-enabled knowledge, IQs may even increase to 1 000. Furthermore, human-enhancement technologies and brain–computer interfacing technologies could increase the brain's capacity to store and process information (Barrenechea 2018). The use of increasingly advanced algorithms (which use data from a sequential series of actions to perform calculations, process data, solve problems and draw conclusions) is regarded by some as a threat because they might rob us of our human qualities of intelligence and creativity. Algorithms are the engines behind AI that control, for example, the processes behind self-driving cars, blockchain technologies and cryptocurrencies, and play a major role in social media, smartphones and tablets and in online services such as Uber and Airbnb (Mokwena 2018). Are they the sinister 'big brother' that George Orwell warned us about?

The headlong progress in robotics, AI, nanotechnology and biotechnology has made many humans wary of the future, especially of the so-called Moment of Singularity when computers theoretically become more intelligent than us. The European Patent Office recently turned down a patent application for a new food container, not because it was not innovative but because it was created by AI. By law, inventors must be actual people, but will it always be that way? Machines have already created many new products, including musical scores, new manufacturing materials and books (Schweisfurth & Goduscheit 2020). Innovation is a process of search and combination. We start with one piece of knowledge and connect it with another into something that is new and useful. In principle, this is something that a machine could also do. In fact, machines excel at storing, processing and making connections, but this does not necessarily mean that they will steal the spark of human creativity any time soon. Innovation is a problem-solving process and for it to happen, problems need to be identified and then combined with solutions. Humans can go in either direction: they can start with a problem and find a solution or they can start with a solution and find a problem that it can solve. Machines cannot do that – yet. An example of the latter is the laser, a unidirectional beam of light that had no obvious use when invented by Theodore H. Maiman in 1960 – a solution looking for a problem. Today, lasers have revolutionised science, medicine, telecommunications and entertainment.

Problem finding is, however, difficult for machines because problems are often beyond the boundaries of the datapool, fed to them by humans, from which they can innovate. What is more, innovation is often based on needs that we did not know we had – think of the Walkman and smartphone – and latent needs are unlikely to find their way into the datapool on which machines base their innovations. Humans also have the advantage over machines in that they can draw on a lifetime of experiences from which to generate ideas. Furthermore, breakthrough innovations often result from connecting disciplines that are far apart and usually unconnected, for example the digital laser (invented by South African Sandile Ngcobo) that connects the worlds of lasers and holograms. Also, creativity is not only about novelty but also about usefulness. While machines may be able to create something that is new, it might not be useful. However, even if machines cannot replace humans in the creative domain – at least for the foreseeable future – they are very useful as aids to human creativity, but their usefulness will be determined largely by the quality and quantity of data that we feed into them (Schweisfurth & Goduscheit 2020).

Instead of fearing technology, we need to understand how it can improve our lives and enable us to live more sustainably, in both urban and rural locations. For example, by 're-designing' our DNA with the Crispr-Cas9 tool, we could increase disease resistance and remedy undesirable hereditary conditions. We could use genome editing to introduce one strand of DNA into another at an exact point to augment who we (or our crops) are at a genetic level. And, using neural lace (an ultra-thin mesh developed by one of Elon Musk's companies), we may one day be able to upload new skills and information directly into our brains and become superhuman. Eventually we could be connected to pseudo-sentient machines that will make us exponentially more intelligent, and hopefully also wiser, than we are today. And because artificial neural networks mimic the way in which the human brain operates, we would not be trying to connect organic minds with mechanical machines but rather meeting one another somewhere in the middle (Sanei 2019b). Of course, there are also many lower-tech solutions that will be created by robots, AI and other modern technologies – as illustrated in this book.

But the biggest changes will be in the workforce. As Baby Boomers retire, Generation X will turn its attention to knowledge transfer, millennials will become the dominant demographic, and Generation Z will enter and totally transform the workplace. They will represent the greatest generational shift the workplace has ever seen, as they take hyperconnectivity to a new level and adopt digital technologies that are second nature to them. In their gig economy, experience will top education, market viability will be more important than marketing, and making a difference will be more important than making a dollar.

But it all depends on how we lay the foundation today. Do we continue to bicker with other members of the human race; fight military, water and trade wars; destroy biodiversity; and disrupt essential ecological processes – or do we learn to work together as a species for the long-term good of all the inhabitants of the planet (human and non-human, plant and animal) as we have done, to a certain extent, during the Covid-19 pandemic? Our potential is unlimited, but what role will African innovators play in this great human adventure?

26

General discussion

'As more Africans get the Foresight fever, the better for
the continent and its people. What is important is that we
must not forget to do what is right for the "public".'

David Lefutso, futurist at Kayamandi Informatics (Lefutso 2021)

Joe Kobuthi, philosopher and curator of the new Kenyan media platform The
Elephant, has firm views on the role that Africa can and should play in future:

> The scientific and technological advancements, the military-industrial complex,
> the sophisticated economic, social and political arrangements of the Western
> hegemonic model now appear futile in the face of the coronavirus pandemic.
> Humanity and nature have been groaning with eager expectation for something
> other than this five-hundred-year European experiment that has revealed
> itself to be rapacious and genocidal where the world is concerned. The West
> now lives at such a crazy, reckless speed that it has lost all reason and moral
> authority as it sinks into the abyss.
>
> Here in the Global South, we must no longer benchmark with this edifice
> as the standard for human advancement. No, we do not want to catch up to
> anyone. What we want is to move forward in the company of all men. It is
> now time for the peoples at the periphery to begin a new history for mankind.
> After Covid-19, if our desire is human progress, we must create other ways of
> being. For Mother Nature and for humanity's sake, we must rebuild from the
> ruins, think anew and attempt to set afoot a new philosophy of man.

Joe Kobuthi, philosopher and curator at the new media platform
The Elephant (The Elephant 2020)

But do the facts support Kobuthi's assertion? Other commentators have suggested that the reaction to the Covid-19 crisis, and the ability of nations to ensure the well-being of their citizens, has exploded the myth that 'First World' countries are more capable of dealing with crisis than 'Third World' ones – one of the most deeply held and offensive prejudices in the modern world. This view suggests that 'First World' implies competence, sophistication, and higher ethical and moral standards. In response, some African countries aspire to have a better material standard of living by becoming 'First World'. This outlook suggests that 'First World' countries would take the Covid-19 pandemic in their stride, leaving the 'Third World' to buckle under the horror of failing health systems and economic chaos. At the time that this was written (November 2020), this had not happened.

One of Africa's main colonisers, Great Britain, and the English-speaking superpower the USA have been among the least successful nations in containing the pandemic – as has another former coloniser, France, and other European countries such as Italy and Spain. In contrast, some of the countries that have been most successful in containing the pandemic have been in Asia and Africa, including South Korea, Singapore and Senegal. The latter has been the stand-out example in Africa because it has not only 'flattened the curve' but has also used 3D printers to produce cheap ventilators and developed a cheap Covid-19 testing kit using its experience in dealing with HIV/Aids and EVD (Friedman 2020).

> But the Covid-19 experience may just trigger new thinking in the 'Third World'. The most basic function of a government is to protect the safety of its citizens. Ensuring that people remain healthy is at least as important a guarantee of safety as protecting them against violence. Reasonable people would surely much rather be living in Kerala or Senegal (or East Asia) right now than in Europe and North America, raising obvious questions about who really does offer a better life. That should inspire Africans and others in the 'Third World' to ask themselves whether it makes sense to want to be America, Britain or France.
>
> *Steven Friedman, professor of political studies at the University of Johannesburg*
> (Friedman 2020)

As businesses transition from profit to philanthropy during the Covid-19 pandemic, the true spirit of ubuntu is being experienced in Africa. According to a report by the Centre for Science, Technology and Innovation Indicators of the HSRC in South Africa, the pandemic has catalysed innovation. Kruss and Sithole (2020) report that nearly 70 per cent of South African companies are 'innovation active' (i.e. they have taken some scientific, technological, organisational, financial or commercial steps towards implementing an innovation). This proportion compares favourably with trends in Organisation for Economic Cooperation and Development countries. These companies are adjusting their practices and strategies; introducing new technologies, products and designs; and determining how they can use digital and automated technologies. The survey also found that innovation is pervasive across all sectors, but is particularly strong in engineering, manufacturing and trade. Many innovation-active businesses reported the use or development of advanced new technologies such as computerised design, material handling, supply chain and logistics technologies, business intelligence technologies, and green technologies (Kruss & Sithole 2020).

* * *

There have been many visionaries in Africa's postcolonial history who have changed the course of events for the better. One visionary pan-African thinker who has a relatively low profile because he has mainly worked behind the scenes is Ghanaian K.Y. Amaoka, who has played a pivotal role in shaping many of Africa's institutions and initiatives. Amaoka served in the World Bank, then as the executive secretary of Uneca and finally as the founder and president of the non-profit think-tank the African Centre for Economic Transformation (ACET) in Accra. His vision has always been to transform the continent's economy from reliance on subsistence agriculture, primary commodities and fuel exports to manufacturing and services. As he gained experience and respect, his intimate understanding of Africa's problems and the perspectives of its leaders, and his championing of the importance of government ownership of policies, led him to become a highly effective interlocutor between global financial institutions and African governments (Michalopoulos 2020).

Under Amaoka's leadership, Uneca provided the technical support that underpinned and paved the way for the biggest pan-African initiative ever – the New Partnership for Africa's Development (Nepad 2018). In 2015, in support of Nepad, the AU adopted Agenda 2063, a long-term African strategy for inclusive growth and sustainable development. In the context of Nepad, Amaoka promoted the establishment of Africa's Peer Review Mechanism, whereby the effectiveness of individual African government policies can be reviewed by other African governments with the support of a secretariat – a unique initiative not undertaken by any other country grouping in the world (Michalopoulos 2020).

In 2007, Amaoka founded ACET to continue his work on Africa's transformation, and he and his staff produced the influential 2014 African

Transformation Report that defines seven sets of actions necessary for the effective transformation of the continent: developing a shared vision, state focus on core functions, coordination, public–private partnerships, trade integration, resource mobilisation and effective leadership. On the latter topic, he noted that notwithstanding the remarkable progress of Rwanda under President Paul Kagame's autocratic rule, transformation in Africa takes place best under a multiparty democracy, provided that those in authority are visionary, honest and capable, and focus on empowering and supporting civil society, traditional communities and families. In his autobiography *Know the beginning well: An inside journey through five decades of African development* (Amaoka 2019), he argues that if this were to happen, Africa could indeed realise its full potential and that the 21st century could still be Africa's century.

The title of his autobiography is based on the African proverb 'If you know the beginning well, the end shall not trouble you'. Wise counsel indeed from this great African visionary. However, what I find missing from his vision and – dare one say it – also from the visions of other great African leaders such as Nelson Mandela, Haile Selassie, Kwame Nkrumah, Julius Nyerere, Patrice Lumumba, Thomas Sankara and Jomo Kenyatta (but not Kofi Annan) is that they all placed too little emphasis on the importance of techpreneurs in driving economic growth, creating employment opportunities and using the innate spirit of innovation that is so evident in Africa to its full potential. If we interpret the proverb in the literal sense, Africa's historic contributions to innovation have been immense and there is no good reason why they should not continue to be significant in future. The transformation that has taken place from the Third Industrial Revolution to this First Digital Revolution (with hyperconnectivity, smartphone technology and the democratisation of innovation taking centre stage) has created a situation where virtually anyone can now contribute to the technological revolution – which bodes well for Africa.

* * *

Some argue that Africa missed the industrial revolutions, but arrived just in time for the digital revolution. If this is the case, maybe it was an advantage because many African countries have sidestepped the environmental despoliation that wholesale industrial development has wrought. To an extent, many African countries also avoided some of the other unexpected consequences of the first three industrial revolutions, such as conspicuous consumption, gross materialism and unsustainable lifestyles, which in turn led to increased incidences of obesity and other lifestyle diseases. They also averted environmentally costly trade practices. However, some of these are now starting to bedevil the African continent as well. Perhaps we should introduce new terminology that distinguishes between 'appropriately developed' countries and those that are 'overdeveloped'? So-called developed countries have much to learn from so-called undeveloped countries about sustainable living and happiness, which will in future become more important imperatives than GDP.

We also need to consider whether Klaus Schwab's (2016) designation of this as the 'Fourth Industrial Revolution' is accurate. I do not think so. I prefer to refer to the three eras of recent human industrial evolution simply as pre-industrial, industrial and post-industrial, with considerable overlap between them. In this scheme, the pre-industrial era includes wisdom derived from indigenous knowledge as well as the many advances made by pre-industrial societies around the world; the industrial era includes the first three industrial revolutions as conventionally defined; and the Post-Industrial or First Digital Revolution includes the so-called 4IR that we are experiencing now.

The naming of the post-industrial era is not just an exercise in semantics but an important distinction, as it signifies a major change of mindset in our approach to what 'industry' is and how it impacts our lives. This post-industrial era is an opportunity for us to recognise that the first three industrial revolutions did a lot of good, but they also did a great deal of harm. The Post-Industrial Revolution equips us with the tools, connectivity and mindset to correct past wrongs and redress past imbalances, especially in Africa. This can be achieved because the fuel of this Post-Industrial Revolution is not steam, coal, gas, oil or nuclear power, but data and information. The symbol of this revolution is not smoke billowing from chimneys, but the interconnected community of human brains and the IoT which have created a world community that is more interlinked than ever. Furthermore, we are not just connected linearly but we are irreversibly entangled, not only with a vast network comprising nearly half the world's population but also with over 10 billion things. This unprecedented connectivity – combined with robotics; AI; quantum computing; 5G wireless technology; telemedicine; 3D printing; and major advances in nanotechnology, biotechnology, alternative energy, battery power and other fields – creates problem-solving opportunities that we have never had before.

What is most exciting about the post-industrial era is that it is not confined to an educated elite, as in past industrial revolutions. Everyone can and does participate, male and female, rich and poor, young and old, 'First- and Third-Worlders'. It especially creates opportunities for digitally competent, globally connected and ambitious young people to contribute ideas, innovations and solutions – and this is where Africa comes to the fore as the 'youngest' continent. Furthermore, many post-industrial developments facilitate technology leapfrogs that allow historically disadvantaged people to quickly enter the information age and contribute to, and benefit from, its services.

Post-industrial inventions have, in effect, created a multibrained, multigenerational superorganism – a kind of 'collective genius' – that has the potential, more than in any previous era, to co-create solutions to some of the world's most intractable problems such as overpopulation, climate change, urbanisation, biodiversity loss, human displacement, poverty and pandemics. These new technologies also teach us to be humble, to recognise our shortcomings and to acknowledge our ignorance. Humans are now by far the most numerous large animals that have ever existed, and we are also the first species to domesticate itself and to lose its ecological niche. It is estimated that we need 3.6 Earth planets to sustain our present rates of resource usage. The Covid-19 crisis is an urgent short-term problem for us to solve, but the crisis of global climate change will haunt us for much longer. Africa has the potential to be part of the solution.

* * *

So, what is the nature of African innovation? The continent has a personality, like ubuntu, that is difficult to define in English. That definition probably also applies to the nature of its innovations. The spirit of Africa is unlike that of any other continent – raw, untamed, impish, dynamic, always on edge. It is at times frustratingly unpredictable but also bold and fearless; and its lingering inferiority complex, stoked by colonialism, is rapidly evaporating. The continent has survived physical and socioeconomic calamities, yet it continues to survive and thrive. These events have shaped a resilient and highly determined class of entrepreneurial people who are motivated rather than discouraged by adversity and who thrive on challenges. They also share a unique bond of togetherness, a shared soul and connectedness with one another that is epitomised by ubuntu. This is the character of Africa.

> That is *ubuntu* – the feeling with each other, the feeling for each other. It is not a mystery; it is something plain and simple … That I must do to other people what I want other people to do to me. That I am a *muntu*, a human being, because of other human beings around me. If there were no other human beings on this planet, I would no longer be a '*muntu*', no longer be a child of *ubuntu*.
>
> *Vusamazulu Credo Mutwa, South African philosopher*
> (Bell-Roberts & Jamal 2014)

The key quality of African innovations is that they are expressed at all three levels, from high- to middle- to low-tech simultaneously, as the needs and wants of its people/community/society dictate. High-tech innovations blend seamlessly into mid-tech innovations, while the underbelly of low-tech innovations continues to proliferate. Some innovations (such as Oil of Olay, DryBath and fly farming) are relatively low tech but require high-level business management and marketing skills to be successful. The different levels feed into one another to create a unique mix of innovations which generate solutions that solve, or strive to solve, the problems on the continent. The 'stepladder' of innovations also promotes the upskilling of African innovators so that they can eventually become global players even if they have had humble beginnings.

Africa has the prospect of becoming the brightest continent in the 21st century if it develops its human and natural resources to their full potential, but we need to recognise that African innovation expresses itself in a different way. It is more akin to a role-swapping, improvisational jazz band that has a high level of alertness to new opportunities (riffs) and co-creates something new, rather than to a formal Western symphony orchestra that is conducted according to a strict formula. The improvisational style of African innovation is epitomised by the hybrid, culturally-mobile ethio-jazz of legendary Ethiopian musician Mulatu Astatke, whose signature piece *'Yekermo Sew'* ('A man of experience and wisdom') emerged from the rhythms and sounds he had absorbed from traditional Ethiopian music combined with the new sounds he had experienced during his travels in Europe and North America. More than anything, African innovation is characterised by its exuberance and spontaneity, its flouting of rules, its home-grown agenda, and its ability to develop at all levels simultaneously. If we play to its strengths, Africa will be *the* crucible of innovation in the 21st century.

In an interview with Werewere Liking of the pan-African arts cooperative in Côte d'Ivoire, published as the article 'Aesthetics of necessity', Michelle Mielly (2003) discusses Liking's concept that practical creativity is often sparked in highly constrained, resource-strapped environments. Liking supports Plato's age-old maxim that 'necessity is the mother of invention' and argues that need is what spurs people, especially women, into creative action. The tension between African 'poor yet hopeful' and Western 'wealthy yet depressed' may be explained by the extent to which deprivation promotes innovation. In the West, the more wealth and technology people have, the more they log on to the internet but the less happy they seem to be.

In developed countries, there is also a problem of young people believing that if they work or practise hard enough, they will succeed because they feel entitled to be in control of the outcomes of their lives by virtue of their sweat equity.

When they lose or fail, they are wracked with self-doubt and self-blame because their parents have protected them from discomfort and disorder (Simmons 2019). This is not a problem in Africa, where many – if not most – young people (especially girls) have experienced failure as well as discrimination and inequality during their short lives, and learned to cope with it. In fact, failing often and quickly, and getting up and starting again, is a far more beneficial lesson in life than (nearly always) succeeding.

Triumph over adversity by resilient, resourceful people at all levels of society: that is the nature of African innovation. It is something worth celebrating.

Author biography

Mike Bruton originally trained as an ichthyologist at Rhodes University in Grahamstown (now Makhanda), South Africa, and carried out his master's and doctoral research on the fishes of coastal lakes in northern Zululand. After a postdoctoral year at the Natural History Museum in London (UK), he was appointed as senior lecturer and then professor and head of the Department of Ichthyology and Fisheries Science at Rhodes University. He then served as the director of the JLB Smith Institute of Ichthyology (now the South African Institute for Aquatic Biodiversity) before being appointed as the education director at the new Two Oceans Aquarium in Cape Town and then as the founding director of the MTN ScienCentre (now the Cape Town Science Centre). Thereafter, he worked in South Africa, Dubai, Bahrain and Saudi Arabia for the company MTN Studios (which designs and installs interactive science exhibitions, science centres and museums) before retiring in 2015 and setting up the consultancy company Mike Bruton Imagineering in Cape Town, where he has been active providing guidance to science and technology centres and writing popular science books.

Throughout his career, he has recognised the importance of innovation in science and technology while acknowledging that creativity in these fields is expressed differently than in the arts. Whereas artists use their imagination to create new artworks, scientists use it to discover new ways of looking at old things, to develop new techniques and technologies in their search for the truth, to derive new relationships and contexts, and to understand the complex mechanisms of nature. Great scientists and technologists, like great artists, are restless and curious individuals who are ill at ease with the status quo. They are individuals who are disruptive risk takers, easily bored, make lots of mistakes, work across disciplines, hate rules, have a reputation for eccentricity, and dream big. In contrast, faithful formalists make incremental contributions but do not develop novel ideas, discoveries or inventions.

The idea for this book arose from the interactive exhibition 'Inventions that Changed the World', which was mounted in the MTN ScienCentre (now the Cape Town Science Centre) in 2002. Following an enlightening comment by the then president of the National Research Foundation in South Africa, Dr Khotso Mokhele, another exhibition ('Great South African Inventions') was held in 2004 to celebrate the 10th anniversary of South Africa's democracy. An eponymous book published in 2010 further stimulated interest in this topic and led to a further, more comprehensive tome *What a great idea! Awesome South African inventions* (2017). Following the success of this book, and interest in the topic from correspondents throughout the continent, the author was encouraged to spread his net wider and write about African inventions – a much more daunting task. This book is the outcome of that daring endeavour.

The end of the beginning.

Bibliography

Abbott A (2019) KZN scores a home run, *Country Life*, March

Abdulai A-R, Duncan E & Fraser E (2019) How digital technologies can help Africa's smallholder farmers, *The Conversation*, 11 July

Adams R (2019) The fourth industrial revolution risks leaving women behind, *The Conversation*, 5 August

Addison G (2004) *Science and technology in South Africa: Progress and achievements in the first decade of democracy, 1994–2004*. Pretoria: Department of Science and Technology

Addison G (2005a) *The competitive edge: Making innovation happen*. Meyersdal: The Engineering Association

Addison G (2005b) *The leading edge: Dividends of democracy*. Meyersdal: The Engineering Association

Adegbite O, Mkandawire E & Machethe L (2020) Nigeria needs to close the financial inclusion gap for women smallholder farmers, *The Conversation*, 6 March

Adeshokan O (2020) Commuting conundrums, *The Africa Report*

Adeyemo O (2020) Appointment breaks proverbial glass ceiling in science in Nigeria, *The Conversation*, 1 April

Africa.com (2018) African Women in Tech: Temie Giwa-Tubosun (March 10). Accessed 29 July 2021, https://www.africa.com/african-women-in-tech-temie-giwa-tubosun/?amp?pr=167410&lang=fr

Africa.com (2018) African Women in Tech: Lilian Makoi (26 August). Accessed 29 July 2021, https://www.africa.com/african-women-in-tech-lilian-makoi/

Africa.com (2020) African Women in Tech: Hilda Moraa (13 January) Accessed 29 July 2021, https://www.africa.com/african-women-tech-hilda-moraa/

Agriprotein (2017) Kingdom to host region's first fly farm. Accessed 8 October 2021, https://www.agriprotein.com/press-articles/kingdom-to-host-regions-first-fly-farm/

Aitchison J (2018) South Africa's reading crisis is a cognitive catastrophe, *The Conversation*, 26 February

Akindele EO (2020) Destroying Nigeria's riverside forests is bad for the freshwater ecosystem, *The Conversation*, 3 March

Akinwande B (2020) The best time to harvest yams? Science says when the lower leaves turn yellow, *The Conversation*, 25 February

Akombi B (2020) Why many Nigerian children still aren't getting proper nutrition, *The Conversation*, 10 February

Akyeampong E, Bates RH, Nunn N & Robinson JA (Eds) (2014) *Africa's development in historical perspective*. New York: Cambridge University Press

Al-Hassani STS (Ed.) (2007) *1001 inventions: Muslim heritage in our world*. Manchester: Foundation for Science, Technology and Civilisation

Allison S (2018) Whose injera is it anyway? Accessed 25 April 2021, https://mg.co.za/article/2018-06-29-00-whose-injera-is-it-anyway/

Allison S (2019) After 45 years, Africa has a new tallest building. Accessed 15 April 2021, https://mg.co.za/article/2019-10-11-00-after-45-years-africa-has-a-new-tallest-building/

Allison S (2020a) The year of the locust. Accessed 15 April 2021, https://mg.co.za/article/2020-02-21-the-year-of-the-locust-2/

Allison S (2020b) In East Africa, locusts come for more. Accessed 15 April 2021, https://mg.co.za/article/2020-04-01-in-east-africa-the-locusts-are-coming-back-for-more/

Allison S (2020c) The face of Africa's COVID-19 fight, *Mail & Guardian*, 9 April

Alobo M (2020) Africa's scientists set out their COVID-19 research priorities, *The Conversation*, 1 July

Amaoka KY (2019) *Know the beginning well – An inside journey through five decades of African development*. Accra: Africa World Press

Anderson S (2011) I want to put a ding in the universe – Steve Jobs quote. Accessed 15 May 2021, www.hobo-web.co.uk/the-best-steve-jobs-quotes-putting-a-ding-in-the-universe/#:~:text=I%20Want%20To%20Put%20A%20Ding%20In%20The%20Universe%20%E2%80%93%20Steve%20Jobs%20Quote

Andi S & Painter J (2020) How much do people around the world care about climate change? We surveyed 80,000 people in 40 countries to find out, *The Conversation*, 16 June

Annan K (2015) It's time to reform Africa's food system. Accessed 15 April 2021, www.linkedin.com/pulse/its-time-reform-africas-food-system-kofi-annan/

Anon (2016) Pair hit high road with transport app, *Pretoria News*, 16 March

Anon (2018a) Study to explore potential benefits of rooibos tea, *Cape Times*, 26 July

Anon (2020a) Spekboom: The wonder plant! Accessed 31 March 2021, www.starkeayresgc.co.za/arbor-week-2019-spekboom-the-wonder-plant/

Anon (2020b) Khoi and San to share in the benefits of rooibos, *SABI*, December 2019/January 2020

Cape Argus (2020) Internet in the sky for Kenya, 10 July

Anon (2020c) UWC celebrates 60 years of hope, action and knowledge with virtual anniversary launch. Accessed 15 May 2021, www.uwc.ac.za/news-

and-announcements/announcements/uwc-celebrates-60-years-of-hope-action-and-knowledge-with-virtual-anniversary-launch

Forbes (2015) Forbes quotes on the business of life. Accessed 15 May 2021, www.forbes.com/quotes/11396/

Anon (2021) *Imagine* (John Lennon song). Accessed 15 May 2021, www.en.wikipedia.org/wiki/Imagine_(John_Lennon_song)

Appel R (2019) Cedric Mizero is bringing Rwandan villages to fashion's global stage. Accessed 21 March 2019, edition.cnn.com/style/article/cedric-mizero-london-fashion-week/index.html

Arnesen J (2020) High data prices remain an obstacle for African smartphone users. *Black Business Quarterly* 83: 58–59

Aswad J (2020) Def Jam Records launches in Africa, *Variety*, 26 May

Awosanya Y (2020) Nigerian startups raised $55.37m in Q1 2020, with over 99% from foreign sources. Accessed 15 April 2021, https://techpoint.africa/2021/05/04/nigerian-startups-funding-report-q1-2021/

Ayalew YE (2020) How internet shutdowns have affected the lives of millions of Ethiopians, *The Conversation*, 2 April

Ayene T, Conroy E, Culliford A & Norbrook N (2020) The top 50 African disruptors (46–50). Accessed 15 April 2021, https://www.theafricareport.com/35662/the-top-50-african-disruptors-46-50/

Ayres G (2020) Sustainable waste. Biogas, *SABI*, December 2019/January 2020

Babalola AB (2020) How we found the earliest glass production south of the Sahara, and what it means, *The Conversation*, 14 July

Bailey S (2020) Why this South African gin is made with elephant dung. Accessed 15 April 2021, https://edition.cnn.com/2020/01/22/business/elephant-dung-gin-south-africa-intl/index.html

Bain K (2017) The shaper. Accessed 15 May 2021, www.kulula.com

Baker A (2017) Linda Mugaruka: Queen of beans, *Time*, 23 October

Baker A (2018) In Madagascar, soaring vanilla prices have a bittersweet cost, *Time*, 2 July

Baker A (2019a) The great green wall of Africa, *Time*, 23 September

Baker A (2019b) Delivering hope, *Time*, 25 November

Baker A (2019c) Meet the man whose bloodless test is revolutionizing the fight against malaria: Brian Gitta, Uganda, *Time*, 10 October

Baker A (2019d) Art, restored: Museum of Black Civilizations, *Time*, 2–9 September

Baker A (2020a) Nawal El Saadawi: For a more equal Egypt, *Time*, 16–23 March

Baker A (2020b) Ellen Johnson Sirleaf: A first for Africa, *Time*, 16–23 March

Ball EB (2018) *Founding women: African women who are defying the odds to build businesses in tech*. Nairobi: Africa Technology Business Network

Banik D & Gresko KE (2020) Why is used clothing popular across Africa? We found out in Malawi, *The Conversation*, 21 April

Barnett P (2006) *Beer: Facts, figures and fun*. Johannesburg: Trinity Books

Barrenechea M (2018) Intelligent enterprise: Top tech predictions. *Leadership* 391: 24–27

Barrero JM & Higgins TJ (2020) Nigeria has given a new GM cowpea variety the go ahead: Why it matters, *The Conversation*, 27 January

Bateman M (2019) African countries should turn to lower risk solutions to fight fall armyworm, *The Conversation*, 9 January

Bates O (1917) Ancient Egyptian fishing. *Harvard African Studies* 1: 199–272

Baxter W & Allison S (2020) 4.6-million people, one psychologist, *Mail & Guardian*, 6 March

BBC (2020a) Chad repaying $100 m debt to Angola with cattle. Accessed 15 April 2021, https://www.bbc.com/news/world-africa-51925035

BBC (2020b) Tanzanite: Tanzanian miner becomes overnight millionaire. BBC News, 24 June. Accessed 26 July 2021 https://www.bbc.com/news/world-africa-53148612

BBC (2020c) World population census, *BBC Online*, 15 July

Beinart W (2020) The story of the pharma giant and the African yam, *The Conversation*, 9 February

Bello TK (2020) How nutrition education can make a difference to people with HIV in Nigeria, *The Conversation*, 12 February

Bell-Roberts B & Jamal A (2014) *100 good ideas: Celebrating 20 years of democracy*. Cape Town: Struik

Berners-Lee T (2020) A letter to Nigeria's STEM students, *Time*, 3 February

Bhogal N (2018) How the SKA telescope is boosting South Africa's knowledge economy, *The Conversation*, 30 May

Bladergroen BJ (2018) Explainer: Why lithium ion batteries could be a game changer in Africa, *The Conversation*, 18 September

Blain L (2019) BST's wild Hypertek: A new aesthetic standard for electric motorcycles. Accessed 31 March 2021, https://newatlas.com/motorcycles/bst-hypertek-crazy-electric-motorcycle/

Booysen J (2018a) Demand for insect pollination to double, *Cape Argus*, 5 June

Booysen J (2018b) Hortgro and Tru-Cape bring back good-looking first apple variety, *Cape Times*, 26 March

Borrill A (2017) National treasure. Accessed 15 May 2021, www.kulula.com

Braczkowski AR, O'Bryan C, Biggs D & Jansen R (2020) Back from extinction: A world first effort to return threatened pangolins to the wild, *The Conversation*, 4 June

Bradfield J, Reynard J, Lombard M & Wurz S (2020) What a bone arrowhead from South Africa reveals about ancient human cognition, *The Conversation*, 17 May

Brain CK & Sillen A (1988). Evidence from the Swartkrans Cave for the earliest use of fire. *Nature* 336(6198): 464–466

Brewer DJ & Friedman RF (1989) *Fish and fishing in Ancient Egypt* Cairo: American University in Cairo Press

Bright C (2020) Wits is tackling the Covid-19 pandemic head-on. *Mail & Guardian*, 24 April. Accessed 15 May 2021, https://mg.co.za/special-reports/2020-04-24-wits-is-tackling-the-covid-19-pandemic-head-on/

Brümmer M (2019) Counting on hemp to save Morocco's High Central Rif, *Hemp Today*, December

Bruton CM (2020) How Refresh can help your business in a dramatically changed COVID-19 world. Accessed 31 March 2021, www.refresh.co.za/our-blog/

Bruton MN (2010) *Great South African inventions*. Cape Town: Cambridge University Press

Bruton MN (2016) *Traditional fishing methods of Africa*. Cape Town: Cambridge University Press

Bruton MN (2017a) *What a great idea! Awesome South African inventions*. Cape Town: Jacana Media

Bruton MN (2017b) *The annotated Old Fourlegs: The* updated *story of the coelacanth*. Cape Town: Struik Nature

Bruton MN (2018) *The fishy Smiths: A biography of JLB and Margaret Smith*. Cape Town: Penguin

Bruton MN, Cooke A, Ravololoharinjara M & Ravelo C (2020) *Latimeria chalumnae*, coelacanth, *fiandolo*. In S Goodman (Ed.) *The ecology of Madagascar*. Chicago: Chicago University Press

Bruton MN & Haacke WD (1980) The reptiles of Maputaland. In MN Bruton & KH Cooper (Eds) *Studies on the ecology of Maputaland*. Durban: Rhodes University and the Wildlife Society of Southern Africa

Buckley D, Lombard M, Lomberg M, Meiring K & Theron R (2005) *Africa's giant eye: Building the Southern African Large Telescope*. Cape Town: SALT Foundation

Byaruhanga C (2020) How do you fight a locust invasion amid coronavirus? Accessed 15 April 2021, https://www.bbc.com/news/world-africa-52394888

Calatayud P-A & Subramarian S (2020) New bugs, found in Kenya, can help control maize pests, *The Conversation*, 29 March

Campbell C & Gunia A (2020) A deadly new virus goes global, *Time*, 3 February

Cape Times (2018) Bitcoinhub, 16 April 2018

Chahl J (2020) Learning from nature: A new flapping drone can take off, hover and swoop like a bird, *The Conversation*, 27 July

Champoux-Paillé L & Croteau A-M (2020) The reason why female leaders are excelling at managing the coronavirus, *The Conversation*, 18 May

Chaves I, Engerman SL & Robinson JA (2014) Reinventing the wheel: The economic benefits of wheeled transportation in early colonial British West Africa. In E Akyeampong, RH Bates, N Nunn & JA Robinson (Eds) *Africa's development in historical perspective*. New York: Cambridge University Press

Chinyamurindi WT (2020) Learning in the time of coronavirus, *Mail & Guardian*, 20–26 March

Chirikure S (2010) *Indigenous mining and metallurgy in Africa*. Cape Town: Cambridge University Press

Chirikure S (2020) Archaeology shows how ancient African societies managed pandemics, *The Conversation*, 14 May

The Citizen (2018) Facebook plans to make Africa net safer, 7 February

The Citizen (2019) App lets townies be farmers, 30 July

CNN (2020) Job recovery after Covid-19. *CNN Online*, 28 May

Coleman D (Ed.) (2016) *Spinoff: Bringing NASA technology down to Earth*. Washington, DC: NASA

Cooke AJ, Bruton MN & Ravololoharinjara M (2020) Coelacanth discoveries in Madagascar, with recommendations on research and conservation. *South African Journal of Science* 117(3/4): 1–11

Courie E-L (2020) *#StartupStory: David Phume, an African technologist on a mission*. Accessed 15 May 2021, www.bizcommunity.com/Article/196/842/204549.html

Creamer M (2018) New 'disruptive' mine borer heads for Indaba unveiling, *Mining Weekly*, 2–8 February

Daly N (2020) More Chinese push to end wildlife markets as WHO declares coronavirus emergency, *National Geographic Online*, 30 January

Dasygupta S (2017) 30 years of protecting the mysterious okapi, *Mongabay Online*, 13 June

Daudi L (2020) Kenya's coast is losing huge amounts of seagrass. But all isn't lost, *The Conversation*, 10 August

Davenport J (2020) We sacrifice for Covid, so we can fight global warming, *Mail & Guardian*, 2 July

Davidson DB (2018) Potential technological spin-offs from MeerKAT and the South African Square Kilometre Array bid. *South African Journal of Science* 108: 1–3

Davie K (2019) Free up energy before the budget, *Mail & Guardian*, 17 January

DeGeorges A (2012) African traditional hunters and their weapons. *African Hunter Magazine* 18(1): 1–4

De Greef K (2019) *500 vultures killed in Botswana by poachers' poison, government says.* Accessed 15 April 2021, https://www.nytimes.com/2019/06/21/world/africa/vultures-poisoned-botswana-poachers-elephants.html

Demaria F & Todt M (2020) How waste pickers in the Global South are being sidelined by new policies, *The Conversation*, 1 March

Desalegn H (2020) To silence guns, restore nature, *Mail & Guardian*, 14 February

Diamond J (1987) *The worst mistake in the history of the human race.* Accessed 31 March 2021, http://discovermagazine.com/planet-earth/the-worst-mistake-in-the-history-of-the-human-race

Diamond J (1992) *The third chimpanzee: The evolution and future of the human animal.* New York: HarperCollins

Diamond J (1998) *Guns, germs and steel: A short history of everybody for the last 13,000 years.* London: Vintage

Diphoko W (2019) Gaming is not just for fun, it can be a billion-dollar dream for Africans, *The Star*, 2 August

Djunga H (2020) An incredible journey, *Weekend Argus*, 25 January

Dudhla S (2020). Pledge of $160bn for sub-Saharan Africa region, *Cape Times*, 20 April

Duma S (2020) Bid to settle Mars could solve Earth's crisis, *Mail & Guardian*, 7 February

Du Plessis M (2020) Grass roots of ice-age extinctions, *Mail & Guardian*, 14 February

Dweck CS (2008) *Mindset: The new psychology of success.* New York: Ballantine Books

Dzakwa Z (2020) Boy a force against global warming, *Weekend Argus*, 25 January

Ebrahim Z (2020) This is what the locally built, non-invasive Covid-19 ventilator looks like. Accessed 15 April 2021, https://www.news24.com/health24/Medical/Infectious-diseases/Coronavirus/pics-this-is-what-the-locally-built-non-invasive-covid-19-ventilator-looks-like-20200803

The Economist (2015) Why does Kenya lead the world in mobile money? 2 March

Egbejule E (2020) Lagos is a country. *The Africa Report,* 18 May

Ehret C (2014) Africa in world history before ca. 1440. In E Akyeampong, RHG Bates, N Nunn & JA Robinson (Eds) *Africa's development in historical perspective*. New York: Cambridge University Press

The Elephant (2020) Elephant series: Journaling a pandemic. Accessed 15 May 2021, https://www.theelephant.info/op-eds/2020/04/22/elephant-series-journaling-a-pandemic/

Elliott H (2020) China and AI: What the world can learn and what it should be wary of, *The Conversation*, 2 July

El Saadawi N (1980). *The hidden face of Eve: Women in the Arab world*. London: Zed Press

Embashu W, Lileka O & Nantanga KM. (2019) Namibia opaque beer: A review. *Journal of the Institute of Brewing* 125(1): 4–9

Embashu W & Nantanga KM. (2019) Pearl millet grain: A mini-review of the milling, fermentation and brewing of *ontaku*, a non-alcoholic traditional beverage in Namibia. *Transactions of the Royal Society of South Africa* 74(3): 276–282

Endolyn O (2019) Tunde Wey: Food for thought, *Time*, 27 May

English A (2020) The right to read, *The Big Issue*, January/February

Evans S (2020) Coronavirus has finally made us recognise the illegal wildlife trade is a public health issue, *The Conversation*, 17 March

Fairlady (2016) Grind Coffee Company, October

Faith JT, Rowan J, Du A & Koch PL (2018) Plio-pleistocene decline of African megaherbivores: No evidence for ancient hominin impacts. *Science* 362(6417): 938–941

Fanon F (1963) *The wretched of the Earth*. Paris: Grove Press

Fatunla Y (2018) The right tools to build Africa. *Leadership* 392: 56–58

Fayehun F, Omigbodun A & Owoaje ET (2020) Mobile technology can improve access to healthcare in Nigeria – if it is regulated, *The Conversation*, 6 May

Fourie L (2019) Graphene: 21st century wonder material, *The Star*, 2 August

Friedman M (1967) *Capitalism and freedom*. Chicago: University of Chicago Press

Friedman S (2020) COVID-19 has blown away the myth about 'First' and 'Third' world competence, *The Conversation*, 13 May. Accessed 28 July 2021, https://theconversation.com/covid-19-has-blown-away-the-myth-about-first-and-third-world-competence-138464

Gaboué S (2020) The pioneering Cameroonian designer taking on haute couture. Accessed 31 March 2021, https://edition.cnn.com/style/article/imane-ayissi-haute-couture/index.html

Gabriel O, Lange K, Dahm E & Wendt T (2005) *Von Brandt's fish catching methods of the world*. Oxford: Blackwell Publishing

Gallo C (2010) *The innovation secrets of Steve Jobs*. London: McGraw-Hill Education

Garbett B (2020) What's behind hundreds of vulture deaths in Guinea-Bissau: And what can be done? *The Conversation*, 14 June

Garrod B (2020) Bushmeat could cause the next global pandemic, *The Conversation*, 18 May

Gates B & Gates M (2021) Gavi, the Vaccine Alliance, helps vaccinate almost half the world's children against deadly and debilitating infectious diseases. Accessed 15 May 2021, www.gavi.org/our-alliance/about

Gates M (2020) Tu Youyou. Curing malaria, *Time*, 16–23 March

Gaye D (2018) Female entrepreneurs: The future of the African continent, *Le Monde*, 29 November

Gettleman J (2015) Meant to fight malaria, nets are used to fish instead, *International New York Times*, 26 January

Gibbons A (2020) Primatologists work to keep great apes safe from coronavirus. Accessed 15 April 2021, https://www.sciencemag.org/news/2020/05/primatologists-work-keep-great-apes-safe-coronavirus

Githahu M (2020) Say 'ahh' for Quintin, the new hospital ward robot, *Mail & Guardian*, 15 April

Gladwell (2008) *Outliers: The story of success*. New York: Little, Brown and Company

Goh B & Sun Y (2020) Tesla on the road to driving 'autonomy', *Cape Argus*, 10 July

Goitsemodimo GA (2019) Weapons of Southern Africa, *National Museum Reports*, 29 March

Goodwin AJH (1946) Prehistoric fishing methods in South Africa. *Antiquity* 20: 134–141

Gordon DM (2006) *Nachituti's gift: Economy, society, and environment in Central Africa*. Madison: University of Wisconsin Press

Gourvenec S (2020) Floating wind farms: How to make them the future of green electricity, *The Conversation*, 20 July

Goyanes C (2017) These badass women are taking on poachers – and winning, *National Geographic online*, 12 October

Gregory S (2019) South African track star Caster Semenya won't stop fighting for her right to run, just as she is, *Time*, 29 July

Grehan K (2020) Bats are hosts to a range of viruses but don't get sick – why? *The Conversation*, 8 July

Grigson C (1991) An African origin for African cattle? Some archaeological evidence. *The African Archaeological Review* 9: 119–144

Haag S (2020) Energy sector is the new colonial project, *Mail & Guardian*, 21 February

Haas ARN (2020) Shaping Africa's urban areas to withstand future pandemics, *The Conversation*, 1 April

Hagger A (2017) Saving the great white, *Sawubona*, December

Hancock G, Pankhurst R & Willets D (1997) *Under Ethiopian skies*. Nairobi: Camerapix Publishers

Hannan S, Reddy V & Juan A (2016) *Science engagement framework and youth into science strategy: Science centre capacity building project evaluation 2016*. Pretoria: HSRC Press

Harari YN (2014) *Sapiens: A brief history of humankind*. London: Harvill Seker

Harper P (2020a) No legal sales in new cannabis Bill, *Mail & Guardian*, 14 February

Harper P (2020b) The powers of crocodile meat, *Mail & Guardian*, 6 March

Harper P (2020c) Meet South Africa's own online medical myth buster, *Mail & Guardian*, 9 April

Harrington J (2018) Kenya's struggle to modernize traditional medicine is far from won, *The Conversation*, 30 August

Harris T (2020) Greenpeace Africa response to Mauritius oil spill. Accessed 15 April 2021, https://www.greenpeace.org/africa/en/press/11864/greenpeace -africa-response-to-mauritius-oil-spill/

Harvey R (2020) What is the wildlife trade? And what are the answers to managing it? *The Conversation*, 23 April

Hayes P (2020) Here's how scientists know the coronavirus came from bats and wasn't made in a laboratory, *The Conversation*, 13 July

Haynes B (2020) *Art@Africa. So much talent in our country*. Cape Town: Art@ Africa

Haynes S (2019) Selly Raby Kane: Celebrating African design, *Time*, 21–28 October

Henderson E (2018) Veganism not always best for our planet, *Cape Times*, 30 January

Henshilwood C & Van Niekerk KL (2018) South Africa's Blombos cave is home to the earliest drawing by a human, *The Conversation*, 12 September

Herren JK, Mbaisi L, Mararo E, Makhulu EE, Mobegi VA, Butungi H, Mancini MV, Oundo JW, Teal ET, Pinaud S, Lawniczak MKN, Jabara J, Nattoh G & Sinkins SP (2020) A microsporidian impairs *Plasmodium falciparum* transmission in *Anopheles arabiensis* mosquitoes. *Nature Communication,* 11: 2187

Himmelman N, Sarmiento N & Thipe T (2020) A tribute to poet and professor Harry Garuba: We continue to learn with you, *The Conversation,* 31 March

Hissmann K, Fricke H, Schauer J, Ribbink AJ, Roberts MJ, Sink K & Heemstra PC (2006) The South African coelacanths: An account of what is known after three submersible expeditions. *South African Journal of Science* 102: 491–500

Hlabangane S (2017) Locally developed asthma grid wins international award, *eHealth News,* 27 January

Hodges J, Lombrana LM & Pogkas D (2020) Alarmed cities answering call to report on emissions, *Business Day,* 19 February

Hollands G (2012) *The reef: A legacy of surfing in East London.* East London: Glenn Hollands

Hurum HJ (1976) *A history of the fish hook and the story of Mustad, the hook maker.* London: Adam & Charles Black

Ihekweazu C (2020) Steps Nigeria should take to prepare for cases of coronavirus, *The Conversation,* 28 January

Independent On-line (2019) Having raised over R50m, SweepSouth's Aisha Pandor gives her tips to success. © Independent On-line 2021. All rights reserved. Accessed 15 May 2021, www.iol.co.za/business-report/entrepreneurs/watch-having-raised-over-r50m-sweepsouths-aisha-pandor-gives-her-tips-to-success-36400728

Iversen K (2021) Women in leadership: A Q&A with President Sahle-Work Zewde. Accessed 15 May 2021, www.deliverforgood.org/women-in-leadership-qa-with-president-sahle-work-zewde/

Izugbara C & Obiyan MO (2020). Why more must be done to fight bogus COVID-19 cure claims, *The Conversation,* 17 May

Jackson T (2020) i4Policy launches a pan-African consultation on innovation policy. Accessed 15 April 2021, https://disrupt-africa.com/2020/02/17/i4policy-launches-pan-african-consultation-on-innovation-policy/

Jaruzelski B (2015) Innovation's New World Order. *Forbes,* 16 November. Accessed 26 July 2021, https://www.forbes.com/sites/strategyand/2015/11/16/innovations-new-world-order/?sh=56a84be53ba0_

Javad R (2020) Tunisia deploys police robot on lockdown patrol. Accessed 31 March 2021, www.bbc.com/news/world-africa-52148639

Jobs S (2005) Commencement speech, Standford University

Jones T (2020) Slice of pi: African memory techniques. *InterpNEWS* 9(4): 8

Jordan NR, Radford C & Rogers T (2020) Lions are less likely to attack cattle with eyes painted on their backsides, *The Conversation*, 7 August

Joubert M & Mkansi S (2020) South Africa: Science communication throughout turbulent times. In T Gascoigne (Ed.) *Communicating science: A global perspective*. Brisbane: Australian National University Press

Kangbai JB (2020) Sierra Leone is using lessons from Ebola to prepare for coronavirus, *The Conversation*, 27 February

Kaya H (2020) African multilingualism in teaching and learning African indigenous languages and home-grown philosophies for domestication of Sustainable Development Goals (SDGs), *DSI-NRF Centre in Indigenous Knowledge Systems newsletter*, June

Kayuni H, Banik D & Chunga J (2019) Malawi's dream of a waterway to the Indian Ocean may yet come true, *The Conversation,* 5 November

Kedward K (2020) The commons, not the market, *Mail & Guardian*, 29 February–5 March

Kgomoeswana V (2014) *Africa is open for business: Ten years of game-changing headlines*. Johannesburg: Pan Macmillan South Africa

Kgoroeadira K (2014) Small-scale farms grow African women's income. Accessed 15 May 2021, www.brandsouthafrica.com/people-culture/people/small-scale-farms-grow-african-women-s-income-2

Kiruga M (2019) Ethiopia joins the space race, *The Africa Report*, 20 December

Kiruga M (2020) Fueling flames: Nile dam still raging, despite global pause for COVID-19. Accessed 15 May 2021, www.theafricareport.com/25874/nile-dam-still-raging-despite-global-pause-for-covid-19/

Kluger (2020) This disability rights activist wants to be the first wheelchair user in space, *Time*, 24 January

Knight T (2020) *Chimp rescue in Liberia: Safeguarding orphans and saving a species*. Accessed 15 April 2021, https://www.fauna-flora.org/news/chimp-rescue-liberia-safeguarding-orphans-saving-species

Kretzmann S (2018) Tech start-up could save agriculture millions, *City Press*, 15 July

Kruger L & Ramukhithi V (2019) Indigenous sheep and goat breeds in South Africa, *Agricultural Research Council leaflet*

Kruss G & Sithole M (2020) Survey shows South African firms in some sectors are highly innovative, *The Conversation*, 20 July

Kumbani J (2020) What archaeology tells us about the music and sounds made by Africa's ancestors, *The Conversation*, 24 August

Kwayu AC (2020) Tanzania's COVID-19 response puts Magufuli's leadership style in sharp relief, *The Conversation*, 31 May

Larson K (2020) Veteran AP video, photo-journalist in Congo dies of COVID-19, *APNews online*, 22 June

Lategan A (2020) Strengthening international ties in conservation, *Safari News*, Summer

Lawanson T (2020) Lagos's size and slums will make stopping the spread of COVID-19 a tough task, *The Conversation*, 1 April

Lawson G (2017a) *No pavement special: A salute to AfriCanis, Southern Africa's truly indigenous dog.* Accessed 15 May 2021, www.kulula.com

Lawson G (2017b) *Around the world in 80 games.* Accessed 15 May 2021, www.kulula.com

Lawson G (2017c) *Hey bru, what's brewing.* www.kulula.com (April)

Lazareva I (2017) Joburg's rooftop hydroponics provide promising future for the unemployed, *Cape Times*, 4 December

Lazareva I (2018) Widows and rebel wives rebuild war-scarred Uganda, *City Press*, 20 May

Lefutso D (2021) *David Lefutso: Futurist profile.* Accessed 16 May 2021, www.foresightfordevelopment.org/profile/david-lefutso

Le Quéré A & Wade TK (2020) Insights from Senegal: Involving farmers in research is key to boosting agriculture, *The Conversation*, 6 February

Le Roux H (2020) Architecture: Four ideas from history that offer healthier design, *The Conversation*, 31 March

Levy B (2018) Active citizens for better schooling: What Kenya's history can teach South Africa, *The Conversation*, 1 March. Accessed 23 July 2021, https://theconversation.com/active-citizens-for-better-schooling-what-kenyas-history-can-teach-south-africa-92534

Lewton T (2018) In South Africa, 'decolonizing' mathematics. Accessed 31 March 2021, undark.org/2018/12/31/in-south-africa-decolonizing-mathematics

Liem S (2014) A soap to fight malaria, *Reader's Digest*, April

Lind PL (2016) Madagascar's £152m vanilla industry soured by child labour and poverty. Accessed 15 April 2021, https://www.theguardian.com/global-development/2016/dec/08/madagascar-152m-vanilla-industry-soured-child-labour-poverty

Lindow, M (2005) Girth of a nation, *Time*, 8 August

Lockwood G (2020a) Donkey disaster on the African continent, *Safari News*, Summer

Lockwood G (2020b) The snake whisperer, *Safari News*, Summer

Lockwood G (2020c) African manatees are omnivorous, *Safari News*, Summer

Logan S (2017) How mobile money benefited Kenya, *IOL online*, 5 January

Lombela J (2020) The digital African economy: Inclusive, transparent and accountable. *Black Business Quarterly* 83: 48–51

Losh J (2020) Beloved silverback gorilla killed by poachers in Uganda. Accessed 15 May 2021, www.nationalgeographic.com/animals/article/silverback-gorilla-killed-poachers-uganda

Lubele G (2018) Africa's most prominent marketplace is online, or is it? *Leadership* 391: 62–63

Lundahl P (2020) Africa's climate activists in the spotlight, *Weekend Argus,* 1 February

Lynn B. 2020, 'Coronavirus hairstyle' gains popularity in Kenya. Accessed 15 April 2021, https://learningenglish.voanews.com/a/coronavirus-hairstyle-gains-popularity-in-kenya/5415346.html

Maathai W (2010) *The challenge for Africa.* London: Arrow Books

MacAllister C (2019) Eve rising, *Juice*, December

Machel G (2019) Africa can be the launchpad for a green-energy revolution, *Time*, 23 September

Maclaren PIR (1958) The fishing devices of central and southern Africa. *Occasional Papers, Rhodes-Livingstone Museum* 12: 1–48

Maditla N (2020) How architect Mariam Kamara is masterminding a sustainable future for Niger. Accessed 16 March 2020, https://edition.cnn.com/style/article/mariam-kamara-architect/index.html

Maeko T (2020) Sanitiser feeds family, *Mail & Guardian*, 20 March

Maggs J (2018) *Win! Compelling conversations with 20 successful South Africans.* Auckland Park: Jacana Media

Maharaj V (2011) From plant to production: The *hoodia* story. *Quest* 7(1): 25–27

Maichomo M (2020) Why Kenya has banned the commercial slaughter of donkeys, *The Conversation*, 1 March

Mail & Guardian (online) (2020a) Digital innovation in the Covid-19 era, 17 April. Accessed 15 May 2021, https://mg.co.za/special-reports/2020-04-17-digital-innovation-in-the-covid-19-era/

Mail & Guardian (online) (2020b) New energy mix on the cards, 3 April. Accessed 15 April 2021, https://mg.co.za/special-reports/2020-04-03-new-energy-mix-on-the-cards/

Makwa DDB (2020) Uganda's musicians are fighting COVID-19: Why government should work with them, *The Conversation*, 7 May

Malinga S (2018) SolarTurtle's power plan for rural women. Accessed 15 May 2021, www.itweb.co.za/content/o1Jr5MxEXbJqKdWL

Mallinson T (2020) Accra's one-woman library, *Mail & Guardian*, 3 July

Mandela N (1994) *Long Walk to Freedom*. London: Little, Brown and Company

Mann, J (2019) South Africa now has twenty new marine protected areas. *African Wildlife & Environment* 74: 8–12

Manyonga E (2017a) Keeping it clean, *Fast Company South Africa*, October

Manyonga E (2017b) Saving water, saving the earth, *Fast Company South Africa*, October

Manyonga E (2017c) Cooking goes solar, *Fast Company South Africa*, October

Manyonga E (2017d) Ecological engineering is key, *Fast Company South Africa*, October

Manyonga E (2017e) Getting to the hot spot, *Fast Company South Africa*, October

Maritz B & Maritz R (2020) How we tracked the eating habits of snakes in Africa with the help of Facebook, *The Conversation*, 5 August

Mark M (2012) Nigeria's love of pidgin dey scatter my brain yet ginger my swagger, *The Guardian*, 24 September. Accessed 27 July 2021, https://www.theguardian.com/world/2012/sep/24/nigeria-pidgin-scatter-brain-swagger

Markotter W (2020a) Scientists are still searching for the source of COVID-19: Why it matters, *The Conversation*, 12 March

Markotter W (2020b) Why it's important to study coronaviruses in African bats, *The Conversation*, 19 February

Martin PS (1967) Africa and pleistocene overkill. *Nature* 212: 339–342

Mashaba H & Morris I (2012) *Black like you: Herman Mashaba, An autobiography*. Johannesburg: MME Media

Mathe T (2019) Uber cool firms won't always work, *Mail & Guardian*, 11 October

Mathe T (2020a) African free trade is vital for growth, *Mail & Guardian*, 6 March

Mathe T (2020b) The age-old stokvel moves into the digital era, *Mail & Guardian*, 17 January

Matsipa M (2020) On bioclimatic architecture: 'We have our own science, but we have forgotten how to transmit it', *Mail & Guardian*, 24 July

Matthews C (2019) Paying it forward: Gorongosa National Park, *Time*, 2–9 September

Matumi M (2019) Farmworkers get to grips with 3G devices, *The Star*, 25 October

Mazzucato M (2019) The platform economy needs laws, *Mail & Guardian*, 11 October

McCabe MF, Aragon B, Houborg R & Mascaro J (2017) CubeSats in hydrology: Ultra-high-resolution insights into vegetation dynamics and terrestrial evaporation. *Water Resources Research* 53: 10017–10024

McCormick TL (2019) How Cape Town's 'Gayle' has endured – and been adopted by straight people, *The Conversation*, 21 May

McDowell M (2006) Lasers developed in South Africa. South African electronics companies profile: Denel Eloptro, *Electronics News Digest*, 19 April

McHarg L (1969) *Design with nature*. New York: Natural History Press

Medupe T (2010) *Astronomy of Timbuktu*. Cape Town: Cambridge University Press

Meier JD (2019) Best lessons from Bill Gates. Acccessed 15 May 2021, www.medium.com/@jdmeier/25-amazing-lessons-from-the-ever-brilliant-bill-gates-e9dc6eee8ead

Meijaard E (2020) Coconut oil production threatens five times more species than palm oil: New findings, *The Conversation*, 6 July

Michalopoulos C (2020) Review: A masterful look at five decades of African development, *Mail & Guardian*, 28 April

Mielly M (2003) The aesthetics of necessity: An interview with Werewere Liking. *World Literature Today* 2(3): 52–56

The Millenial Source, Zimbabwe doctors to end strike after billionaire's offer to pay salaries, 24 January 2020, accessed 29 July 2021 https://themilsource.com/2020/01/24/zimbabwe-doctors-strike-billionaire-offer-salaries-2020

Mikia A (2020) As climate crisis bites, cricket-eating goes global, *Mail & Guardian*, 31 January

Minnaar A de V (1989) Nagana, big-game drives and the Zululand game reserves (1890s–1950s). *Contree* 25: 12–20

Mittermeier RA & Mittermeier CG (2019) *Megadiversity: Earth's biologically wealthiest nations*. Toronto: Cemex

Mokwena TO (2018) The soul of algorithms. *Leadership* 392: 8

Montgomery K (2020) Wondrous wetlands, *Weekend Argus*, 1 February

Moodley K (2018) New telescope chases the mysteries of radio flashes and dark energy, *The Conversation*, 17 August

Mpinganjira M, Brett A & Maiden LFN (2018) E-commerce puts Africa in the frame, *Africa Voice*, 5 June

Msiza I, Mistry J & Nelwamondo, FV (2011). Towards secondary fingerprint classification. International Conference on Computer Engineering and Applications (ICCEA), Haikou (China), 15–17 July

Mtunzi-Hairwadzi K (2019) MTN drives launch of award-winning literacy app, *Mail & Guardian*, 11 October

Muganda N (2020) Kenya's internet balloons could help to bridge the digital divide, *The Conversation*, 16 July

Murphy RR, Adams J & Gadudi VBM (2020) Robots are playing many roles in the coronavirus crisis – and offering lessons for future disasters, *The Conversation*, 27 April

Nagaoka L, Rick T & Wolverton S (2018) The overkill model and its impact on environmental research. *Ecology and Evolution* 8(19): 9683–9696

Naidoo D (2018) Harnessing the power of water, *Cape Argus*, 31 December

Naidoo J (2017) *Change: Organising tomorrow, today*. Cape Town: Penguin

Naidu D (2020) Lockdown risks for continent, *Sunday Independent* (19 April)

Nair N (2020) SA masks change lives, cheer up Belgians, *Sunday Times*, 16 August

Nash J (2020) Digital innovation in the COVID-19 era, *Mail & Guardian advertising supplement*, 17–23 April

Ndabezitha T (2019) Ikhaya Garden. *Sawubona*, 14 June

Nepad (New Partnership for Africa's Development) (2018) *An integrated, connected Africa: The official NEPAD yearbook*. Rivonia: Nepad Business Foundation

New Frontier Data (2018) *African Regional Hemp and Cannabis Report: 2019 Industry Outlook*. Washington, DC: New Frontier Data

News24 (2020) 'We can get it done here': Africa's tech scene tackles coronavirus. Accessed 14 May 2020, www.m.news24.com/Africa/2020519?isapp=true

Ng M et al. (2014) Global, regional and national prevalence of overweight and obesity in children and adults 1980–2013: A systematic analysis. *The Lancet* 384(9945): 766–781

Ngonyama T (2020) GeoPoint Africa is dominating the anthracite market. *Black Business Quarterly* 83: 54–55

Ngumbi EN (2020) Lessons on how to effectively tackle insect invasions, *The Conversation* 17 January

Niassy S (2018) Exploring the best tactics to combat fall army worm outbreaks in Africa, *The Conversation*, 7 May

Nicolon T (2020) Snake bites are 'neglected' health crisis in Africa. Accessed 15 April 2021, https://www.nationalgeographic.co.uk/photography/2020/05/snakebites-are-neglected-health-crisis-in-africa

Nkosi P (2020a) Revolutionary ways to read: Unconventional public libraries. *InFlight* 1:53–58

Nkosi P (2020b) How much are you worth? Calculating *lobola* the modern way. *InFlight* 1:47–50

Noakes T & Sboros M (2017) *Lore of nutrition: Challenging conventional dietary beliefs*. Cape Town: Penguin

Nogry S & Varly P (2018) How kids in a low-income country use laptops: Lessons from Madagascar, *The Conversation*, 19 March

Norbrook N (2020a) Satellite city solutions, *The Africa Report*, April-May-June

Norbrook N (2020b). US/Africa: Team Trump steps on the gas as China looms, *The Africa Report*, 1 May

Norton B (2020) Tales from Zulu to Arabic shared online, *Weekend Argus*, 25 January

Nubukpo K (2020) The Eco: A real test for West African vision and governance, *The Africa Report*, 24 April

Nugent C (2020) A revolution's evolution, *Time*, 20–27 July

Nulens R, Scott L & Herbin M (2011) An updated inventory of all known specimens of the coelacanth, *Latimeria* spp. *Smithiana Special Publication* 3: 1–52

Nuwer R (2018) In Chad, the elephants (so many elephants) are back. Accessed 15 April 2021, https://www.nytimes.com/2018/05/14/travel/chad-elephants-zakouma-park.html

Obura D (2020) We're coming up with a new set of targets to protect the natural world: Here's how, *The Conversation*, 26 February

O'Carroll S (2021) How you can be part of the solution to our literacy crisis, *Mail & Guardian*, 26 November.. Accessed 15 May 2021, www.mg.co.za/article/2018-11-26-how-you-can-be-part-of-the-solution-to-our-literacy-crisis/

Odendaal N (2020a) Pratley unveils new cable gland. *Engineering News & Mining Weekly* 40(8): 61

Odendaal N (2020b) Baddies' target. Africa in cybercriminals' cross hairs as internet connectivity doubles. *Engineering News & Mining Weekly* 40(8): 61

Odubanjo D (2020) Lassa fever: Why there's a call to declare a health emergency, *The Conversation*, 13 February

Odume N & Slaughter A (2018) Invest more in water experts, *The New Age*, 27 February

Okafor-Yarwood I (2019) EU targets fragile West African fish stocks, despite protection laws, *The Conversation*, 5 November

Olopade D (2014) *The bright continent: Breaking rules and making change in modern Africa*. London: Duckworth Overlook

Olurounbi R (2020) Canaries in digital coal mines, *The Africa Report*

Oosthuyse J (2020) Effective elephant contraceptive, *Safari News*, Summer

Oudenhuijsen L (2020) The covert social effects of Covid-19, *Mail & Guardian*, 29 April

Owino EA (2020) Heavy rains put Kenya at risk of mosquito-borne diseases, *The Conversation*, 22 January

Owuor VO (2018) The uneasy relationship between online betting and mobile money transactions, *The Conversation*, 23 April

Paquette D (2020) The sudden rise of the coronavirus grim reapers: Ghana's dancing pallbearers, *The Washington Post Online*, 25 April

Park A & Campbell C (2020) Containing a crisis: Are we doomed to a future of relentless viral outbreaks? *Time*, 10 February

Paweska J (2020) Seven factors that turned the DRC's Ebola outbreak around, *The Conversation*, 8 March

Payne J & George L (2018) Billionaire's huge Nigerian oil refinery likely delayed until 2022. Accessed 15 May 2021, www.reuters.com/article/us-nigeria-dangote-refinery-idUSKBN1KV0WA

Peter Z (2018) Doing the Kasi hustle, *City Press*, 27 May

Philander R (2018) MeerKAT captures heart of Milky Way, *Cape Argus*, 18 July

Pierce J (2017) Djibouti Telecom and TESubCom DARE contract comes into force. Accessed 15 April 2021, https://www.capacitymedia.com/articles/3777078/Djibouti-Telecom-and-TESubCom-DARE-contract-comes-into-force

Pilane P (2015) Lobola calculator: The thin line between humour and misogyny, *The Daily Vox*, 3 February

Piot P (2020) Lessons from Ebola, *Time*, 17 February

Piou C (2020) Locust invasions are cyclical: African states shouldn't be caught napping, *The Conversation*, 3 February

Piroth L, Cottenet J, Mariet A-S, Bonniaud P, Blot M, Tubert-Bitter P & Quantin C (2021) Comparison of the characteristics, morbidity, and mortality of COVID-19 and seasonal influenza: A nationwide, population-based retrospective cohort study. *Lancet Respiratory Medicine* 9(3): 251–259

Platter E & Friedman C (2019) *Durban curry: Up2Date*. Durban: Pawpaw Publishing

Potgieter L (2020) The unsung heroes making (city) waves, *InFlight*, January

Preston-Mafham K (1991) *Madagascar: A natural history*. Cape Town: Struik

Prins C (2019) Coding dreams, *Juice*, July

Quansah HN-B (2019) Lamugin: Hausa beer 'brewed' in Zongo, *Daily Graphic*, 16 August

Quérnet J-F (2020) A history of Africans in the Tour de France, *CyclingNews*, 24 June

Rahman A (2019) *What is 3DIMO? Sport tech start-up using AI to predict injuries when athletes train.* Accessed 15 May 2021, www.nsmedicaldevices.com/news/3dimo-startup-predict-injuries-athletes/#

Raynard L (Ed.) (2015) *Journeys of discovery: Stories of human innovation in Africa.* Cape Town: SKA South Africa

Reid C (2020) Make way for the Tutu Tester! Accessed 31 March 2021, http://desmondtutuhivfoundation.org.za/blog_post/make-way-tutu-tester

Reid S (2019) Sweet success in peanut butter venture, *Tatler*, 5 September

Rhodes CJ (2014) Mycoremediation (bioremediation with fungi): Growing mushrooms to clean the earth. *Chemical Speciation & Bioavailability* 26(3): 196–198

Riché M (2019) Isatou Ceesay, queen of recycling in The Gambia. Accessed 31 March 2021, http://climateheroes.org/heroes/isatou-ceesay-queen-plastic-recycling-gambia/

Ridovhona M (2019) Netshidzati creates gloves to help deaf people, *Limpopo Mirror*, 16 September

Robinson JM (2018) Kenya's struggle to modernize traditional medicine is far from won, *The Conversation*, 30 August

Robinson JM (2020) Biodiversity loss could be making us sick – here's why, *The Conversation*, 4 August

Roby L, Erickson L & Nagaisha C (2016) Education for children in sub-Saharan Africa: Predictors impacting school attendance. *Children and Youth Services Review* 64: 110–116

Rockwood, K (2018) A new home of African art, *Time*, 3–10 September

Rodriguez I & Inturias M (2020) Bolivia: Contribution of indigenous people to fighting climate change is hanging by a thread, *The Conversation*, 11 February

Rogan J (2018) *Elon Musk told about AI, tunnels and dancing Tesla.* Accessed 15 May 2021, www.medium.com/predict/elon-musk-told-about-ai-tunnels-and-dancing-tesla-5f3f81ff9cd3

Root T (2019) Octopus' garden: The race to build the world's first octopus farm, *Time*, 2–9 September

Rupiah K (2020) The dance app that can diagnose Parkinson's, *Mail & Guardian*, 6 March

Rusch N (2020) Does alcohol have an undisclosed African heritage? *The Conversation*, 9 June

Ruzicka A (2018) Getting banked in minutes, *City Press*, 29 April

Ryan E (2020) The global crisis is just starting to affect Africa, *Mail & Guardian supplement*, 17 April

Sabeti P (2020) Early detection is a key defense, *Time*, 17 February

Sadr K (2018) How we recreated a lost African city with laser technology, *The Conversation*, 15 March

Saldivia G (2020) Some good news: An 'elephant baby boom' in one Kenyan national park. Accessed 15 April 2021, https://www.npr.org/2020/08/14/902177466/some-good-news-an-elephant-baby-boom-in-one-kenyan-national-park

Samson L (2018) Inspired by the dung beetle, this reactor turns plastic into gas. Accessed 31 March 2021, https://www.news24.com/citypress/News/inspired-by-the-dung-beetle-this-reactor-turns-plastic-into-usable-fuel-20180626

Sanei J (2019a) Rise of the humachines, *GQ*, November

Sanei J (2019b) *FOREsight: Awaken curiosity. Cultivate wisdom. Discover the abundant future.* Johannesburg: Burnet Media

Sanral (South African National Roads Agency) (2017) N2 Wild Coast update. *SANRAL Bulletin* 3: 1–8

Schreiber M & González LL (2020). Rethinking the mosquito net, *Mail & Guardian*, 28 February–5 March 2020

Schwab K (2016) *The Fourth Industrial Revolution*. Geneva: World Economic Forum

Schwartz I (1957) Protea. The story of an African car, *Car*, March

Schweisfurth T & Goduscheit RC (2020) From the pyramids to Apollo 11: Can AI ever rival human creativity? *The Conversation*, 5 February

Scigliano E (2020) The world wants to eat more octopus: Is farming them ethical? Accessed 15 April 2021, https://www.nationalgeographic.co.uk/animals/2020/02/world-wants-eat-more-octopus-farming-them-ethical

Scroll.in (2019) Olympic champion Eliud Kipchoge becomes first man to complete marathon in under two hours, 12 October

Semuels A (2020) Fewer jobs, more machines, *Time*, 17–24 August

Sha R (2017) Taking the pulse of the natural world, *Fast Company South Africa*, October

Shaban ARA (2020) COVID-Organics: Madagascar launches Africa's first cure for virus. Accessed 15 April 2021, https://www.africanews.com/2020/04/22/covid-organics-madagascar-launches-africa-s-first-cure-for-virus//

Shapshak T (2017) How Kenya's SupaBRCK aims to solve Africa's internet problems. Accessed 15 May 2021, www.forbes.com/sites/tobyshapshak/2017/03/07/how-kenyas-supabrck-aims-to-solve-africas-internet-problems/?sh=2bd9bb0f17f3

Shaw B (2020) CT company creates proteins in plants to be used in rapid diagnostic test kit for COVID-19. Accessed 31 March 2021, www.news24.

com/Video/SouthAfrica/News/watch-ct-company-uses-plants-to-create-antigens-to-test-for-covid-19-20200512

Sigenu Z (2020) Use African languages to promote learning, *Mail & Guardian*, 20 March

Simmons R (2019) Tell kids the truth: Hard work doesn't always pay off, *Time*, 1 July

Skupien, S (2019) *Elizabeth Rasekoala on science and ownership*. Accessed 15 May 2021, www.sureco-review.net/2019/01/17/interview-elizabeth-rasakoka/

Smith C (2018) The future of work in digital world, *The Mercury*, 3 August

Smith LN (2019) Why Zimbabwe's female rangers are better at stopping poaching. Accessed 15 April 2021, https://www.nationalgeographic.com/magazine/article/akashinga-women-rangers-fight-poaching-in-zimbabwe-phundundu-wildlife-area

Smith P (2020) The contradictions of capital, *The Africa Report*, 27 February

Sokanyile A (2020) SA's first cannabis college on cards for Eastern Cape, *Weekend Argus*, 15 February

Sondo G (2020a) Cameroon's language barriers, *Mail & Guardian*, 24 April

Sondo G (2020b) *Mail & Guardian* (24–29 April)

Sosibo K & Allison S (2020) Soundtrack to a pandemic: Africa's best coronavirus songs, *Mail & Guardian*, 2 April

Soumarè M, Velluet Q, Galtier M & Norbrook N (2020) Tech hubs across Africa to incubate the next generation, *The Africa Report*, 14 February

Space in Africa (2019) *Mandla Maseko, first black African scheduled to travel to space, passes on*. Accessed 15 May 2021, www.africanews.space/mandla-maseko-first-black-african-meant-to-travel-to-space-dies/

Standaert M (2020) Coronavirus closures reveal vast scale of China's secretive wildlife farm industry, *The Guardian online*, 28 February

Steel G (2018) Sudanese women are using social media to trade – and break gender barriers, *The Conversation*, 5 May

Stobbs RE & Bruton MN (1991) The fishery of the Comoros, with comments on its possible impact on coelacanth survival. *Environmental Biology of Fishes* 32: 341–359

Stuart-Findlay D (2019) John Myers and the Protea, *The Crankhandle Chronicle*, November

Sunderland J (2020) Spend less on arms and more on science, *Mail & Guardian*, 14 February

Sunter C (1987) *The world and South Africa in the 1990s* Johannesburg: Tafelberg

Swain S (2020) *BioWise: Taking a cue from nature*. Accessed 15 May 2021, www.pezulahotel.com/2020/01/11/biowise-taking-a-cue-from-nature

Swart J-M (2019) By women for women, *Juice*, December

Swilling M (2019) Gloom the theme of Davos conversations. *New Agenda* 73: 17–20

Sydow F & Nordien GS (2019) *Cape, curry and koeksisters*. Cape Town: Human & Rosseau

Tade O (2020) Nigeria's tradition of matching outfits at events has a downside, *The Conversation*, 9 February

Tebape N (2019) From the co-founder of Nubian Seed: The life of a 'mompreneur', *Botswana Family Gems*, 24 April

Tella S (2020) What Nigeria needs to do to avoid World Bank's 'worst case scenario', *The Conversation*, 19 January

Terrefe B (2020) Megaprojects in Addis Ababa raise questions about spatial justice, *The Conversation*, 23 June

Thornton C (2018) Building for blockchain, *SkyWays*, June

Thunberg G (2020) No 'green deal' will be ambitious enough to save the planet, *Time*, 20–27 July

Tiedeu B (2020) Moves are afoot in Africa to keep more women in science careers, *The Conversation*, 9 June

Timothy S (2018) More conservation efforts needed for Uganda's long-horned Ankole cows, *CCTV News*, 7 September

Tinley KL (1964) Fishing methods of the Tonga tribe in north-eastern Zululand and southern Mozambique. *Lammergeyer* 3(1): 9–39

Tobias PV (1969) *Man's past and future*. Johannesburg: Wits University Press

Tobin D (2019) The future of education. *Black Business Quarterly*, fourth quarter: 34–37

Tofield-Pasche N (2020) Rwanda extracts methane from Lake Kivu for electricity. How it works, *The Conversation*, 15 July

Top Women (2020) Getting Africa into top tech condition. Standard Bank *Top Women*, 14 January

Trautmann C & Monjero K (2019) Science centers in Africa. *Informal Learning Review* 154: 1–10

Tsarwe S (2020) How apps on mobile phones are changing Zimbabwe's talk radio, *The Conversation*, 2 August

Tsekleves E (2020) New ways to clean homes may help in Ghana's fight against bacterial disease, *The Conversation*, 11 February

Tucker P (1997) *Just the ticket! My 50 years in show business*. Johannesburg: Jonathan Ball

Tyrell P (2019) Kenyan wildlife policies must extend beyond protected areas, *The Conversation*, 19 December

Ugwuanyi K (2020) Giving back to English: How Nigerian words made it into the Oxford English dictionary, *The Conversation*, 16 February

Uneca (2019) COVID-19 in Africa: The implications for macroeconomic and socioeconomic dimensions. Accessed 15 May 2021, www.uneca.org/sites/default/files/AEC/2020/presentations/covid-19_in_africa_-_the_implications_for_macroeconomic_and_socioeconomic_dimensions1_.pdf

Unesco (United Nations Educational, Scientific and Cultural Organization) (2017) *Cracking the code: Girls and women's education in science, technology, engineering and mathematics* (STEM). Geneva: Unesco

Unesco (2020) COVID-19: UNESCO and ICOM concerned about the situation faced by the world's museums. Accessed 15 April 2021, https://en.unesco.org/news/covid-19-unesco-and-icom-concerned-about-situation-faced-worlds-museums

United Nations (2006) *United Nations Fact Sheet on Climate Change*. UN Climate Change Conference Nairobi

Urama JO & Holbrook JC (2009) The African Cultural Astronomy Project. *Proceedings of the International Astronomical Union* 5(S260): 48–53

Urrego MR (2019) Message from Martha Rojas Urrego Secretary General of the Ramsar Convention on Wetlands on the occasion of the International Day for Biological Diversity. Accessed 15 April 2021, https://www.ramsar.org/news/message-from-martha-rojas-urrego-secretary-general-of-the-ramsar-convention-on-wetlands-on-4

Vance A (2015) *Elon Musk: How the billionaire CEO of SpaceX and Tesla is shaping our future*. London: Virgin Books

Van Itterbeeck J (2020) Swarming locusts: People used to eat them, but shouldn't anymore, *The Conversation*, 24 April

Van Velden J, Moho B, Biggs D & Travers H (2020) Malawi study shows how dependency on bushmeat hunting can be reduced, *The Conversation*, 13 July

Van Wyngaardt M (2017) Transnet unveils Trans-Africa locomotive. Accessed 15 April 2021, https://m.engineeringnews.co.za/article/transnet-unveils-trans-africa-locomotive-2017-04-04

Vaughan CL (2008) *Imagining the elephant: A biography of Allan Macleod Cormack*. London: Imperial College Press

Venter J (2017) Keeping it clean and simple, *Fast Company South Africa*, October

Verhoef AH & Kruger HA (2020) What connects Shaka Zulu, decolonisation and mathematical models, *The Conversation*, 16 February

Vick K (2020) Wangari Maathai: Seeding a movement, *Time*, 16–23 March

Viljoen A (2018) A budding revolution. *Leadership* 391: 42–44

Wadley L & Sievers C (2020) How we deduced that our ancestors liked roast vegetables too, *The Conversation*, 15 January

Ward D (2005) Do we understand the causes of bush encroachment in African savannas? *African Journal of Range and Forage Science* 22(2): 101–105

Waterworth T (2020) Bright future for sunnies brand, *Weekend Argus*, 25 January

Webb J (2019) African mathematics. Cape Town: University of Cape Town Summer School

Weber A, Kalema-Zikusoka G & Stevens NJ (2020) Lack of rule-adherence during mountain gorilla tourism encounters in Bwindi Impenetrable National Park, Uganda, places gorillas at risk from human disease. *Frontiers in Public Health* 8: 1. DOI: 10.3389/fpubh 2020 00001

WeeTracker (2020) Nigerian ventures secured highest investment in 2019: WeeTracker Report. Accessed 15 April 2021, https://weetracker.com/2020/01/08/nigerian-ventures-highest-funding-2019-weetracker-report/

Weldon C (2020) We found a way to trap stable flies: Their dung preferences helped us, *The Conversation*, 2 February

Wentzel L (2019) Biotech entrepreneurs pitch ideas, *Tatler*, 5 September

Wentzel L (2020) UCT alumni develop corona app to counter fake news, *Southern Suburbs Tatler*, 26 March

Whitby N (2019) Health benefits of surfing for body and mind. Accessed 15 April 2021, https://www.thewave.com/blog/health-benefits-surfing-body-and-mind/

White T (2018) Economic crime spirals out of control. *Leadership* 391: 20–22

WHO (World Health Organisation) (2018) *2018 World malaria report*. Geneva: WHO

Wicander S & Coad L (2018) Can the provision of alternative livelihoods reduce the impact of wild meat hunting in West and Central Africa? *Conservation and Society* 16(4): 441–458

Wild S (2013) *Searching African skies: The square kilometre array and South Africa's quest to hear the songs of the stars*. Auckland Park: Jacana Media

Wild S (2015) *Innovation: Shaping South Africa through science*. Johannesburg: Pan Macmillan

Wodinsky S (2018) World's first brick made of urine requires a lot of it. Accessed 31 March 2021, https://www.nbcnews.com/mach/science/world-s-first-brick-made-urine-requires-lot-it-ncna941356

Worland J (2020) The defining year, *Time*, 20–27 July

Wright MG, Spencer C, Cook RM, Henley MP, North W & Mafra-Neto A (2018) African bush elephants respond to honeybee alarm pheromone blend. *Current Biology* 28(14): 778–780

Yende F (2019) Failings turned to success. *Black Business Quarterly*, fourth quarter: 46–47

Yunus M (2020) Social entrepreneur means business, *Mail & Guardian advertising supplement*, 17 February

Yuora BA (2019) Kaolin clay touted for its many uses, craved by many, *Daily Graphic*, 23 August

Zewde S-W & Azoulay A (2020) Rethinking education is an urgent matter, *Mail & Guardian*, 14 February

Zhang M (2020) This guy made an aerial 'street view' of his island nation's coastline. Accessed 15 May 2021, www.petapixel.com/2020/06/25/this-guy-made-an-aerial-street-view-of-his-island-nations-coastline/

Ziraba A & Quashie PK (2020) Perspectives from Kenya and Ghana on coronavirus preparations, *The Conversation*, 28 January

Zorthian, J (2017) Flying jet taxis, *Time*, 15 May

Index